James E. Keeler
Pioneer American Astrophysicist

James E. Keeler, pioneer American astrophysicist (1895).

(Reproduced by kind permission of the Mary Lea Shane Archives
of Lick Observatory.)

James E. Keeler

PIONEER AMERICAN ASTROPHYSICIST
AND THE EARLY DEVELOPMENT OF
AMERICAN ASTROPHYSICS

DONALD E. OSTERBROCK

Lick Observatory, University of California

The right of the
University of Cambridge
to print and sell
all manner of books
was granted by
Henry VIII in 1534.
The University has printed
and published continuously
since 1584.

CAMBRIDGE UNIVERSITY PRESS

Cambridge

London New York New Rochelle

Melbourne Sydney

PUBLISHED BY THE PRESS SYNDICATE OF THE UNIVERSITY OF CAMBRIDGE
The Pitt Building, Trumpington Street, Cambridge, United Kingdom

CAMBRIDGE UNIVERSITY PRESS
The Edinburgh Building, Cambridge CB2 2RU, UK
40 West 20th Street, New York NY 10011–4211, USA
477 Williamstown Road, Port Melbourne, VIC 3207, Australia
Ruiz de Alarcón 13, 28014 Madrid, Spain
Dock House, The Waterfront, Cape Town 8001, South Africa

http://www.cambridge.org

First published 1984
First paperback edition 2002

A catalogue record for this book is available from the British Library

Library of Congress catalogue card number: 84-7727

ISBN 0 521 26582 7 hardback
ISBN 0 521 52480 6 paperback

To Irene

Contents

Preface

James E. Keeler was an outstanding American scientist. His fellow astronomers and physicists at the end of the last century considered him the leading astronomical spectroscopist of his generation. He was a pioneer of the new field of astrophysics, the application of the methods of physics to understanding the nature of stars, planets, nebulae, comets and the other objects that populate the universe. One of the first students at Johns Hopkins, America's first research university, he had accomplished much in the two decades after his graduation, and was just reaching the peak of his career in 1900. His contemporaries all saw a long and productive future before him. Then unexpectedly he died, at the age of forty-two.

Over the years, his fame evaporated, and we today know little of him, although his discoveries are still taught in every astronomy course in the United States. We no longer associate them with his name. I first became aware of the extent of Keeler's work when I went to Lick Observatory as its director in 1973. There, in the Archives assembled and organized by Mary Lea Shane, I began to learn of Keeler's career from his letters, each of them so modern in tone and content. Then I started to read his published scientific papers. Keeler was much more like us today in the problems he studied and the methods he used than any of his contemporaries. The more I read of Keeler, the more fascinated I became by his life and career that took him back and forth across America three times in his scientific career. Then I began tracking down his correspondence in other archives and other observatories. This book is the record of his life in science.

It is not only about Keeler, although it is focussed strongly on him. Through him it also tells the story of the little group of pioneers who followed in the footsteps of Samuel P. Langley and Charles A. Young and made "the new astronomy" into the science of astrophysics. Keeler's career was closely intertwined with those of George Ellery Hale and W.W. Campbell, and his story cannot be told without telling theirs. Laboratory spectroscopists like Henry A. Rowland, Henry Crew, and Joseph S. Ames were part of his life too. Astrophysics was (and is) immersed in the world of astronomy, and Keeler was as closely associated with Edward S. Holden and E.E. Barnard as with Hale and Campbell. Simon Newcomb was the acknowledged leader of American astronomy during Keeler's entire life as a research scientist. All of these men's lives and careers are part of this book.

Keeler was a good man, a great scientist, a careful diplomat, a supportive director and a poor fund-raiser. These attributes are hard if not impossible to convey in a vacuum. But by telling his story against the background of these men's lives, and of the lives of other scientists like George Davidson and T.J.J. See, and of university administrators like Daniel Coit Gilman and William Rainey Harper, Keeler's strengths become clearer, as do his weaknesses.

One of the main themes of the book is the increasing professionalization of American science at the end of the nineteenth century. Keeler was intimately involved in the building and early days of the two large "big-science" observatories, Lick and Yerkes, each constructed around a large instrument, designed to be used for research by a team of specialized observers. He was a key participant in the first international astronomical meeting held in the United States, the World Congress of Astronomy and Astro-Physics of 1893, and gave the main invited address at the first national meeting, the scientific conferences at the Yerkes dedication in 1897. With Hale, Keeler was one of the founding editors of the *Astrophysical Journal*, the first professional journal of the new field he helped create. At the University of California Keeler started the first regular graduate program, built around the Lick Observatory fellowships, to produce theoretically trained, observationally oriented professional researchers in astronomy and astrophysics. In the last two years of his life he became the first American astronomer to use a large reflecting telescope, until then shunned by professional astronomers in the United States, for his research. With it he began the systematic study of nebulae, including planetary nebulae, diffuse nebulae, and the "spiral nebulae" that he recognized as an important constituent of the universe and which we now know as galaxies, remote stellar systems. All these are the stories of this book.

It is a pleasure indeed to record my indebtedness to the many people who helped me over the years that went into the preparation of this volume. First and foremost I would like to thank the late Mary Lea Shane. She conceived, organized, and headed for many years the Archives of Lick Observatory that now bear her name. Her vision, energy and persistence converted the mouldering old files of the observatory into archives that are a treasure-house of documentary records of the early days of American astronomy. She was able to add to them, through her personal contacts, many collections of scientific and personal correspondence of early Lick astronomers. This material provided the primary source for my book. From the beginning, Mary Shane encouraged me in my study of Keeler, shared with me her unrivalled knowledge of the history of Lick Observatory, and helped my research in every way she could. I am most grateful to her memory.

Dorothy Schaumberg, who followed her as the curator of the Mary Lea Shane Archives of Lick Observatory, has also helped far above the call of duty, locating material for me and sending me many needed dates, names, pictures and identifications. To her and to David Heron and Allan J. Dyson, who as successive university librarians at the University of California, Santa Cruz, supported the archives and personally encouraged my work, I owe special debts of gratitude.

I consulted many other university and observatory archives in the preparation of this book. They are all listed at the beginning of the reference section at the back of the book. I am grateful to all of them for preserving, and providing for my use, valuable historical documents that enable us all to explore our astronomical roots. I am grateful to these archives for permission to quote from their material in this book. The few translations from German are my own.

I am particularly grateful to Judith A. Lola of the Yerkes Observatory Archives, to Frank A. Zabrovsky and Charles E. Aston, Jr. of the Archives of Industrial Society, in the Hillman Library of the University of Pittsburgh, and to J.R.K. Kantor of the University of California Archives in Berkeley. I returned to these three archives frequently, often at times very inconvenient for these individuals, but they always were pleasant and supportive, helping me track down the facts that went into this book. Other librarians, historians, and archivists who were most helpful to me, and whom I wish I could thank individually, were M.C. Beecheno, Gwen M. Cain, Marie T. Capps, J. Frank Cook, Kenneth C. Cramer, Sue Crosby, Catherine T. Engel, Agnes T. Gates, Janice F. Goldblum, Frances R. Greeby, Josephine Harper, Joan Hodgson, Dorothy Michelson Livingston, Donald Marion, Daniel Meyer, Julia Morgan, Ruth Neuendorffer, Roxanne Nilan, Julia O'Keefe, Howard Plotkin, Mary J. Pugh, Virginia J. Renner, Joseph L. Rishel, Glenora E. Rossell, Michael

T. Ryan, Bernard Schermetzler, Len Smith, Sid Squibb, Vivian Sykes and Joan N. Warnow. Likewise three university officials, W.L. Rogers, Marjorie J. Woolman, and Elaine Wright, provided important source materials from their files, for which I am very grateful.

I am very grateful to the editors of the *Astronomy Quarterly, Chicago Magazine of History, Mercury, Physics Today* and *Sky and Telescope* for permission to use material that appeared in different form in articles that I wrote for those magazines.

I am extremely grateful to the late Elizabeth Day Breneiser, the niece of James E. Keeler. She encouraged me in my study of her uncle, and presented to the Mary Lea Shane Archives of Lick Observatory several letters, a scrapbook that belonged to her grandmother, Anna D. Keeler, and two journals kept by her grandfather, William F. Keeler, in the years 1878-86. These journals, with the newspaper clippings they contain, are the main source of contemporary information we have on her "Uncle Edden's" early life. To Babs, as all her friends knew her, I owe a very special debt of thanks.

I am also most grateful to Lucile T. Clark, Isabel T. Marx, and Jane T. Blew, who allowed me to consult the papers of their father, Sidney D. Townley. These papers include letters and a diary that provide an invaluable picture of Lick Observatory in the year 1892-93.

In addition, many of my friends in the astronomical community have helped greatly by providing specific information or contacts. Some of the most significant, but unexpected, facts came from these sources. In this connection I would like particularly to thank Dennis Butler, Eugene R. Capriotti, Dale P. Cruikshank, John S. Gallagher, Eugene A. Harlan, Arnold M. Heiser, W. Albert Hiltner, Philip Keenan, Arnold Klemola, Katherine Gordon Kron, Leonard V. Kuhi, Brian G. Marsden, Nicholas U. Mayall, Orren C. Mohler, Wallace L.W. Sargent, Remington P.S. Stone, Daniel W. Weedman, Kingsley Wightman, and the late Paul Herget, C. Donald Shane and Joseph Tapscott. Two other friends who helped me with especially significant information were Robert A. Alberty and Alfred G. Swan.

After Mary Shane, I owe the greatest debt of gratitude to Helen Wright. Her biography of George Ellery Hale is the standard to which we all aim, no matter how far below it we fall. Over the years I worked on this book, she continually encouraged me to go on with it. She shared with me many interpretations and facts from her very great knowledge of American astronomy, not only of Hale and his colleagues, but also especially of Richard S. Floyd, Thomas E. Fraser, the Lick Trust, and the early days of Lick Observatory. She does not agree with all my interpretations, but her comments, advice and criticism have greatly strengthened this book.

Also my friend since graduate-student days, Wallace R. Beardsley, helped me greatly in the research for this work. When I began collecting material, he was a faculty member at the Allegheny Observatory of the University of Pittsburgh. Himself the biographer of Samuel P. Langley, he was able to locate for me many important letters, memoranda, minute-books, plans and daily records that would otherwise have been lost in the observatory's files, cupboards and drawers. He helped me on many visits to Pittsburgh, and by copying material for me when I could not come. His knowledge of the early history of Allegheny Observatory, shared freely with me, was essential in the preparation of the early chapters of the book.

Three other historian friends, Owen Gingerich, William G. Hoyt, and John Lankford were also of great help to me. They, together with Helen Wright, read the entire book in draft form. Their comments, criticisms, suggestions and insights greatly strengthened the book; whatever errors or false interpretations remain are my responsibility alone. I thank them all for their efforts, freely given in spite of their very busy schedules.

I began writing this book during two sabbatical quarters at the University of Minnesota in 1977-78, made good progress on it during a sabbatical quarter at the University of Chicago in 1980, and completed it during a sabbatical year at the Institute for Advanced Study in 1982-83. I am grateful to those institutions and to Lewis M. Hobbs, Edward P. Ney, David N. Schramm, Harry Woolf, and especially John N. Bahcall for their encouragement and hospitality in making available to me pleasant, stimulating environments in which to work. John Bahcall in addition took time from his busy schedule to read and comment on several of the chapters in draft form, providing additional insights from the viewpoint of a working astrophysicist. Also, I am grateful to the Hill Family Foundation for partial support under a Visiting Professorship at the University of Minnesota, to the John Simon Guggenheim Memorial Foundation for partial support under a Guggenheim Fellowship at the Institute for Advanced Study, and of course to the University of California for its continuing support.

I am also grateful to the National Academy of Sciences, the National Science Foundation, the University of California, and the Universities Research Association which, by paying my travel expenses to scientific meetings and conferences in the East and Midwest, made it possible for me also to visit distant archives and collect valuable information on James E. Keeler.

At Lick Observatory I am grateful to Sue Robinson, who typed the first drafts of several of the early chapters of the book, from often barely legible and always heavily corrected written copy. I am especially grateful to Barbara Pinkham who, at the Institute for Advanced Study, typed on the word processor second drafts and further revised drafts of all the chapters, as well as the entire final manuscript. Their intelligence, skill and dedication contributed greatly to this book.

My greatest debt of all is to my wife, Irene H. Osterbrock. As a volunteer at the Mary Lea Shane Archives of Lick Observatory she found many important letters, clippings and references that would otherwise have been overlooked. Her insights helped me understand the true meanings of several letters and incidents. She typed the first drafts of most of the chapters. Most importantly, she always encouraged me to go on with the book to the end. To her it is dedicated.

1

A raw country boy from Florida

On August 12, 1900, James E. Keeler lay dying in a San Francisco hospital. An outstanding scientist, he had been felled by a stroke at the age of forty-two. The director of the Lick Observatory of the University of California, in two short years he had photographed with its reflecting telescope hundreds of spiral "nebulae" and had recognized them as important constituents of the universe. His work was the beginning of the scientific study of the galaxies.

Keeler was a pioneer astrophysicist, whose work and scientific techniques underlie much of our modern view of the universe. He was among the first American scientists to apply the methods of physics to investigate the physical nature of planets, stars, nebulae, and the objects that we today call galaxies. In Keeler's day, the new word "astrophysics" meant the use of spectroscopes and spectrographs on telescopes to analyze the light from celestial objects that reaches the surface of the earth. In our day it has grown to include measurements of X-ray, ultraviolet, far infrared and radio radiation as well, many of them made from rockets or artificial satellites. The principles have not changed one iota.

Keeler was born just before the American Civil War, in the same decade as Theodore Roosevelt and Emperor William II of Germany, Max Planck, Pierre Curie and Henry Ford. He was trained at the first American research university, Johns Hopkins in Baltimore. He worked as an assistant, the equivalent of a modern "postdoc", for Samuel P. Langley, one of the first astrophysicists in America and indeed in the world. Keeler had one year of postgraduate study in Germany, then the acknowledged world center of physics. Like many of the leading American scientists of his time, he attended the lectures of Hermann von Helmholtz, Robert Bunsen and Heinrich Kayser, learning from the men

1

who were laying the groundwork of electrodynamics, optics and spectroscopy.

In 1886 Keeler was the first professional astronomer hired at Lick Observatory, then still in the building stage. The mechanical and technological developments of post-Civil War America were being applied to scientific research, especially in astronomy, the study of the whole universe. The country's tradition had been tinkering and invention, but the astronomer Simon Newcomb, and the physicist Henry A. Rowland were two of the first and most influential voices to speak up for pure research. A very few wealthy men and women, like James Lick, William Thaw, Charles T. Yerkes and Catherine W. Bruce, personally financed the building of successive monster telescopes and auxiliary spectroscopes and spectrographs to unlock the secrets of the universe. American money and technology, applied at fine observing sites in the favorable climate of California, enabled the United States to overtake Germany and Great Britain, and become the world leader in observational astronomy. American physics emphasized increasingly accurate quantitative measurements, and the physical conclusions that could be drawn from them.

Keeler was intimately involved in every stage of this development. Lick Observatory was the first large mountain-top astronomical observatory, and the first large research institution in California. There with the 36-inch refracting telescope, the largest in the world, Keeler studied the spectra of the planets, stars, and nebulae. Most of his work was based on extremely accurate wavelength measurements, for which he set new standards of astronomical precision. At Lick and later at Allegheny Observatory near Pittsburgh, Keeler made many fundamental new discoveries on the nature of the objects that make up the universe.

Especially during his years in the East, Keeler's career was very closely associated with that of his younger friend George Ellery Hale, the founder of Yerkes Observatory and later of Mount Wilson as well. Keeler was intimately involved in the beginnings of Yerkes Observatory, the second high-technology, big astronomy research institution that started its life at the end of the last century. Hale and Keeler founded the *Astrophysical Journal*, the first and for many years the only research journal of its field.

By his contemporaries, Keeler was considered the outstanding American astrophysicist of his time. He was admired and respected by astronomers all over the world, as much for his personality as for his research attainments. His contemporaries gave him nearly every award and honor to which an astronomer could aspire. They expected even greater things of him, but like Adonis, the darling of the gods, he was struck down before his time.

Through Keeler's life we can see not only his scientific discoveries, but the whole little world of American astronomy and spectroscopy at the end of the last century. His career was intimately linked not only with Hale's but with those of Edward S. Holden, W.W. Campbell and E.E. Barnard. Three future directors of Lick Observatory worked under Keeler in the years 1898-1900. He knew practically every important American astronomer, corresponded with many of them, helped scores of them in their research. He started the first serious graduate program in observational astronomy at Lick Observatory in 1898, a program that over the years turned out many of America's research pioneers. Every person who works in astrophysics today is, in some sense or other, one of James E. Keeler's scientific heirs.

James E. Keeler's wife always considered him a Connecticut Yankee,[1] and at least one of his fellow students at Johns Hopkins thought of him as "a raw country boy from Florida,"[2] but he was born in La Salle, Illinois on September 10, 1857. La Salle is on the Illinois River, in the rich and fertile northern part of that state, settled by New England pioneers in the first half of the nineteenth century. Both Keeler's parents could trace their ancestry through Revolutionary War soldiers back to the Pilgrim settlers.[3] His father, William F. Keeler, was descended from Ralph Keeler, who was born in England and emigrated to Hartford, Connecticut in 1637. Ralph Keeler's great great grandson, Samuel Keeler, was an officer in the Continental army during the Revolution. Samuel Keeler's grandson Roswell married Mary Plant, daughter of a wealthy family in upstate New York, and their son William was born in Utica on June 21, 1821. He in turn married Anna Dutton, daughter of Henry Dutton, of Bridgeport, Connecticut, on October 5, 1846.

Henry Dutton was an honor graduate of Yale, a lawyer who served several terms in the Connecticut legislature and Senate. In 1847 he was appointed a professor of law at Yale and moved to New Haven, where he continued to practice until he became a county judge in 1852. In 1854 he was elected governor of Connecticut and served one full term; a few years later he was appointed a judge of the Supreme Court of Connecticut, a post he held until he retired in 1866 at the age of seventy. During all this time he had continued to teach at the Yale Law School, and after his retirement he devoted his main efforts to it until his death in 1869. As a judge he was not considered particularly learned, but practical, with a good knowledge of precedents, quick to grasp a legal point, and advanced in his ideas.

Henry Dutton's elder brother Matthew was another Yale product, who had graduated with highest honors, and stayed at New Haven as a tutor until he was ordained as a minister in 1814. He served a congregation at Stratford, Connecticut for seven years, but then returned to Yale as professor of mathematics and natural philosophy, as science was then called. He was considered a remarkable mathematician, and was a very popular teacher, but he died at the age of 42, "from general debility brought on by intense application to his work."[4] James E. Keeler combined in his career many of the characteristics of these two brothers. Henry Dutton's father and grandfather had both served in the Revolution, and through them Keeler was descended from John Punderson, one of the founders of the first church in New Haven.

William and Anna Keeler's first son, Henry Dutton Keeler, was born on November 3, 1847. A few years later William, with his brothers James and Edward, sailed from New York around Cape Horn to San Francisco during the California Gold Rush, to seek their fortunes. James died on the voyage, and Edward in California. William did not make his fortune, but he certainly did see the gold fields, and probably worked in them for a time.[5] From California he took passage on another ship for China, and then continued around the world via the Cape of Good Hope and then home again. By 1853 he was in business as a watchmaker in La Salle. The Keelers' first daughter Minnie was born there in 1854, but she died at the age of three, the same year James Edward, named for his two dead uncles, was born. By this time his father had become a senior partner in the La Salle Iron Works, which did all kinds of foundry and machine work, steam engines, casting and general machinery.[6]

The Keelers' last child, Elizabeth Eliot, was born in 1860. A few months later Abraham Lincoln was elected president, on a platform of stopping the expansion of slavery. The Civil War soon broke out, and by the end of 1861 Willliam Keeler obtained, with the aid of his congressman, Owen Lovejoy, a commission as acting assistant paymaster in the United States Navy. In January 1862 he was assigned to the crew of U.S.S *Monitor,* the famous Union ironclad steamship. During her voyage from New York to Hampton Roads the *Monitor*'s engine-room ventilating system failed and the engineers were overcome; Paymaster Keeler was able to draw on his La Salle Iron Works experience and operate the engines until the engineers were revived and returned to duty. During the *Monitor*'s battle with the Confederate ironclad *Merrimack,* he was in the pilot house and when Captain John L. Worden was blinded by a shell that smashed into the armor, Keeler gave him first aid and notified the executive officer to take over command. As the "saviors of the North," the crew of the *Monitor* were instant celebrities, and in the months following the battle their ship was visited by President Lincoln, Vice President Hannibal Hamlin, several cabinet officers and senators, and, by Keeler's count, at least 19 generals.[6]

That summer, while the *Monitor* was supporting the Union army on the James River, south of Richmond, Anna's brother Henry M. Dutton, a lieutenant in the 5th Connecticut Volunteers, was killed in the fighting at Cedar Mountain.[7] The carnage was fearful as the Confederate and Union Armies blasted away at each other in the dark woods, and after the battle over half the company officers in the 5th Connecticut were reported killed or missing in action, the latter meaning that their corpses were so mutilated by gunfire or grapeshot as to be unidentifiable. Judge Dutton hurried to the battlefield, but was unable to find the body of his namesake and only son.

Keeler was able to go home on leave for a month when the *Monitor* was sent to the Washington Navy Yard for repairs in October. He continued to serve on her until she went down at sea in a gale off Cape Hatteras at the end of December 1862. Keeler tried to save the ship's money, but the safe in his cabin was underwater by the time he got to it, and he had to go over the side on a rope. The only possessions he had when he was picked up by a boat from the U.S.S. *Rhode Island* were the clothes on his back.[6]

Keeler was then transferred to the U.S.S *Florida,* assigned to blockade duty off Southern ports until the end of the war. He was wounded twice, and received a pension of $10 a month for the rest of his life. In March 1865, Keeler was able to bring his teen-age son Henry aboard the *Florida,* where he helped his father buy stores and deliver payrolls, and lived in the cabin with him. They saw many Florida and Gulf Coast ports together, and after General Robert E. Lee's surrender and the defeat of the Confederacy, they enjoyed a cruise to Panama, when the *Florida* carried men and money destined for the Pacific Squadron to the Isthmus. Keeler was finally allowed to go home in November 1865, and was released from the Navy a few months later.[8]

Meanwhile, young Eddie, who throughout his life was always known by his middle name to his family and close friends,[9] was growing up back in Illinois. During and after the Civil War he attended school at an institution he later described as La Salle High School,[10] perhaps with some exaggeration, for the family moved to Florida in November 1869, when he was only 12 years old. Just a few months before they left, Keeler had seen the solar eclipse which swept across the United States in 1869, a vivid sight that he never forgot.[11]

William Keeler had not much liked Key West or the other Florida ports he saw or visited from his ship during the War, but perhaps a few winters back on the snowy prairies of northern Illinois convinced him of the advantages of living in a more moderate climate. His sister and brother-in-law, David and Frances Brown, had moved from Chicago to Lake Worth, in southern Florida near Palm Beach, and their reports probably helped convert him. Anna's father, Henry Dutton, died in New Haven in April 1869, and no doubt whatever inheritance she received from his estate helped the Keelers buy land and build a house near Mayport, at the mouth of the St. John's River, eighteen miles downstream from Jacksonville.

Mayport was a very small town. The Keelers' house, which they named Thalassa, was a large mansion, surrounded by a palmetto forest that opened out to a fine view of the sea.[12] William Keeler promoted himself and was known as "Major Keeler" to the locals; under this name he operated a general repair business with his own small shop and forge, and planted and cultivated orange, pecan, and date-palm trees, as well as beans, corn, potatoes, and other garden vegetables.[13] He wrote a weekly Mayport letter for the Florida Times-Union, signed with the pseudonym "Silex", or later, "Monitor", in which he chronicled the comings and goings of ships from the sea and vacationers from Jacksonville and points north. He often wrote of the healthful effects of an outdoor Florida life, as when he described a physician who "worn down by the cares and anxieties of his professional business, has for a few days laid aside pills, powder and physic, and has prescribed for himself the sea air and salt water of our Mayport beach. His prescription is a wise one, and if followed will result in his complete restoration to health."[14] Although the Major welcomed visitors from Jacksonville and urged them to enjoy their holidays and vacations in Mayport, he felt it his duty to warn them not to bathe in the nude in front of the summer cottages, but rather to go further away up the beach for the sake of modesty.[15]

There was no school remotely matching Eddie Keeler's interests or abilities in tiny Mayport, and all his further education after the family left Illinois was at home with his parents. He helped out with work on the house and in the shop, and had plenty of time for tramps on the beach with his brother Henry, and games and visits with his friends in and around the little community of Mayport.[16] He liked boats and sailing.[17] But he studied too, with his father, who subscribed to the *Scientific American,* and his mother, who subscribed to *Harper's Magazine,* and somehow, between them, they managed not only to give the boy a solid grounding in reading, writing and arithmetic, but also to help him make himself an expert sketch artist, and a master of tact who got along well with people all his life.[9]

Ed Keeler developed his interest in astronomy from the practical side of surveying,[3] which he learned from his father. He ordered a two-inch achromatic lens from Queens, an optical dealer in Philadelphia who advertised in *Scientific American*, and two smaller lenses for an eyepiece. Within a week after he got them, he had assembled his first telescope, and although the cloudy winter weather prevented him from observing stars, he tested it on distant ships and the Mayport lighthouse. Soon he was observing the moon, double stars, Saturn, Jupiter, the Orion nebula, and the Ring nebula in Lyra.[16] He kept a careful record of the objects he observed, and drew sketches of lunar craters and planets. Astronomers who saw these drawings in later years pronounced them excellent.[18] In 1877 Keeler built a meridian circle from the telescope, using a pasted-on protractor scale and a small kitchen clock which "kept execrable time" to convert it into an instrument of at least semiprecision.[19]

At last near the end of 1877 he got his chance to go north for an education. Earlier that same year his brother, Henry, who had been working as a clerk in a trust company in New York, died unexpectedly at the age of 29, while on vacation at Branford Point, Connecticut.[20] Probably he had contracted some kind of fever, for the cause of death was given as "colic."[6] His father thought of Henry as a "young and promising botanist,"[12] but there is no indication that he had ever entered college. Ed's chance came through his sister Lizzie, who was attending a private school in Tarrytown, New York. She was taken with her class to visit the private observatory of Charles H. Rockwell, a wealthy amateur astronomer. When they looked at Saturn through the telescope, she said she had often seen it in her brother's telescope down home in Florida. Her remark attracted Rockwell's attention, who wanted to learn more about this unusual brother, and ended up bringing him north and helping finance his education.[9]

Rockwell, five years younger than Keeler's father, had been born to a wealthy family in Norwich, Connecticut, and studied chemistry at Yale. In 1850 he sailed around Cape Horn to California, and may conceivably have met William Keeler at that time, although no record of such a meeting survives. During the Civil War he served as a captain in the Quartermaster Corps at Port Hudson, Vicksburg and St. Louis, purchasing ammunition and stores for the Army and forwarding them to New Orleans. At the end of the War he was on General Phil Sheridan's staff and distinguished himself at Cedar Creek, the battle from which Sheridan was absent in Winchester, twenty miles away.[21] Rockwell lived in Chicago for a few years after the War, and then in 1869 moved to Tarrytown, in

Westchester County, New York, the millionaire colony where he counted among his neighbors John D. Rockefeller, William Rockefeller, and Jay Gould. Here at his "Italian Villa" Rockwell built his own observatory, and devoted himself to astronomy for the last thirty-five years of his life.[22]

Somewhere along the line Rockwell and Keeler's father had met and become friends.[23] Both their families go back to the early days of Connecticut, and were heavily involved in the New England migration to northern Illinois.[3] At any rate, Rockwell, impressed by what he could learn of young Keeler, arranged for him to come north and visit Yale, Harvard, and Johns Hopkins. In writing to these universities, Rockwell described Keeler's knowledge of mathematics and science, emphasizing that he was brilliant and hard working but relatively untrained. He wrote of Keeler's practical skill and ability, demonstrated by the instruments he had constructed himself "with only the resources of a poverty stricken plantation."[24] Another by no means negligible qualification of Keeler's was that, as the grandson of the late Governor Dutton, he had "good stuff in him."[25] When Rockwell met Keeler on December 5, 1877, he liked the "lankey green country boy" who had worked his passage north from Mayport to New London, helping the captain navigate the schooner by the sun and stars.[26] No doubt he did; yet Keeler's free trip was probably not completely independent of the fact that his father was a deputy collector of customs at Mayport and a close personal friend of all the captains who visited that port frequently. George J. Brush, the professor of physics who interviewed Keeler at Yale, was William Keeler's cousin.[3] Keeler's own idea was to become a scientific workman in Alvan Clark's optical establishment in Cambridgeport, Massachusetts, well known to him from the pages of the *Scientific American*, but luckily for astronomy there were no job openings there at the time.[27] All the professors who met Keeler agreed that he should become a scientist, and that although his level of training would not qualify him to enter Yale or Harvard, Johns Hopkins was just the place for him.[28]

Johns Hopkins University was very young when Keeler arrived in Baltimore late in December 1877; it had admitted its first students in the fall of the previous year, and Keeler's class was the second in the history of the institution. It had been founded by the will of Johns Hopkins, a wealthy Baltimore financier who left half his seven-million-dollar estate, consisting largely of Baltimore and Ohio railroad stock, to build a university. The trustees lured Daniel Coit Gilman away from the University of California to be Hopkins' first president, and he built it as America's first research university. Gilman had been trained at Yale, and had been deeply involved in the founding of its Sheffield Scientific School, where he was librarian, professor of geography and secretary of the governing board for many years before he left for California. At Johns Hopkins, with a large endowment and complete freedom from church or state control, he was determined to promote scholarship of the highest order. He modelled Johns Hopkins on the German universities which then dominated so much of world science. It was primarily a university for advanced work by graduate students, but also had an undergraduate department where students were supposed to receive the kind of instruction that other universities gave only in their post-graduate courses. Gilman wanted to leave the kind of work usually done by undergraduates to other institutions. During Keeler's years at Johns Hopkins it was a growing institution, with on the average about eighty graduate students and only thirty undergraduates in the whole university.[29]

The faculty numbered twelve, all personally selected by Gilman. The first professor appointed had been young Henry A. Rowland, the outstanding physicist, whose laboratory was the first in the country with an instrument shop in which the apparatus needed for physical research could actually be produced. He rarely lectured, and never taught a systematic course. On one occasion, when asked what he would do with the undergraduates, Rowland replied, "Do with them? Do with them? *I shall ignore them!*" In fact, he stimulated them by his example, rather than by lecturing to them. With his vigorous personality, keen pursuit of science, and incisive intelligence, Rowland provided an inspiring model for the budding young physicists. A few years later he was to deliver a stirring "Plea for Pure Science".[30] Keeler took most of his physics courses from Charles S. Hastings, a young Yale Ph.D. who had studied abroad at Berlin, Heidelberg and the Sorbonne. He specialized in optics and had a keen interest in astronomy.[31] The star of the faculty was J.J. Sylvester, the English mathematician, who, like Rowland, did not teach undergraduates. Another outstanding scientist was Ira Remsen, the professor of chemistry, who had received his Ph.D. at Göttingen, and followed the German scheme of involving his graduate students fully in his research. There was no astronomy professor at Hopkins, but Simon Newcomb, the outstanding astronomer of the day, came over twice a week from the Nautical Almanac Office in nearby Washington to teach a course in the winter of 1877, and frequently thereafter.[32] Samuel P. Langley, the pioneer astrophysicist from Allegheny Observatory, near Pittsburgh, gave a series of lectures on the sun and radiant energy a few years later.[29]

When Keeler entered Johns Hopkins he was twenty years old. He was nearly six feet tall, with light hair, blue-gray eyes and a fair complexion.[33] The father of one of his classmates described him as a backwoods character straight from a Mark Twain novel, speaking with a pronounced "Cracker drawl."[34] Yet his air of quiet competence impressed everyone with whom he came in contact, and he made many close friends among his classmates and the graduate students at Johns Hopkins. Gilman had introduced an undergraduate group system in place of the traditional college schedule in which all the students took essentially the same courses. Instead, students at Hopkins could push forward as rapidly as they chose, or stretch out their course of studies over a longer period if they preferred. No fixed time was required to complete any of the seven possible groups of courses; usually it took three years, rarely four, and students who came with better preparation than required for matriculation could finish in as little as two years. Keeler, who did not complete his matriculation requirements until shortly before he graduated, took a little less than four years to complete his program in Group 3, "for one who prefers Mathematical studies, with reference to Engineering, Astronomy, Teaching, etc." He majored in physics and German, and took minors in mathematics, chemistry and astronomy. In addition, he worked as a part-time laboratory assistant and did some numerical computing for the *Baltimore Sun Almanac* to help pay his expenses.[35]

It was hard work, but Keeler enjoyed it. One fellow student testified to the spirit of those early times at Johns Hopkins, calling them "a dawn wherein 'twas bliss to be alive".[29] In later years, Keeler always felt extremely positive about his alma mater and the education that he and his classmates had received there.[36]

At the end of his first year at Hopkins, Keeler had an unusual summer opportunity. Through his patron Rockwell, he got the chance to join a scientific expedition to Colorado to observe the total solar eclipse of July 29, 1878. Solar eclipses provide the only conditions under which the faint solar corona and the regions of space immediately surrounding the sun can be observed. They happen only infrequently, and the path of totality is a narrow corridor that may be located in any part of the world. Rockwell, a skilled amateur astronomer, was going with the party sent out by the Naval Observatory, under the leadership of Professor Edward S. Holden, to make scientific measurements of the eclipse. Holden's group included Hastings, Keeler's teacher at Johns Hopkins, Lieutenant Samuel W. Very, a scientifically trained Navy officer, and Edgar W. Bass, a mathematics professor from West Point. Rockwell and Keeler accompanied them, as volunteers, at Rockwell's expense.[37] Keeler had already met Holden soon after his arrival at Hopkins. He had, at the Naval Observatory astronomer's request, made a drawing of the Orion

Nebula, based on his observations with a small telescope, for comparison with an earlier drawing by Sir William Herschel.[38]

The West was still wild in 1878. Holden, on the basis of available climatological data, originally planned to go to Virginia City, a mining camp located on the eclipse path in the Montana Territory. However, knowledgeable Army officers warned the superintendent of the Naval Observatory that the Indian hostiles might be out on the warpath that summer, and Holden was ordered to stay a little closer to civilization. The site he picked, Central City, Colorado, is less than 450 miles from the spot on the Little Big Horn where Crazy Horse, Gall and their Sioux and Cheyenne warriors had annihilated Custer and five troops of the Seventh Cavalry just two summers before.

Keeler traveled alone from Baltimore to Denver, getting up at 3 a.m. on a hot July morning to catch the 6:55 train to Harrisburg. Two of his closest friends at Hopkins, David T. and William C. Day, got him to the station with two minutes to spare. En route to Denver he had to change trains five times, at Harrisburg, Pittsburgh, Indianapolis, St. Louis, and Kansas City. An avid sightseer, Keeler admired the iron and steel works he saw in Harrisburg and in Pittsburgh, but he admired a pretty waitress in the station restaurant at Topeka even more. The rest of the passengers behaved like hogs, according to Keeler's account, eating with their hats on, drinking whiskey, and otherwise acting in ungentlemanly fashion, so when Keeler "smashed" (flirted with) the waitress, she ignored the other customers and devoted all her efforts to him. But the train soon left Topeka, and he was again bored by the dry flat prairies of Kansas and eastern Colorado. Finally he reached Denver, more than three days after his departure from Baltimore. He took the horse-drawn bus to the American House, as instructed, but found that Rockwell, who had arrived there two or three days previously, had gone off on a trip. Keeler decided to take advantage of the situation to improve his education and "went down 16th street . . . into a sort of rum shop theatre, more to see what kind of rowdies Colorado furnished than to see the play." But it was pretty tame stuff by Eastern standards, "nothing compared to the New Central in Baltimore, and the girls even wore high-necked dresses, but they were bold and cheeky." Disappointed, he left after the first act and went back to the hotel.

The next day he made a sightseeing trip around the city, and was entranced by the sight of the Rocky Mountains, rising abruptly out of the plain just west of Denver. Back at the hotel he met Very and Rockwell, who had returned from his trip, and the three of them drove out to visit the Princeton eclipse party, which was under the direction of Charles A.

Young, the pioneer astronomical spectroscopist. He and his group, which totaled thirteen observers, including Princeton physicist Cyrus F. Brackett and seven graduating seniors, were preparing to observe the eclipse from a site near Denver. Arthur C. Ranyard, an English spectroscopist who was Secretary of the Royal Astronomical Society, was with them.[39] Keeler was introduced to Young and was impressed by the fine instruments the Princeton astronomers had brought to Colorado, particularly the first astronomical spectroscope he had ever seen. That day, with Rockwell in town, Keeler confined his theatrical activities to watching a performance by an educated pig, which could count and do simple arithmetic, and, in the evening, seeing the famous actor Joseph Jefferson star as Rip Van Winkle at the Opera House.

The next day Hastings and Holden arrived in Denver, discussed their plans for observing the eclipse with Rockwell, Very and Keeler, and then went on to Central City.[40] The other three followed the next day, and Keeler was greatly impressed by the mountain scenery, and by the little narrow-gauge railroad that zigzagged up the mountain from Golden to Central City, 8,400 feet above sea level. They settled in the Teller House, a three-story, 150-room hotel, and Holden arranged with the manager for his party to observe the eclipse from its flat roof. While waiting, they all went sightseeing in Central City's Bob-tail Gold and Silver Mine, at that time one of the largest and richest mines in Colorado.[41] The weather was perfect, cool and comfortable, with clear skies every day.

On July 29, the date of the eclipse, a few light clouds appeared before noon, but soon dissipated. As the moon covered the sun, the sky darkened, and Keeler heard roosters crowing and dogs howling, and saw nighthawks wheeling around the hotel. One local wag reported that a few well-known citizens went into their favorite bar for a nightcap,[42] but Keeler does not mention this episode. Just before totality he saw the shadow bands, an alternating dark and light eclipse phenomenon, sweep across the landscape. At totality he rapidly sketched the corona, observing with a two-inch aperture telescope which had a large field of view. He took one hasty glance with his naked eye and saw the grandest sight he ever expected to see.[41] As soon as the eclipse was over he made a crayon drawing on black board of his sketch, from his memory, which he had trained by observing scenes for one or two minutes and then drawing them without looking at them again. Keeler's drawing, published in his first scientific paper as part of the Naval Observatory report on the eclipse, shows coronal streamers and the flattened, equatorial shape of the corona characteristic of times of sunspot minimum.[37]

Holden, using a larger telescope, searched the darkened sky near the sun during the eclipse, looking for a planet inside the orbit of Mercury, which the French astronomer Urbain LeVerrier thought he had observed at an earlier eclipse. Holden found no such object, and from his observations was able to rule out the existence of any intramercurial planet as bright as a third-magnitude star. Hastings measured the polarization of the light from various places in the corona quantitatively, but found to his consternation that it was tangential, rather than radial as all previous reliable observers had found, and as he expected it to be. He suspected something was wrong with his instrument. Nevertheless, when he checked it he found everything in order, so he published the results, but neither he nor anyone else afterward ever really believed the polarization was tangential.[43]

The eclipse over, that evening the astronomers invited the people of Central City up to the roof of the Teller House and showed them the planet Jupiter, double stars and clusters through their telescopes. Keeler exhibited his drawing of the corona, and received congratulations on it. Everyone in the group felt satisfied with their results as they headed back to the East. Keeler stopped in Illinois for his first visit with friends and relatives in La Salle since his departure with his family nine years before.[44] A few weeks later, on his twenty-first birthday, he was back in Baltimore, beginning his second year at Johns Hopkins.

Along with physics and mathematics, Keeler took one or more courses in German nearly every term.[45] Many of the Johns Hopkins fellows had studied in German universities, and they founded an informal "Kneipe" or club, at which they drank beer and discussed the problems of the world in German.[29] Keeler must have taken a lively part in these proceedings. Yet all was not peaches and cream; at least one unreconstructed rebel, the Virginian Walter Hines Page, was unhappy with the "Germanism" of Johns Hopkins, which he considered "at least unnatural" in a Southern city like Baltimore. He described one fellow graduate student whom he did not like as a horrible example of a grind: "He can make dictionaries but can no more appreciate the soul beauties of literature than a piano manufacturer can appreciate Wagner. He is a native of Connecticut, and Connecticut, I suppose, is capable of producing any phenomenon".[29]

During Keeler's sophomore year his 18-year-old sister Lizzie came up from Florida by schooner for a month-long visit with him in Baltimore in May 1879. There she first met his friend and classmate David Day, her future husband. A year later, in the summer of 1880, Day accompanied Keeler on his first trip back home to Florida since he had entered Johns Hopkins. The two of them spent most of July and August at Mayport, dividing their time among picnics, trips, visits, and helping Keeler's father with his work in the shop. The next winter Lizzie went north again and this time she stayed in Baltimore nearly four months. Day by this time was considered nearly one of the family, and Keeler's father took pride in sending him huge oranges from his orchard, and in receiving a shipment of goldfish from him in return.[13]

At the university, Keeler was a good student who was liked by all. Twenty years later one of his college chums, by then himself a physics professor at Western Reserve University in Cleveland, remembered him for "[h]is kindness, his cordiality, his boyish high spirits, his earnestness of purpose, and his quickness as a student".[46] Finally, in June 1881, having completed all his requirements Keeler received his bachelor's degree from Johns Hopkins. He had learned a lot in his courses as well as in his laboratory work, and was considered a great success by all his teachers.[9] Keeler had also learned, from observing Gilman, how to harmonize and keep the peace among professors and scientists who were often prima donnas. The Johns Hopkins president was always friendly and positive about the achievements of others, keeping his true feelings well hidden under a mask of bland and courteous phraseology.[29] Keeler followed his example all the rest of his life. Johns Hopkins granted nine Ph.D. degrees that June, and twelve B.A.'s, including David Day's as well as Keeler's. But even before the graduation, Keeler had left Baltimore to begin work at Allegheny Observatory, near Pittsburgh, as an assistant to Samuel P. Langley.[47]

2

I shall be glad to keep him here for the present

Samuel P. Langley, the director of Allegheny Observatory, was a self-taught scientist and a pioneer astrophysicist of the generation before James E. Keeler's. Born in Roxbury, Massachusetts in 1834, Langley had been educated in the Boston Latin School and Boston High School. He had been interested in astronomy from childhood, and indeed in nearly every scientific and mechanical subject. His father could not afford to send him to a university, so Langley decided to become a civil engineer, which he thought was as close to astronomy as any profession in which he might be able to find a job. For several years he practiced as an architect, mostly in Chicago and St. Louis. Next he became the manager of the collection department of a large mercantile agency, and then headed in turn its Milwaukee and Philadelphia offices.[1] By 1864 Langley had saved enough money to leave the business world, and he returned to New England. He and his brother John built a telescope, and then toured through Europe together for a year, visiting observatories, scientific institutions, and art galleries. Back in Boston again Langley was hired as an assistant at the Harvard College Observatory. He was then 30 years old, and just beginning his career in astronomy.

At Harvard Langley sharpened his eye, and his observing skills, but after only two years he left for a better job as assistant professor of mathematics at the United States Naval Academy at Annapolis, in charge of its small observatory. However, just a year later he moved on once more, to a position as director of the Allegheny Observatory of the Western University of Pennsylvania, now the University of Pittsburgh, where he stayed for over twenty years.[2]

Allegheny Observatory was located in Allegheny, now a part of Pittsburgh, but then a separate city on the north bank of the Allegheny River. The observatory dates back to 1860, when a group of leading citizens of

15

Pittsburgh and Allegheny chartered the Allegheny Telescope Association, bought a 13-inch telescope, and built an observatory to house it. At first the observatory was strictly an amateur operation, and the telescope was used only for looking at stars, but in 1865 the Western University of Pennsylvania took it over, and in 1867 Langley was appointed as its director.[3]

When he arrived, the observatory consisted of the small building containing the telescope, a table, three chairs, and not much else. Everything was needed, but there was no money for anything. The university administration expected Langley to devote his time to teaching, while he wanted to do research, and he spent two years of bitter discouragement, years he hated to look back on, torn between classroom duties and astronomical science. Then he had his great money-raising idea, of setting up a time service and selling highly accurate time signals to the railroads, and to jewelers and watchmakers. In the days before standard time zones every large city had its own time. No regular time signals were sent out by the Naval Observatory or any other government agency, so each city was on its own in regulating its clocks. The railroads were rapidly expanding after the Civil War, and confusion over time was becoming intolerable. Pittsburgh was one of the largest cities in the United States, a manufacturing center and a railroad hub. Langley saw his opportunity and seized it. He organized a time service in which accurate observations, made at the observatory, were used to regulate master clocks whose signals were sent out over the regular telegraph wires to a wide network of paying customers. He drew on his astronomical knowledge, and also on his experience as a store manager and bill collector, to make it a paying business. The income, several thousand dollars a year, provided an operating fund for the observatory.[1]

Freed of his teaching duties, Langley began astronomical research, chiefly on the sun. The old astronomy, going back to Galileo and Newton, was concerned with the positions of the stars, and planets, considered as mass points moving under gravitation. Langley was one of the little band of pioneers who tried to use the concepts and methods of physics to analyze the nature of astronomical objects – their compositions, temperatures, energy sources, how they radiated, how they evolved. Thus he was one of the first practitioners of the infant science of astrophysics, which he always called "the new astronomy." He and his contemporaries, William Huggins and J. Norman Lockyer in England, Hermann Vogel in Germany, Angelo Secchi in Rome, and Charles A. Young and Edward C. Pickering in America, were following in the footsteps of Gustav Kirchhoff, who had studied the strongest absorption lines in the solar spec-

Fig.1. Samuel P. Langley, Keeler's research supervisor at Mount Whitney and at Allegheny Observatory (1881). (Reproduced by kind permission of the Mary Lea Shane Archives of Lick Observatory.)

trum and recognized the signatures of the same elements as on the earth. All of the astrophysicists together numbered only a very small minority of the working astronomers in the 1870's and 1880's.

At Allegheny, Langley studied sunspots, the small dark areas on the sun first seen with a telescope by Galileo. Langley began by drawing them carefully, and his skill as a visual observer and draftsman provided some of the best representations of sunspots ever made, until the Princeton Stratoscope I balloon photographs, taken from above most of the distorting effects of the earth's atmosphere 80 years later, superseded them. In 1878, when Keeler was with Edward S. Holden's party at Central City, Langley was observing the same solar eclipse from Pike's Peak, with a group financed by the U.S. Army Signal Service. In the transparent mountain air, he was able to measure radiation from the solar corona out to twelve solar diameters, further out than anyone had previously realized it extended.[4]

From visual observations of the sun, Langley's physical interest naturally led him to try to measure quantitatively the radiation from sunspots and from the sun itself. He soon found that the radiation detectors then available were not nearly sensitive enough for his needs, so he invented the bolometer.[2] It is essentially a resistance bridge, or balanced electrical circuit, in which one of the resistors, a blackened platinum strip, is unusually sensitive to slight temperature changes. This strip is placed at the focus of the telescope, and changes in its resistance, due to heating by the radiation from the sun, moon, sunspot, planet or other object at which the telescope is pointed, result in changes in the electric current through the bridge. Thus measurement of the current, made in Langley's day with a galvanometer, a sensitive electrical meter, gives a measurement of the energy from the source falling on the telescope. Together with a spectroscope, the bolometer could be used to measure the amount of energy emitted by the source, from the visible range of colors to the invisible, long wavelength, infrared spectral region.

These studies required money for apparatus and for an assistant to carry out the measurements, and Langley was fortunate in finding a wealthy patron, William Thaw, who financed his research.[5] Thaw, born in Pittsburgh in 1818, had begun his financial career at the age of 16, traveling through the Ohio Valley on horseback, making collections for the Bank of the United States during the presidency of Andrew Jackson. He quickly grasped the importance of communications in an expanding, pioneer nation. In the 1840's, in partnership with his brother-in-law, Thaw took over the Pennsylvania and Ohio Canal Line, a company in the business of shipping merchandise by canal and river boats. By 1855 they had interests in over 150 steamboats operating on many lines, including the famous Pittsburgh and Cincinnati Packet Line. But they saw that railroads were the wave of the future and sold out their canal line that year, putting the money into coal lands and railroads running west from Pittsburgh. By 1881 Thaw was a director of the Pennsylvania Railroad, and owned huge blocks of stock in various other railroad and shipping lines. He was soon one of the richest men in Pittsburgh, with a fortune estimated at between $8 and $12 million.

Thaw became a philanthropist, who avoided publicity and ostentation while trying to build up Pittsburgh as an educational center. As a young man he had attended the Western University of Pennsylvania, and now he could afford to devote part of his time and money to supporting it. He met Langley in 1868, and soon formed a close personal relationship with him. From that time until his death in 1889, Thaw supported astronomical research at Allegheny Observatory, providing on many occasions the money to buy new instruments that Langley needed,[5] and even once personally guaranteeing his salary.[6] Thaw was one of the first of the small group of wealthy American capitalists who, for one reason or another, financed astronomical research and enabled the United States to become the world leader in observational astrophysics.

By 1880 Langley had made significant progress in measuring the solar radiation. He found that, contrary to the generally accepted opinion of the time, the absorption of the sun's rays by the earth's atmosphere decreased toward longer wavelength, that is, toward the infrared, or, as Langley called it, the "dark heat" region. Both these names mean the spectral region invisible to the human eye which nevertheless contains a good deal of the energy flux from the sun, at longer wavelengths than the red spectral region. To check this point, to measure the transmission of air in detail, and to measure the sun's radiation with as little absorption as possible, Langley decided he needed to get to a high mountaintop, above as much of the disturbing effects of the atmosphere as possible. Then, by comparing measurements of the solar radiation made at the mountaintop and at a much lower elevation, ideally at the same time, he could determine quantitatively the amount of radiation removed by the intervening air. Thus he could measure the absorption of the atmosphere and, at least in principle, correct the measurements for this absorption and thus determine the total amount of solar radiation incident at the top of the atmosphere. Aided by the advice of John Wesley Powell and Clarence King, two of the leaders of the U.S. Geological Survey, Langley picked Mount Whitney in the Sierra Nevada range in California, about 200 miles north of Los Angeles, as the place to make these measurements. At

14,495 feet it is the highest peak in the continental United States, but even more important, it rises steeply 11,000 feet from the floor of the Owens Valley, making possible measurements from two stations with this difference of altitude but only 12 miles apart. Langley also knew that the skies are practically always clear there between June and September. Thus he proposed to go to Mount Whitney in the summer of 1881 to measure the solar radiation, and when Thaw agreed to finance it, the expedition was on.[7]

During the previous winter, Langley had been a visiting professor at Johns Hopkins, where he gave a three-week series of lectures on astrophysics.[8] Keeler of course attended, and no doubt was stimulated by the new ideas and insights on the marriage of physics and astronomy. When Langley decided he needed an assistant to take along on the expedition to help with making the measurements, and reducing them afterward back at Allegheny Observatory, Henry A. Rowland and Charles S. Hastings naturally recommended their star student and laboratory assistant who was about to graduate. One of the reasons Hastings recommended him was because "it was always easy to find out what he didn't know." By this Hastings meant that Keeler did not act as if he knew everything, but was intellectually modest and willing to learn.[9] The kind of assistant Langley had in mind is indicated by his letter to Pickering, the director of Harvard College Observatory:

> I am getting into a very interesting line of work, and I want some additional assistance, perhaps temporary, possibly for some time. Do you know of any quite young man with a turn for physical research . . . who could assist me & Very and can be had cheap? . . . Let him have a touch of Faraday – a good deal of Melloni, a strong flavor of both Herschels, the best points of Humboldt (I may take him on a scientific journey) and feel these modest gifts overpaid with 5 or 600 a year! It is a hard case if I can't have at least as much as that for my money![10]

When Pickering also recommended Keeler, whom he remembered from his visit to Harvard in December 1877, Langley was convinced and hired him. Although he received special mention from President Daniel Coit Gilman at the Johns Hopkins graduation on June 6, 1881, Keeler was not there himself, for he had already begun work at Allegheny Observatory.[11] His first task was to familiarize himself with Langley's instruments and methods; soon he was packing the telescopes, mirrors, gratings, prisms, bolometers, thermometers, and all the other apparatus they would need, over five thousand pounds in all. In addition to Langley and Keeler, their party included William C. Day, his friend from Johns Hopkins, who would be returning to Baltimore as a fellow after the summer, and Captain O.E. Michaelis, of the U.S. Army, for the expedition was partly financed by the Signal Service, as part of its mission to study meteorology.[7] They left Allegheny on July 7, traveling in a private Pullman car provided by the Pennsylvania Railroad, through the courtesy of Thaw and his friend Frank Thomson, the vice president of the railroad. Langley was a well-read, highly cultured scientist, and a vivid conversationalist with his friends, but he was somewhat reserved and highly conscious of the social gulf between professor and assistant. The trip to California was a long one, and although the Isaak Walton, their private car, was comfortably furnished, they soon became weary of travel. Luckily for Keeler and Day, a former attorney general of the United States and his young niece were in another car just ahead of them in the train, and they struck up a friendship with her.[12]

On July 13 Langley's party reached San Francisco, where they met George Davidson, who helped them with last-minute preparations for the final leg of the journey to Mount Whitney. Davidson, whose career intersected Keeler's at several crucial times in the next nineteen years, was then fifty-six, nine years older than Langley. Born in England, Davidson had emigrated to Philadelphia with his parents in 1832. At its Central High School, he had studied under Alexander Bache, who guided him toward science. The school had a telescope and a magnetic observatory; Davidson became an expert with both and for three years averaged only three hours' sleep a night, devoting the rest of his time to studying, observing, and reducing data. Soon after graduation from high school Davidson went to work for the U.S. Coast Survey, of which Bache was by then superintendent. In 1850, only two years after California had been ceded to the United States by Mexico, Davidson was sent out to make a much needed survey, in which he mapped the most important harbors along the west coast, from San Diego to the mouth of the Columbia River. Except for eight years in the East, during and just after the Civil War, he spent the rest of his long life working out of San Francisco, and was in charge of the Pacific branch of the Coast and Geodetic Survey from 1868 until 1895. Davidson was keenly interested in astronomy, and had his own private observatory, with a 6-inch telescope, on Holladay's Hill (now Lafayette Square) in San Francisco, located in the Western Addition, between the downtown area and the Presidio. He usually observed every clear night, either early in the evening or before dawn, and worked a full day besides. Most of his astronomical research was concerned with accurate measurements of positions of stars, and the conclusions on the shape of the earth, the motion of the earth's axis of rotation, the constant of aberration and similar results that could be drawn from them.[13] As president of the California Academy of Sciences from 1871 until 1887, Davidson had invaluable contacts with many of the wealthy civic leaders of San Francisco, and he was personally acquainted with just about every important American astronomer, meteorologist and geodesist.[14] Langley was glad to have his help during the nine days the party spent in San Francisco, and afterward in forwarding apparatus and mail to their field headquarters.[15]

By the time the Langley expedition left San Francisco, it had grown by the addition of an escort of soldiers from the Eighth Infantry, two Signal Service sergeants, and one of Davidson's sons, called George F. Davidson to distinguish him from his father. Young George was nineteen years old, five years younger than Keeler, and had joined the party as a volunteer. The railroad took them south through the Central Valley and up into the foothills of the Tehachapi Mountains. Langley, Keeler and part of the escort left the train at the little station of Caliente, and continued their trip on horseback, 120 miles northeast up the Kern River, and across the Inyo desert to Lone Pine, their base. The rest of the group stayed with the train and crossed Tehachapi Pass to Mojave, and then brought the instruments and provisions to Lone Pine in wagons by a level but longer and slower route, taking a week to make the journey.

Lone Pine was a little mining town in the Owens Valley at the north end of Owens Lake, at an elevation of nearly four thousand feet. Mount Whitney juts up from the desert thirteen miles to the west, but the clear desert air and the abrupt rise of the mountain combine to make it seem much closer. Langley and Keeler remained in Lone Pine for two weeks, setting up their instruments and making measurements of the solar radiation. It was very hot, dry and dusty, with temperatures often as high as 110 °F inside their tents. Keeler demonstrated his ingenuity in setting up a double tent to use as a dark room so that, although the sunlight was screened out, there was sufficient ventilation to keep the inside cool enough so that they could observe effectively. Under Langley's supervision, he handled the spectrobolometer, the most important instrument.

Then the time came to push on up Mount Whitney to make the measurements above as much of the atmosphere as possible. They hired a local guide, William Crapo, loaded their instruments on mules, and headed up the trail, leaving the two sergeants in Lone Pine to continue barometric measurements for comparison with their measurements to be made on the peak. Their route went around to the west side of the mountain, from which the ascent was not so steep. By the time they reached nine thousand feet the sky was noticeably darker and more violet than at Lone Pine. Langley, who went ahead with the guides and a small escort, leaving Keeler in charge at Lone Pine with orders to follow with the pack train a few days later, reached the mountain camp which was to be their upper station on August 16. It was at 11,625 feet, just above the timberline, in a beautiful little valley with a small lake in the center of it. According to the guides, it was the best campsite in the whole Sierra Nevada. Captain Michaelis, Keeler, Day and the pack train, which included a milk cow, did not get to the camp until August 25. While waiting for them, Langley had gone up the mountain to an altitude of about thirteen thousand feet on August 19, where he found the sky a deep violet-blue color, and had continued all the way to the summit on August 22, returning each night however to the mountain camp. Everyone in the party was badly sunburned at these high altitudes, Langley himself so much that his face was scarcely recognizable. As soon as the pack train arrived with the instruments, measurements of the solar radiation were begun from the mountain camp. Originally Langley had planned to make these measurements at the summit of Mount Whitney, but when he had reached it he realized it was too cold for them to live there, with the nearest firewood at the timberline three thousand feet below.

In addition to his primary aim of measuring the solar radiation, Langley had also planned to try to photograph the solar corona from the high altitude of Mount Whitney. The corona, the faint outer envelope of the sun, can be seen only at total solar eclipses, when the direct light of the sun is blocked off by the moon. Yet the corona is always there; it is only because of the glare of scattered light from the much brighter sun that we cannot see it all the time. Many astronomers had devoted considerable effort to trying to observe the corona outside of eclipse, all without success up to that time. To observe it would require reducing the scattered sunlight as much as possible. One way to do this was to get above as much of the earth's atmosphere as possible, in a dry site with no dust or other solid particles in the air. Langley had been impressed by how far out from the sun he could see the corona in the clear air of Pike's Peak at the 1878 eclipse; he resolved to try without an eclipse on Mount Whitney. He had had a special very long focal-length telescope, with a long tube to eliminate stray light, especially made for this purpose, and had brought it along broken down into sections, with the other apparatus in the pack train. His experiments while waiting for the mules, blocking out the sun behind rocks, and observing the sky close to the sun with a small portable telescope he had carried up with him, convinced him that he had a good chance for success. He spent three days lining up and working with the telescope, which was mounted horizontally on the ground, fed by a flat mirror, or heliostat, which followed the motion of the sun in the sky.

Unfortunately, however, "just as a possibility of success seemed near, a most disheartening accident robbed us of further hope in this direction."[7] One of the soldiers of the escort, seeing a spot of dust on the especially carefully prepared heliostat mirror, took out his buckskin gloves and wiped it clean, thereby roughening the silver reflecting coating so that it scattered much more sunlight than the atmosphere, making further measurements impossible.[16] Langley's words on this occasion have not been preserved.

On September 4 Keeler and Al Johnson, one of the guides, hiked up to the summit and spent the night ("not a pleasant one") there. Keeler made barometric observations all night long, and took bearings to establish their exact geographical position. They returned to the mountain camp on September 5, and Langley and Keeler continued their spectrobolo-metric measurements of the sun. On September 9 they discovered the great infrared absorption feature in the solar spectrum, called Ω by Langley. By now forest fires were multiplying in the distance all around them, the skies were becoming less astronomically usable, and the air definitely colder. They broke camp and came all the way down to Lone Pine on September 11, the day after Keeler's twenty-fourth birthday. Taking the direct route to the east from their camp, they first ascended slightly for two hours, past snowy cliffs and frozen lakes in the northern shadow of the mountain. They then came out through a gap in the ridge so narrow that only one person could pass through it at a time, and saw spread out, far below them, the desert and the bright green spots that marked its oases. Continuing downward, they passed deep-blue lakes and icy streams. After getting below the snowline, they passed through char-red and burning forests, and reached the desert floor by dark. Day had sprained his ankle on the way down, so they left him with all their coats to keep warm, intending to ride back for him the next day, and pushed on to Lone Pine. They reached it at midnight, seventeen hours after leaving their mountain camp. The next morning, Day hobbled in on his own before sunrise. The pack train arrived a few days later, and they were soon in the Isaak Walton again, on their way to San Francisco, where they dropped young George, and then back to Pittsburgh, where they arrived just before the end of September.[7]

Keeler enjoyed the expedition tremendously. It was an unsurpassable chance to do science in strange, adventurous surroundings. He and Day had climbed out on a peak, just off the trail south of Mount Whitney, so steep that they could drop rocks that would fall over half a mile before they struck anything.[17] Langley named this peak "Keeler's Needle" for his young assistant; another nearby peak was later named Day Needle, and now there are in the vicinity a whole galaxy of peaks named for astronomers, including Mount Langley, Mount Barnard, Mount Hale, and Mount Young.[18] Keeler enjoyed the scorching heat of the sun on Mount Whitney, together with the near freezing temperature in the shade, the deep violet of the sky, and the evenings spent around the campfire, listening to the stories of the old Californians who visited their camp.[19] He remembered the physical aspects of the trip all the rest of his life.[20] Keeler proved himself extremely capable on this expedition, a resourceful and imaginative scientist, and Langley reported that his efforts had meant the difference between success and failure on more than one occasion.[7]

Back at Allegheny, Keeler began reducing the data from Mount Whitney and making new measurements to supplement and extend them. He was expected to work all day, six days a week, with only Sunday off.[21] Keeler worked closely with Frank W. Very, Langley's other assistant, a Massachusetts Institute of Technology graduate five years older than himself. Very was a fussy, excitable individual, [22] who had been working with Langley since 1878, handled much of the time-service routine at Allegheny, and had been left in charge of the observatory when Langley went to Mount Whitney. Very taught Keeler the time-service work and Langley's reduction methods; Keeler did most of the laboratory and astronomical measurements under Langley's direction.

One of the important problems they had to solve in order to use their data from Mount Whitney was to determine a wavelength scale in the infrared spectral region. Spectroscopes break up light into its constituent colors or wavelengths, or technically speaking disperse the light into a spectrum, in either of two ways. Prism spectroscopes use prisms, which refract or bend the light, the amount of refraction depending on the wavelength. Grating spectroscopes use reflection gratings, arrays of equally spaced parallel lines ruled on reflecting surfaces, which diffract the light, the amount of diffraction depending on the wavelength. In Keeler's time, gratings could not be made as efficient as prisms in concentrating faint light, but they have the advantage that the wavelength of the radiation diffracted in a particular direction can be directly calculated, and depends only on the spacing between the rulings on the grating. Furthermore, a grating diffracts the light into several overlapping spectra, or orders, which are related by simple mathematical relationships, so that radiation in the first-order spectrum at wavelength 20,000 Å (in the far infrared) is diffracted by exactly the same angle as radiation in the second-order spectrum at 10,000 Å (in the shorter-wavelength infrared), which also is diffracted by exactly the same angle as radiation in the third-order spectrum at 6,667 Å (in the visible red spectral region), radiation in the fourth-order spectrum at 5,000 Å (the visible green spectral region), the fifth-order spectrum at 4,000 Å (the just barely visible violet spectral region), the sixth-order spectrum at 3,333 Å (the invisible ultraviolet spectral region), and so on. Langley had a large concave (focussing) grating, especially made for him by Rowland at Johns Hopkins, which he and Keeler used as a monochromator, with the sun as the source of radiation.

With this grating they could look at the visible radiation from the sun and recognize the well-known absorption lines (the wavelengths at which the solar spectrum is dark) such as the sodium D line at 5,893 Å in the third order. Allowing this radiation to shine into their prism spectroscope, they knew they were also admitting radiation at 11,786 Å in the second order and at 17,679 Å in the first order. They then inserted into the beam of sunlight a filter that cut out all the shorter-wavelength radiation, and transmitted only the longer-wavelength infrared radiation. In this way they knew that the only radiation in the spectrum formed by the prism spectroscope would be at 17,679 Å, and by moving the bolometer until they detected this radiation, they could measure the angle of refraction of the prism, or technically, its index of refraction for this wavelength. Repeating the process at many wavelengths, they could build up the curve of index of refraction as a function of wavelength for the whole infrared spectral region, and thus map the solar spectrum. Langley stated that he considered Keeler more a collaborator than an assistant in this phase of the work, for he relied heavily on his laboratory skills.[23] He wrote his old friend Holden, now director of the University of Wisconsin's Washburn Observatory, who remembered Keeler from the Colorado eclipse expedition, that he was very pleased with his work and was raising his salary. "I think I must keep Keeler and take care that he doesn't lose by my keeping him. He is very good. . . I shall be glad to keep him here for the present."[24]

Although the Ω band at about 19,000 Å or 1.9 μm was the first detected by Langley and Keeler at Mount Whitney, most of their measurements of the solar spectrum there had only extended out to 1.2 μm. However, with this experience, after they returned to Allegheny they found they could improve the sensitivity of their measurements so that in spite of the greater thickness of the atmosphere above them, they were able to push their measurements further into the infrared, out to 2.7 μm where the solar radiation is very faint. These measurements greatly extended the known spectrum of the sun. It had previously been thought that the transmission of the atmosphere decreased at longer wavelengths. Langley showed, however, by comparison of the Mount Whitney, Lone Pine, and Allegheny measurements, that in fact the transmission increases rather than decreases toward the infrared in the spectral regions between the bands, but that the absorption bands become closer together, and wider, and finally obliterate nearly all the solar radiation beyond 2.8 μm. Langley correctly deduced from the way in which they strengthen at lower altitudes that these absorption bands arise in the earth's atmosphere; we know today that all the major bands which Langley and Keeler observed in the infrared are due to water vapor.

One of the most important quantities Langley and Keeler determined from their Mount Whitney measurements was the "solar constant," the amount of solar energy falling on the top of the earth's atmosphere. The final result they found, after all the reductions were completed, was 3 calories per square centimeter per minute, with a fairly large uncertainty. This value, we now know, is somewhat too large (the correct value is 1.95 calories per square centimeter per minute), mostly as a result of an error in the reduction procedure they used. Nevertheless it was a very good first attempt, and the Mount Whitney result was used in atmospheric studies for many years.

Working at Allegheny Observatory, Keeler soon formed a firm friendship with John A. Brashear, who made nearly all of Langley's astronomical instruments for him. Brashear, six years younger than Langley and seventeen years older than Keeler, was a self-taught optician who loved astronomy and the stars. He had been trained as a machinist and had worked in the Pittsburgh steel mills, but his interest in astronomy led him to make a 5-inch refracting telescope in the early 1870's. He met Langley at Allegheny Observatory in 1876 and showed him the telescope. Langley inspected it and found it very good, but told Brashear about the advantages of reflecting telescopes, in which a parabolic mirror rather than a lens brings the light to a focus. Brashear, assisted by his wife, whom he always called "Ma", then made a 12-inch reflector, and Langley realized he had a find on his hands. He told his wealthy patron about Brashear, and Thaw provided the funds that enabled the budding optician to set up his own shop, in which he made telescopes, spectroscopes and other instruments to order for amateur and professional astronomers and other scientists. Brashear was a hard worker who produced many excellent instruments, but he was never a good businessman, largely because he could not bring himself to charge astronomers realistic prices for the custom jobs they brought him. Thaw, recognizing the importance of first-class optical work to astronomical research in general and Langley's in particular, subsidized Brashear's business for many years and bailed him out financially on several crucial occasions.

After he had demonstrated his competence and all-around skill with optical instruments, Brashear formed a business arrangement with Rowland to manufacture gratings for spectroscopy. To give good spectra and get the most out of the available light, a grating must be ruled with all its grooves very accurately parallel, and exactly evenly spaced. The larger the grating, the more efficient it is, but at the same time the harder it is to make. Rowland had developed in his laboratory at Johns Hopkins, a machine with which he could make gratings far superior to any previously known. Typically, one of his gratings would be several inches in diameter, ruled with 14,436 grooves per inch. These gratings revolutionized spectroscopy, and physicists everywhere were clamoring for them. Rowland could not spend his valuable research time on production, but he hired a mechanic who operated the machine, and Brashear supplied the accurately flat speculum-metal blanks on which the gratings were ruled. Rowland then returned the finished gratings to Brashear, who distributed them to the scientists who could use and pay for them. In the course of the years, several thousand Rowland gratings were made and distributed in this way.[5]

A short, peppery individual, Brashear loved to talk, and would expound at length on science, astronomical instruments, his good fortune in life, famous astronomers he had known, and similar subjects at the drop of a hat. He was always happy to let casual visitors who came by his house at night see stars through his telescope, no matter what he was doing.[5] Brashear recognized and appreciated good workmanship, and when he met Keeler for the first time at Allegheny Observatory, and saw his toolbox, which the younger man had made himself, and the fine set of tools he had brought with him from Johns Hopkins, he knew he had met a kindred soul. Brashear admired Keeler's scientific knowledge, his practical skill, and his craftsmanship; they met frequently at Brashear's shop and at the observatory and soon were close friends. Their friendship endured for the rest of Keeler's life, and Brashear made many of the instruments that Keeler designed and used throughout his career.[25]

Working with Langley, Keeler was getting excellent practical training, much like a modern postdoctoral research associate. But he wanted to go back to Johns Hopkins as a graduate student, and in the spring of 1882 applied for a fellowship, giving Langley's name as a reference.[26] Keeler was awarded the fellowship, but Langley, anxious to keep his capable young assistant, proposed an alternative plan. If Keeler would stay at Allegheny for another year, Langley would arrange for his patron Thaw to pay Keeler's expenses to study in Germany, then the world center of science, the following year.[27] Keeler accepted this offer and politely declined the Johns Hopkins fellowship, writing however that he hoped to return there after his year in Germany.[28]

In the summer of 1882 Keeler got his first vacation from Allegheny, going home to Mayport with David T. Day, now as Lizzie's suitor considered one of "our boys". Soon after their departure at the end of summer two years before, Keeler's father had hurt himself while forging a drill in his shop, blinding one of his eyes. Although he was laid up for a month at that time, the Major had returned to work and could handle most jobs himself, but he was glad to have Ed and Dave around again to help him, if only briefly. Ed headed back to Pittsburgh at the end of August, and Dave returned to Johns Hopkins, where he was now a graduate student in chemistry and geology.[29]

At Allegheny, Keeler continued measurements and reductions of the solar spectrum under Langley's direction, but occasionally there was a welcome break in the routine. On December 6, 1882 a rare transit of Venus occurred, in which the planet crossed the disk of the sun as seen from the earth. Accurate observations of such a transit, made at several stations on the earth, may be used to measure one of the fundamental quantities of astronomy, the distance from the sun to the earth. Hence, all over the world, astronomers were poised to observe it, including Langley, Keeler and Very at Allegheny Observatory. Langley himself observed with the 13-inch refractor, the main telescope of the observatory, stopped down to 6 inches to help cut down the light of the sun, while Keeler and Very used smaller portable telescopes set up outside, near the dome. To measure the distance to the sun, very accurate timing of the moments of first contact, when the disk of the planet first touches the disk of the sun, second contact, when the disk of the planet just gets completely inside the disk of the sun, and third and fourth contacts, the corresponding moments when the planet leaves the disk of the sun, are required. Most astronomers who observed the transit concentrated all their efforts on these measurements of the times of contact. Holden and S.W. Burnham, two astronomers of the old school, whom Keeler was to join on the staff of the Lick Observatory a few years later, had made the long journey to Mount Hamilton, the remote California site chosen for the observatory, to observe a transit of Mercury in this way the previous year.[30] Langley, however, had more astrophysical concerns; he and his assistant were not only timing the transit of Venus but were also looking for signs of its atmosphere, which they hoped would be revealed by sunlight shining through it.

Most of the eastern United States was under clouds at the time of the transit, and at Pittsburgh a rainstorm the night before made the prospects for observing poor. But it cleared before daylight at Allegheny, and they saw the transit. Langley and Very were able to keep their professional images intact, but the youthful Keeler cried out, "There she is" when he saw Venus, up to then invisible, touch the sun's disk just at the predicted moment of first contact.[31] As they watched Venus move onto the sun, Langley and Keeler saw a brighter ring around the dark disk of the planet, caused by the refraction of sunlight in its atmosphere. Furthermore, they both saw that the brightest part of this ring was not opposite the center of the sun, as would be expected from symmetry, but was displaced by about twenty-five degrees from the line between the center of the sun and the center of the planet. Both of them sketched and measured this bright spot, and although their observations were also independently confirmed by Brashear, who was observing with his own telescope at his shop, evidently Very did not see it. Langley allowed Keeler to publish his own observations independently; this was his first published paper from Allegheny Observatory.[32] He wrote Hastings, his Johns Hopkins professor, about the observation, and the optics expert provided a theoretical interpretation of the ring of light in terms of refraction, and showed that the asymmetry must be due to clouds deep in the atmosphere of Venus.[33] This indirect evidence for clouds has been abundantly confirmed by space probes of Venus in our own time. In contrast to most of the astronomers of their era, Langley and Keeler were prepared to make astrophysical observations, and when an unexpected phenomenon occurred, they were able to see it, record it, understand its importance, and get the data to a theoretician who had the physical and mathematical tools at his command to interpret it. They learned something new about Venus; this is the purpose of science.

In the following spring after a few months more solar observing, Keeler finally set out for his year in Germany. Langley felt that Keeler's "remarkable skill" had been invaluable to his researches, and said he did not know where he could find another such assistant.[21] Keeler sailed from Baltimore on the North German Lloyd steamship *Braunschweig*, and got a chance to brush up his German en route.[34] In Bremen he enjoyed the picturesque parks and lakes, the old town, and the moat. He was impressed by the theatre, although in his opinion it did not measure up to the Academy of Music in Baltimore. He was amused by the sight of elegant ladies and uniformed men drinking beer in public, and by seeing dogs harnessed to pull carts. Then he went on to Berlin for a week's visit, and saw the old German Emperor Wilhelm I, and the Royal Museum, which he admitted was better than anything in Baltimore.[35]

By early June Keeler was in Heidelberg, where he attended the lectures of three well-known German professors, Georg Quincke on the wave theory of light, Robert Bunsen on chemistry, and Lazarus Fuchs on calculus.[36] Quincke's ideas on physics were firmly rooted in the past; he admired Faraday's experimental methods and ideas but never accepted their mathematical reformulation by Maxwell. Quincke viewed light as elastic waves in a mechanical medium, the ether, with a whole set of properties which made it unobservable except as a medium for transmitting light waves. Keeler, along with most physicists of his generation, learned to think in these terms, which satisfactorily accounted for nearly all the observed properties of light, rather than in terms of Maxwell's equations, as we do today. Bunsen, the chemist who invented the Bunsen burner, was a pioneer in spectroscopic analysis, and he and Gustav Kirchhoff had made the first spectroscopic discovery of an element, cesium, in 1860. Keeler, already well trained in spectroscopy by Hastings and Langley, undoubtedly learned many new techniques and ideas from Bunsen, in spite of the fact that his research laboratory was primitive by Johns Hopkins standards.[35] Fuchs was an impressive mathematician, who rarely prepared his lectures, but produced on the spot the material he wanted to teach. Keeler and the other students thus saw a first-class mathematical mind in actual operation.[37]

At Heidelberg Keeler lived in a room on one of the main roads leading up the hill to the castle, through which he loved to roam, savoring the pleasant paths and the picturesque towers, the winding staircases and the magnificent views over the Necker River. He reported to his Baltimore friends that in Heidelberg water stood in bottles on the tables in restaurants, untouched for weeks, Rhine wine was weak and sour and could be distinguished from vinegar only by its label, but that high quality beer cost less than two cents a glass. Life was enjoyable in the golden days of Heidelberg, but there were serious moments too, as when Keeler called on Gilman, the president of Johns Hopkins, who was in Germany collecting information on medical schools.[38] Keeler reported on the progress of his studies, and no doubt renewed his request for a graduate fellowship at some later date.

After the summer semester ended, Keeler spent one more month studying in Heidelberg, and then went off on a trip through Germany, Switzerland and Italy.[39] He visited Berlin briefly and saw the edge of the Black Forest, then continued to Strassburg and its inspiring cathedral. Traveling always by train, he went on to Zurich, where the Swiss National

Exposition was in progress. He inspected the collection of scientific instruments, all made in Switzerland, and browsed through the exhibits of wood carvings, gold and silver work, and embroidery.[40] He visited Rome and Naples, and climbed to the very top of Mount Vesuvius, where he took shelter from a sudden rainstorm in a small crater, walked on barely solid lava, and finally looked down into the clouds of sulfur smoke mixed with dust and rocks coming up from below.[41] This was the culmination of the trip. A few days later he was back in Berlin, ready for the start of the winter semester.

There he attended the lectures of Hermann von Helmholtz, the patriarch of German science and its government's chief scientific adviser. Originally trained as a medical doctor, Helmholtz' first work in physics had been in physiological optics and sound, but he soon became an expert in electrodynamics. He had accepted Maxwell's theory of electromagnetic fields and lectured in terms of them, rather than the ether. Keeler also attended the lectures of the young docents Paul Glan, an expert on optics; Heinrich Kayser, Helmholtz' assistant; and Carl Runge, then a mathematician lecturing on differential equations.[42] Several other American students were in Berlin that winter, including Henry Crew, who attended Helmholtz' and Kayser's lectures with Keeler. As a boy in Ohio, Crew received a classical education, emphasizing Latin and Greek, and even more of the same as a physics student at Princeton. In contrast to Keeler, he understood little German, but nevertheless was able to grasp nearly everything Kayser said from his clear diagrams and gestures.[43] Kayser had just completed his textbook on spectrum analysis; a few years later he and Runge moved to Hannover where they systematically studied and analyzed the spectra of the elements.[44] The methods, techniques and ideas Keeler learned in Germany were invaluable to him in his later career.

He was not alone in this; many of the best American scientists of his generation received their graduate training in Germany. For instance, Keeler was following almost exactly in the footsteps of Albert A. Michelson, an Annapolis graduate who was five years his senior. Michelson, in preparation for a career switch from naval officer to research physicist, had attended the lectures of Helmholtz, Quincke and Bunsen in Berlin and Heidelberg in 1880-81, before returning to the United States to accept a faculty appointment at the Case Institute in Cleveland. Twenty-five years later he became the first American to earn the Nobel Prize, for his outstanding experimental work on light.[45]

Keeler concentrated almost entirely on physics, and seldom saw an astronomer or visited even the famous Potsdam Observatory.[46] Besides attending lectures in Berlin, Keeler wanted to do experimental work. He had brought with him a bolometer lent by Langley,[47] and a large reflection grating that Rowland had given him.[5] With these credentials he had no difficulty securing laboratory space in Helmholtz' well-equipped Physical Institute.[39] The research Keeler did was on the absorption of infrared radiation by carbon dioxide, one of the gases in the earth's atmosphere. This problem was undoubtedly suggested by his work with Langley. The experimental layout Keeler devised and the construction of the apparatus were excellent, but the paper that resulted is not very profound. Building and assembling the absorption tube and the associated optics took longer than he had expected, and he was not left with enough time to make the spectroscopic measurements with the grating as he had planned. Instead he had to content himself with measuring the absorption in broad spectral regions, defined by four different light sources. Even with this crude method of spectral discrimination Keeler was able to conclude that the absorption of carbon dioxide is strong in the infrared, that it must be concentrated in one or more broad, strong bands, and that the amount of absorption does not increase uniformly as the amount of carbon dioxide increases, a saturation (or as astronomers call it, a curve-of-growth) effect. He considered it probable that one or more of the infrared bands that Langley discovered in the solar spectrum were due to carbon dioxide (actually this is not correct; all of the bands Langley saw are water-vapor bands), but correctly concluded that another gas must also be involved in the atmospheric absorption.[47]

After writing up the results of his experimental work on carbon dioxide in Berlin and doing a bit more sightseeing, Keeler sailed back to America from Bremen, arriving in New York on June 14, 1884.[36] Always the diplomat, he had sent ahead some thin platinum sheets, made by a new process in Germany, which Langley hoped to use to make even more sensitive bolometers.[48] Keeler stopped in Pittsburgh to report to Langley and Thaw;[49] then a few weeks later he was home in Florida for his first visit in nearly two years, bringing with him a close friend, Miss Jeannie Leamy of Baltimore.[29]

A gay round of suppers, picnics and walks on the beach soon began, and in August David Day, who by now had completed his Ph.D. at Johns Hopkins and was working for the U.S. Geological Survey, arrived to squire Lizzie. But in mid-August Keeler had to break away and go back north to Allegheny, where he again went to work as Langley's assistant. It is impossible to tell now, from the scanty available evidence, whether there was a romantic attachment between Keeler and Jeannie Leamy, although it seems very likely. At any rate, it is certain that in December she married William Day in Baltimore; David was his best man, and Keeler was one of the ushers.[50]

By now Langley was measuring the radiation from the moon and comparing it with the sun; Keeler's main job was to work with Very in making and reducing these measurements. Langley had greatly improved the sensitivity of the bolometer, but even so the moon is much fainter than the sun, and the measurements were difficult with the tricky, temperamental instrument.[51] Since glass lenses absorb large amounts of infrared radiation, Langley and his assistants instead used a fixed horizontal reflecting telescope, fed by a flat mirror, to form a concentrated image of the moon. Various filters, diaphragms and screens could be put into the light beam to compare the lunar and solar radiation. In addition to these infrared measurements, they also made visual comparisons of the sunlight and moonlight, using another, smaller reflecting telescope, together with a grating to disperse the light. In this way they found that the moon is redder than the sun, that is, that the moon's reflectivity increases toward longer wavelengths. The infrared measurements were calibrated by comparison measurements of an infrared source that could be filled with water at various temperatures. Although the properties of radiation were not known in anything like their modern form, Langley realized that the hotter an object is, the shorter the wavelength is at which its emitted spectrum is brightest, while conversely, the cooler the object is, the longer the wavelength is at which it is brightest. Using this principle, Langley was trying to determine the temperature of the surface of the sunlit side of the moon. He found that the moon's temperature is definitely below the boiling point of water, 212 °F or 100 °C, and close to the freezing point, 32 °F or 0 °C, somewhat low (the modern value is about 93 °C) but a very good attempt for his time.[52] He did not realize clearly that at these temperatures the wavelengths at which the spectra are brightest are so far in the infrared that they are completely blocked by the absorption in the earth's atmosphere, and thus the maximum of the lunar radiation cannot be observed from the surface of the earth. Nevertheless, the results were far superior to any previous quantitative measurement.

The lunar measurements did not take up all Keeler's time, and he was involved in various other smaller projects. On September 26, 1885, a very bright meteor was seen in the daytime, at 4 o'clock in the afternoon, in western Pennsylvania not far from Pittsburgh. A loud crash was heard, and many people reported that the object must have struck the earth just over the horizon from wherever they were. Langley sent Keeler to the scene to investigate, and from the reports of eyewitnesses he was able to reconstruct the meteor's path and conclude that it had exploded at an altitude of about two miles, approximately twelve to fifteen miles south of Independence, Pennsylvania, and that no fragments had been found on the ground.[53] Keeler also published short notes on his observations of the unusual color of the eclipsed moon on September 23, 1885, and on a method he had developed for repolishing rock-salt prisms, used for infrared spectroscopy.[54] Neither paper is very important, but they show that Keeler was becoming restless and wanted to do independent research work on his own. His prospects at Allegheny were dim. Like nearly all observatories and astronomy departments in those days, it was a one-professor operation, in which Langley initiated all the research, supervised it closely, participated actively in every phase of it, and published the results as his own. Very, who was older than Keeler and had worked for Langley longer, was the first assistant; he was the one left in charge when Langley went away. Keeler considered returning to Johns Hopkins to complete his graduate studies.[55] He planned to get a job in the Coast and Geodetic Survey or in the Signal Service, the main governmental scientific bureaus, which offered both travel and more chance for advancement.[56] But in the end he was saved for astronomy by a job offer from Lick Observatory in far away California.[57]

3

I could not ask for anything better

Lick Observatory was built with money from the fortune of a wealthy man, a pattern that was followed later at Yerkes Observatory, Mount Wilson Observatory, Palomar Observatory, and McDonald Observatory, the great American research observatories built up to the time of World War II. James Lick, the donor of Lick Observatory, was born in Stumpstown, Pennsylvania (later renamed Fredericksburg) in 1796. He was the oldest of seven children of a poor farmer who also worked as a carpenter, and young James learned the wood-working trade from his father. At the age of twenty-one James Lick got a local girl, the daughter of a prosperous miller, pregnant. He offered to marry her, but her father refused to consider the match, thinking Lick too poor and socially too far beneath his daughter.

A few months after the child was born, Lick left Stumpstown to seek his fortune, resolved to return owning a mill so big it would make his would-be father-in-law's "look like a pigsty". After two years of bare survival as a furniture and piano maker in Alexandria, Virginia and in Baltimore, Lick sailed south to Argentina in 1821 and started a piano-making business of his own in Buenos Aires. Here, in a rich land with little competition, he was wildly successful. The wealthy people begged for the chance to buy his pianos. After twelve years in Buenos Aires he moved on to Valparaiso, Chile for a few years, and then to Lima, Peru, everywhere respected and successful as a master builder of fine pianos. Lick had always planned to return to the United States some day; he decided to make the move after the Mexican War and the Bear Flag rebellion in California. He sold his business in Peru, and on January 7, 1848 sailed through the Golden Gate and landed in San Francisco, with $30,000 in gold in his strongbox. He was fifty-one years old, tall, healthy and confident. San Francisco was a little town of about a thousand inhabitants, clustered in a few blocks near the waterfront. Lick decided there was money to be made in real estate, and began a new career that within a few years made him one of the richest men in the rapidly growing city.

His first purchase was a lot at Montgomery and Pacific Streets, which he bought for $270. Many of the lots he bought went for $16 or one ounce of gold, which were then and for many years afterward equivalent. Only seventeen days after Lick had arrived in San Francisco, gold was discovered at Sutter's mill on the American River. Soon the Gold Rush was on, and nearly everyone in San Francisco headed for the gold fields. They needed money for tools and supplies; Lick, with a large supply of ready cash, was glad to oblige them by buying up their downtown land at cheap prices. He skillfully pyramided his investments, and his knowledge of Spanish and of Latin-American customs, gained in South America, stood him in good stead. He branched out into property in and around San Jose, at the south end of San Francisco Bay. He bought a millsite in Alviso, near San Jose, and although his Stumpstown girl friend had long since married another, he built a beautiful cedar and mahogany mill, known to the locals as "Lick's Folly," far finer than any mill in Pennsylvania.[1] On this site he built his "Mansion," a large wooden, colonial style house, surrounded by beautiful plantings of pepper, olive, palm and fruit trees. It still stands to this day, now in use as the office of a real estate developer.

Lick went further afield and bought 50,000 acres of land on Santa Catalina Island, off southern California, for a little over $27,000. He held it tenaciously, like nearly all his other properties; after his death his executors sold it for $250,000. Everything Lick touched turned to money. During the Civil War, he built the Lick House, California's first great luxury hotel, on Montgomery Street in downtown San Francisco. The dining room of the Lick House, with its great arched ceiling and parquetry floor, was the showplace of the West. Paintings of California scenes in 1849 – the Golden Gate, Yosemite Falls, South Dome and El Capitan, Mount Shasta, the Redwood Forest, the Russian River Valley – mounted in rosewood frames designed by Lick, lined the walls. He himself lived in a small back room on the third floor, where he kept his books, his tools and his workbench. He wore old clothes, had a nonexistent social life, and was generally considered an eccentric miser.[1]

As he grew older, Lick, immensely wealthy, single and with no legitimate heir, decided he wanted to leave a monument to himself. His first will had as his main bequest a million dollars for statues of himself and his parents,[2] and another of his early ideas was to build a pyramid, larger than the Great Pyramid in Egypt, at Fourth and Market Streets in downtown San Francisco.[3] Eventually he was persuaded to build an observatory instead. Lick's earliest recorded contact with astronomy occurred when he was sixty-four years old and met George Madeira, a young amateur astronomer and itinerant lecturer. According to Madeira, Lick first heard him speak in San Jose in 1860 and invited him to his ranch, where he stayed several days, answered the old man's questions about the stars, and showed him the small telescope he owned. He claimed he described to Lick the largest telescopes of the time, and the discoveries that had been made with them. Some years later, when they met again in San Francisco, Madeira told Lick that if he had his money, he would use it to construct an even larger telescope, and thus planted the seed of an idea in his mind.[4]

In 1871, when Joseph Henry, secretary of the Smithsonian Institution in Washington and president of the National Academy of Sciences, visited San Francisco, he stayed at the Lick House. He made it a point to meet Lick and took the opportunity to tell him of the needs of science, and of how the wealthy Englishman, James Smithson, had perpetuated his name in the Smithsonian Institution.[1] But the person more responsible than anyone else for turning Lick's thoughts toward giving money for an observatory was George Davidson, the pioneer West Coast scientist who helped Samuel P. Langley and James E. Keeler with their local arrangements for the Mount Whitney expedition eight years later. Davidson first met Lick in February 1873, when he called on him in his cluttered room in the Lick House, to thank him for his unexpected and unsolicited gift of a lot on Market Street to the California Academy of Sciences, of which the geodesist was president. Lick was interested in what Davidson had to say, and asked him to return.[5] Two months later Lick suffered a stroke, and for almost a year afterward was confined to his room.[1] Davidson came often to see the old man, propped up on his bed in the hotel, and they had long talks, on science, astronomy, the rings of Saturn, the belts of Jupiter, the mountains and craters on the moon, and similar topics. As a result of his growing interest in astronomy, Lick decided to abandon the pyramid scheme and instead provide funds for a telescope "superior to and more powerful than any telescope yet made." Originally he planned to erect it at Fourth and Market Streets, and to surround it with three groups of statues, honoring Francis Scott Key, Thomas Paine, and himself (or possibly his father and mother instead). Davidson, who had traveled all over the West on his mapping expeditions, was convinced of the superior qualities of high mountain sites, with their clear, thin air, for astronomical observing. Eventually he managed to persuade Lick, and on October 20, 1873, Davidson announced to the California Academy of Sciences that James Lick had authorized him to state that he would build an astronomical observatory in the Sierra Nevada at an elevation of ten thousand feet.[5] In his first enthusiasm, Lick was willing to spend a million dollars to build the observatory.[6]

As he recuperated from his stroke, the wealthy old man laid his plans. He wrote to Alvan Clark, the famous Massachusetts telescope maker, to ask for his advice on how large a telescope he could build, and what it would cost.[7] Clark, only a few years younger than Lick, had grown up on a farm and worked as an engraver and a commercial portrait artist for the first forty years of his life. But when one of his sons, an engineering student at Andover, tried to make a small telescope, Clark became interested and was soon helping the boy, experimenting on polishing glass, and learning as much as he could about optics. A Harvard professor who saw the resulting telescope encouraged them to go into business, and the firm of Alvan Clark and Sons was born. Although their first telescope was a reflector, they quickly abandoned this design and concentrated entirely on refracting telescopes. They made many small 6- and 8-inch refractors which had excellent optical properties. Clark and his son Alvan G. Clark had the touch that enabled them, using their empirical methods, to polish the glass to just the right shape to bring the light from a star to a near-perfect focus. Their fame spread, and orders rolled in. They began making 12-inch lenses, then repeating the same optical design over and over again, a series of larger and larger telescopes – a 15 ½-inch refractor for the University of Wisconsin, an 18 ½-inch originally ordered by the University of Mississippi and completed just before the Civil War, which actually went to Chicago and later to Northwestern University, and then a 26-inch for the Naval Observatory, which Clark's sons were erecting in Washington when Lick's letter came.[8]

Clark wrote Lick that he wanted to see how the 26-inch Naval Observatory telescope performed in practice before he committed himself to just how large a telescope he could make next. He warned Lick that a mountaintop site should be tested by a competent astronomer before it was definitely chosen, and also told him that as he was nearly seventy years old "what I do I must do quickly."[7] The Naval Observatory telescope turned out to be a good instrument, and once it had been tested the Clarks wrote Lick that they would build him as large a telescope as they could get glass for, mentioning 35 inches as a possibility.[9] On a trip East, Davidson consulted with astronomers at Harvard and the Naval Observatory to find the cost of an observatory such as Lick wanted, and came back with an estimate of $1,500,000. This was more than Lick expected; by now he had planned to spend at most $500,000, but Davidson convinced him to up the figure to $700,000.[5]

In the summer of 1874 Lick signed and published a deed of trust, setting up a board of trustees consisting of seven powerful San Francisco businessmen and bankers, including the mayor and one of his predecessors in office. They were directed first of all to spend $700,000 of Lick's fortune to build an observatory with a telescope "superior to and more powerful than any telescope yet made" at a site on Lake Tahoe near the Nevada border; this was soon changed to a site to be chosen by Lick. Various other sums were set aside to build a school of mechanical arts, a public bath and an old ladies' home, to erect statues, and to support several existing orphanages and societies. The residue of Lick's estate was to be divided equally between the California Academy of Sciences and the Society of California Pioneers.

The trustees took their responsibilities seriously, and they were worried by the fact that Lick wished to leave only $3,000 to his illegitimate son John H. Lick, and even less to his nephew James W. Lick, both of whom had worked for him in California for some years.[1] The trustees feared that one or the other of them would bring suit and break the trust, but Lick refused to increase their bequests, for he was adamant that they should not profit from his death.[3] He refused to acknowledge that John H. Lick was his son.[10] He became more and more impatient with the trustees he had appointed and hired a lawyer to maneuver them into resigning; Lick then drew up a second deed of trust in September 1875. Captain Richard S. Floyd was appointed president of the new board of trust, Bernard D. Murphy, mayor of San Jose, was one of the members, and John H. Lick was also named but did not accept the appointment and never served on the board.[1] A provision was added that the telescope and observatory, when completed, be handed over to the University of California. Davidson left on a two-year trip around the world in the summer of 1874, and while he was gone Lick decided to give up the Lake Tahoe idea, considered various other sites, and eventually decided on Mount Hamilton, a 4,200-foot peak in the Coast Range just east of San Jose. He announced he would place the observatory there if Santa Clara County, where it is located, would build a road to the top. The county supervisors quickly approved the plan unanimously, and the site decision was settled.[1] On his return, Davidson, disappointed that a higher mountain had not been chosen and that Lick had not provided what he considered adequate funding for the observatory, withdrew from the project and refused to have any further conversations with the old man.[5]

Lick, still unhappy with the slow pace of his trustees and under pressure from elderly members of the Society of California Pioneers and the California Academy of Sciences, who wanted to get the trust wound up so they could divide the residual estate, discharged his second board on September 2, 1876, and appointed a new third board of trust. Captain Floyd was kept on as president; the other members were all new: George Schoenwald, the manager of the Lick House, William Sherman, assistant United States treasurer for San Francisco, Charles Plum, a real estate appraiser, and Edwin B. Mastick, an attorney. It was a far less prestigious group than the first board, but Lick hoped he would get action out of them. To counteract rumors that he was insane, Lick had himself examined by nine doctors who all agreed that his mind was perfectly sound. But the eighty-year-old Lick's strength gradually ebbed away, and on October 1, 1876 he died.[1]

Even while Lick was still alive, his successive boards of trust had worked hard to get the telescope started. Soon after the first board was appointed, Trustee D.O. Mills, one of the richest men in California, went East to discuss the whole project with Simon Newcomb, the head of the Nautical Almanac Office in Washington, and the outstanding American astronomer of his time. Although born in Nova Scotia where his father was a country school teacher, Newcomb was descended from a long line of New Englanders. As a boy, he was largely self-taught, and soon showed he had a very good head for mathematics. At the age of twenty-two he went to work as a computer in the Nautical Almanac Office, without any previous university training. The office was then in Cambridge, Massachusetts, and there in his free hours, he studied the works of the great masters of celestial mechanics, the mathematical predictions of the detailed motions of the planets, satellites, comets and asteroids, and soon was an expert in the field. In 1861 he went to the Naval Observatory, where he worked for several years observing with a transit and mural circle, small specialized telescopes for measuring very accurately the positions of stars in the sky. When the Clarks completed the "great" Naval Observatory 26-inch refractor in 1873, Newcomb was placed in charge of it, and as it was the largest telescope in America, and the largest refractor in the world, it was natural for Mills to seek his advice in 1874. Actually, however, Newcomb, although an outstanding theoretical astronomer, was by no means an expert observer, and it was only after he had gone back to computational work and Asaph Hall took over the 26-inch in 1875 that important new results began to flow from it.[11] Two years later Newcomb became head of the Nautical Almanac Office, by then moved from Cambridge to Washington, where he had a long and distinguished career, improving the calculations of the exact motions of the planets and their satellites.[12]

At their meetings in the summer and fall of 1874, Mills and Newcomb discussed plans for Lick Observatory. Newcomb included in the meetings his young assistant, Edward S. Holden, who was soon taking an active part in their discussions. A general plan for the buildings and observatory was drawn up at that time, and was put into written form by Holden.[13] Newcomb suggested at one of these meetings that a director should be chosen for the observatory well in advance of the start of actual work, and recommended Holden for the position.[12] It is very likely that Mills first offered the directorship to Newcomb, but he turned it down.[14] According to Holden himself the position of director was provisionally offered to him at that time, and he accepted it.[13] Certainly from then on he was one of the two chief astronomical advisers of the Lick Trust, second only to Newcomb,[15] and the succeeding boards regarded him as the coming director.[16]

It was a remarkable appointment, for at that time Holden was only twenty-seven years old, and had worked as a professional astronomer for less than a year. He was born in St. Louis, but his mother died when he was only three years old, and he was brought up by an aunt in Cambridge, where he was educated in private schools until he was sixteen.[17] Then he returned to St. Louis and studied in the Washington University Academy (equivalent to a present-day high school) for two years before he entered the university itself in 1862. He studied under William Chauvenet, a well-known theoretical astronomer and mathematician who was then chancellor of Washington University. After receiving his B.S. in 1866, Holden entered West Point, graduated third in his class in 1870, and was commissioned a second lieutenant. After one year with the artillery, he returned to West Point as an instructor, and married Mary Chauvenet, the daughter of his former professor. In 1873 Holden resigned his Army commission to take a position at the Naval Observatory, where he worked briefly with a meridian circle until he became Newcomb's assistant at the 26-inch refractor when it went into operation. In 1874 he had done practically no astronomical research, but he had a wide knowledge of literature and science, was an excellent writer, and a vivid, entertaining conversationalist. Newcomb was tremendously impressed with Holden's energy and ability.[18]

In the winter of 1874-75 Newcomb went to Europe to investigate and report to the Lick Trust on the possibilities for getting the glass disks from which the lens would be made. He concluded that they could be obtained from either of two firms, one in England and the other in France.[12] He and Holden strongly recommended that no site be definitely selected for the observatory until it had been tested by an experienced astronomer, and proposed S.W. Burnham, the noted double-star expert, for this job.[19]

After Lick discharged Mills and the rest of his first board of trust, Captain Floyd, the new president of the second board, went abroad in 1876 and visited various observatories and telescope makers, seeking advice on the project from many of the leading European astronomers and telescope makers.[19] Floyd, the man who more than anyone else built Lick Observatory, was the scion of a well-known Southern family. He had been an 18-year-old midshipman at the Naval Academy when the Civil War broke out in 1861; when Georgia seceded he immediately resigned his commission and went south. He served as an officer in the Confederate Navy, and then after the war ended, moved on to California, and, beginning as a fireman on a ship, worked his way up to master; ever afterward he was known as Captain Floyd.[20] After only a few years in California he met and married Cora Lyons, the only daughter of a very wealthy pioneer lawyer and judge who had come to San Francisco before the Gold Rush. Judge Lyons left his daughter a fortune, and after the marriage Floyd became a businessman in San Francisco and a landowner in Lake County. There he spent much of his time at Kono Tayee, his estate on Clear Lake, north of San Francisco.[21] He was active, intelligent, hard working and scientifically inclined, and he threw himself into building Lick Observatory with all his heart and soul.[22]

When James Lick died on October 1, 1876, Floyd was still in Europe and the third board of trust, appointed less than a month before, had not been confirmed by the courts. Ultimately the board took office, and after several years of legal wrangles, finally reached a settlement with John H. Lick, who had come to California soon after his father's death and filed suit to be recognized as heir to his fortune. In the compromise he received $533,000 and agreed in return to settle with all other purported relatives. Then the trustees could go ahead and begin to build the observatory.[3]

Burnham was brought out from Chicago with his 6-inch telescope, as Clark, Newcomb and Holden had all recommended, and spent six weeks in the summer and fall of 1879 observing at Mount Hamilton. With the land granted by Congress, and the road built by Santa Clara County, it is hard to imagine what would have happened if his findings had been unfavorable. But he reported the site excellent for astronomical observing, as indeed it is, with many clear nights, and also exceptionally good "seeing," or atmospheric steadiness.[23] In 1880 Floyd and Thomas Fraser, formerly James Lick's foreman and now chosen to be superintendent of construction at Mount Hamilton, visited Newcomb and Holden in Washington and drew up plans for the observatory.[13] A contract for the lens was awarded to the Clarks, and the glass disks were ordered from the Feil firm in Paris.[13]

Work began on the top of Mount Hamilton in the summer of 1880. A spring was found that would be an adequate water source, a telephone line was brought up from San Jose, the peak of the mountain was leveled off to make room for the Main Building, a small dome was erected and a 12-inch refractor was installed in it by the fall of 1881. Bricks were made right on the mountain, and by 1885 all of the Main Building had been completed except the large dome for the 36-inch refractor, and several small wooden cottages had been erected for the workmen.[19]

The major delay in the whole project was getting the glass for the 36-inch lens. All astronomical refracting telescopes have so-called achromatic lenses, made of two separate glass elements nearly in contact, a convex crown-glass lens combined with a concave flint-glass lens. The combination optically is much like a single convex lens, but because of the different refractive powers of the two types of glass, they approximately cancel the prismatic effect that makes it impossible for a single lens to bring light of all colors to the same focus. Although the contract with the Clarks to produce the lens was signed at the end of 1880, the first 36-inch glass disk, or blank, was not obtained from Feil, the Paris glassmaker, until 1882.[13] This was the flint disk; the crown-glass blank was even more difficult to produce. The problem is to get the molten glass well mixed, so that it is homogeneous, and then to cool it slowly and evenly so that no strains occur and that no air bubbles form within the blank. In the spring of 1885 Langley, who was in Paris, reported to Holden that he had visited the glassmaker's establishment. Feil claimed to have just produced a good crown disk, but would not let Langley see it, and Langley was therefore suspicious that it would not be acceptable.[24] However, after nineteen failures the crown disk was finally successfully cast in September 1885 and shipped to the Clarks, who could then begin work on grinding, figuring and polishing the two disks to make the 36-inch lens.[13]

Meanwhile, in 1881, Holden had taken a job at the University of Wisconsin as director of its Washburn Observatory. For his first year he was on leave from the Naval Observatory, but he decided to stay in Madison and resigned his Navy appointment in 1882.[25] The Washburn Observatory position gave him more independence and responsibility, but he always regarded it as a temporary stop on the way to Lick Observatory.[17] His files in Madison are full of letters he wrote or received on behalf of the Lick trustees, mostly about books, clocks, small telescopes, auxiliary measuring devices, and all the minutiae of an observatory he knew so well.[26] Holden had first visited Mount Hamilton when he and Burnham had observed the transit of Mercury with Floyd in 1881, again in 1883 on his return from the Caroline Island eclipse, and each succeeding year after that as the construction progressed.[27]

By 1884 Holden was anxious to come to California as soon as possible, to begin work with the smaller instruments and no doubt also to forestall the appointment of anyone else to the directorship.[28] In 1885 the position of president of the University of California became vacant, and the regents, who had trouble finding and keeping presidents, decided to use Lick Observatory to attract a new one. Judge John S. Hager, the regent given the task of recruiting the next head of the university, first offered the position to Newcomb, whom he was reasonably sure would not take it.[29] After Newcomb had declined, Hager offered the post to Holden, who accepted.[30] He was appointed with the understanding that he would give up the presidency to become director of the Lick Observatory as soon as it was completed.[31] Hager was a close friend of Floyd, who no doubt approved this procedure, which had the effect of getting Holden close to the scene of action, without actually appointing him to the directorship immediately.[20] On January 6, 1886, Holden took up his duties as president in Berkeley, but naturally continued actively in his role as adviser to the Lick Trust.[32]

Only after both the glass blanks for the lens had been delivered to the Clarks was it certain that the Lick refractor would be a 36-inch, for the trustees had been prepared to settle for as small a telescope as 32 inches if no bigger glass could be obtained. Even at that size, it would have been the largest refractor in the world. The 36-inch aperture in turn determined the length of the telescope, approximately fifty-five feet, and hence the size of the dome needed to enclose it. Storm Bull, a University of Wisconsin engineering professor, was hired as a consultant on Holden's recommendation, to provide a detailed design of the dome building.[33] For this work Bull was paid at the rate of seventy-five cents an hour, which he regarded as a compromise between a fair professional rate of one dollar an hour and the fifty cents per hour that Holden wanted to pay him.[34] All the materials for the dome were brought to the summit of Mount Hamilton in 1885, in preparation for building it the next year.[19] Floyd, who had previously supervised the project from his Clear Lake estate, visiting Mount Hamilton as often as needed, now decided to move to the mountain and take personal charge of the final stages of installing the telescope and getting it into operation.

According to the terms of Lick's deed of trust, the trustees were to complete the observatory and then turn it over to the university, and Floyd adhered to this provision rigidly, not even permitting Holden to visit the mountain except on rare occasions.[35] However, they both felt it would be wise to hire an experienced younger astronomer to assist Floyd in all the practical details of getting the observatory ready for operation. Holden naturally thought of Keeler, who had accompanied him to the Colorado eclipse back in 1878, and whom he had heard so much about ever since.

Holden was a close friend of Langley, and they corresponded frequently. In 1881, just a few months after the Mount Whitney expedition, Holden had enquired for the first time whether Keeler was available, but Langley had answered that he planned to keep him at Allegheny.[36] Holden often called on his friend for scientific help; Langley in turn frequently farmed the work out to his capable assistant. Thus in February 1885 Keeler measured quantitatively the optical transmission of several wire gauze screens which Holden had sent to Allegheny, and wrote a careful report, which Langley forwarded to Holden at Madison.[37] Later that spring Keeler checked the design of a spectroscope that had been drawn up by Holden and Langley,[38] and when John A. Brashear had completed building it for Washburn Observatory, Keeler helped him test it, using the burning gas "torch" of a nearby steel mill as a light source.[39]

Holden was thus well aware of Keeler's practical knowledge and skills in astrophysics. Keeler himself made sure that Holden was kept abreast of his qualifications by writing him about the measurements he had made and by sending him a long series of questions about instrumentation.[40] Soon after Holden informed Langley that he had definitely been appointed president of the university and director of Lick,[41] Keeler wrote again and asked for a job at the new observatory. Holden replied immediately offering him a future position on the staff.[42] He said he was not authorized to hire anyone on university funds until the observatory was completed, but that he hoped to raise the money to take on one astronomer sooner, and that if he succeeded, Keeler was the man he wanted. He concluded:

> If I were a young man and in your situation I should accept this if offered. As far as it lies in me I mean to make the Lick Observatory the best obsy in America, and I hope to have the very best men there that will come – and to work for the best things in the best way – and to make my personal & first work to be the giving of the very freest choice to every one of my colleagues.

No one could resist this ringing declaration, and Keeler quickly replied:

> It is almost needless for me to say that the position you speak of would

suit me exactly, and that I consider it worth waiting for and taking the chance of its materializing. Once at the observatory I would do my best to carry out whatever projects you might form. For the present then I shall remain here at Allegheny as you recommend.[43]

Langley was not at all anxious to lose his capable assistant. He recognized that Keeler would have far more opportunities at Lick, and did not intend to stand in his way, but he wanted to keep him just as long as he could, and urged Holden not to send for him until he had an "indispensible need of him."[44] Keeler was eager to start the new job at Lick. He planned to resign his post at Allegheny in mid-March, and then go home for a visit before heading for California.[45] But he was called to Florida more quickly by a sudden turn for the worse of his father, who had been in increasingly poor health as a result of injuries he had suffered in a fall from a wagon at the beginning of 1885. Keeler left Allegheny early in the morning on March 2; not until the next day, in a newspaper he bought while the train was in Charleston, did he learn that his father had died on February 27, and his eighty-five-year-old grandmother, who lived with the family, two days later.[46] They were both buried before Keeler reached home on March 4.[47]

While Keeler was still in Mayport, a letter from Holden arrived for him at Allegheny, and Langley opened it, as Keeler had asked him to do before leaving. It contained the news that Holden had received authority from the Lick trustees to hire Keeler, and definitely offered him the job. Langley kept the original at Allegheny, and relayed the substance of the letter to Keeler in Mayport. The young assistant immediately wrote to Holden, accepting the "handsome offer" and saying that he planned to stay in Florida another week to help put his family affairs in order, would then return to Allegheny for two weeks more work he owed Langley, and would be ready to leave for California in early April.[48] The "family affair" was actually his sister Lizzie's marriage to David Day, who had also come down to Mayport immediately after William Keeler's death. They were married on March 17 and went north with Keeler that same day.[47]

After helping Lizzie and Dave get settled in their new home in Baltimore, Keeler went on to Allegheny where he found to his consternation that he had misread Langley's spidery handwriting. Holden had not offered him a salary of $150 a month, as he thought, but only $50, hardly enough to live on. He immediately wrote Holden explaining his mistake, saying that the amount "would be bare support for me and those left dependent upon me" and that he did not like the thought "of going so far from home and friends" without some assurance of more money.[49]

Holden knew that Keeler was making $75 a month at Allegheny,[50] and had promised him $100 a month once the university began paying the salaries.[42] He considered himself morally bound to that salary from the beginning, but the Lick trustees refused to pay Keeler more than $50 at first.[51] A flurry of telegrams followed, Holden raised the offer to $1,000 a year plus room and board,[52] and on April 11 Keeler left for California.[53] Langley's attitude was that Keeler's father's death was unfortunate, but that Allegheny Observatory should not lose any of its valuable assistant's time as a result of it. In the end, however, he let Keeler go with good grace.[54] He hired in his place James Page, another Johns Hopkins graduate, whom one of his professors recommended as "a very gentlemanly fellow of more than average ability" who would probably make an acceptable assistant.[55]

When Keeler arrived in California at the end of the transcontinental railroad, he undoubtedly reported first to Holden in Berkeley, and then quickly went on to San Jose, at the southern end of San Francisco Bay. From San Jose the last 27 miles was by stagecoach, up the winding mountain road, to the 4,200-foot summit of Mount Hamilton. At the top, the Lick Observatory Main Building, completed except for the dome, dominated the mountain. Just below it stood the three-story Brick House, where Floyd stayed when he was at Mount Hamilton, and still further below the wooden cottages where the workmen lived. Nearby were stables, a windmill, workshops, and pens for pigs and cows. The slopes of the mountain were covered with sparse chapparal, still green but beginning to dry out in late April, mixed with scattered pine groves on the north slopes.

Before the end of the month Keeler was hard at work, getting acquainted with the transit and switchboard. Fraser, the mountain superintendent, was initially suspicious, for as he put it, "if he turns out all right I will be his friend, but if he dont I have no use for Drones here." The industrious new assistant soon convinced him that he was no drone.[56] His first duty was to set up a time service, on the Allegheny model. The transit is an instrument for measuring the precise instant at which a star crosses the meridian, thus determining the exact time, and the switchboard was used to connect the various clocks to the telephone line to the Western Union office in San Jose. Much of Keeler's time was spent checking the lines, connections, batteries, clocks and all the other apparatus needed for this work. The system was thoroughly tested, and went into actual operation on January 1, 1887, supplying telegraphic time signals precisely at noon every day through the Southern Pacific station in San Jose to points all over the West, as far away as El Paso, Texas and Ogden,

Fig.2. Keeler at the time service switchboard at Lick Observatory
(1886). (Reproduced by kind permission of the Mary Lea Shane
Archives of Lick Observatory.)

Utah.[57] Once the time service was working, Keeler only had to check the clocks occasionally, for the observatory watchman could make the necessary connections at the switchboard to send the time out over the wires.[58] In addition Keeler was given the job of helping prepare the manuscript of the first volume of the *Lick Observatory Publications,* containing copies of the various deeds and reports from earlier days, together with descriptions of the instruments.[59]

In June Floyd moved up to the mountain with his family, consisting of his wife, their only daughter, and their niece, Cora Matthews.[60] The Floyds had long wanted to have a son, but when their one and only child was born, she turned out to be a girl. The captain, not to be cheated, named her Harry, and when the Floyds moved to Mount Hamilton, Miss Harry, as she was generally known, was fourteen years old. Cora Matthews, the daughter of Cora Floyd's half sister, was a few years older than Miss Harry, and lived with the family for long periods of time as her companion and friend. Keeler was soon spending much of his time working with the captain in preparing for the arrival of the telescope.[61] He was in a delicate position, for he had been hired by Holden, who would be his director when the observatory was completed, but Floyd, the president of the Lick Trust and the man on the scene, was his immediate supervisor. Keeler managed to keep them both happy with his hard work and tactful personality. He reported by letter to Holden in Berkeley at least once a week, frequently more often, and was in constant personal contact with Floyd. A close relationship developed between the captain and his capable assistant, and an even closer relationship between the assistant and the captain's attractive, Louisiana-born niece. Keeler and Cora Matthews became good friends that summer, and their friendship soon ripened into love.

Later in the summer, however, a potential rival appeared in the person of George C. Comstock, a young astronomer less than a year older than Keeler. He had studied at the University of Michigan under the classical astronomer James C. Watson, and as his assistant had followed Watson to the University of Wisconsin when he became first director of its Washburn Observatory. After Watson's sudden death in 1880 Comstock had stayed on as Holden's assistant at Wisconsin until 1885, when he became a faculty member at Ohio State University.[62] Holden hoped to hire Comstock on the Lick staff eventually, and with Floyd's approval invited him to come to Lick during his summer vacation as a volunteer and find out for himself what it was like.[63] An expert in the old astronomy, Comstock relieved Keeler on the transit observing program while he was on Mount Hamilton. He also became a close friend of Floyd and an admirer of Cora

Matthews, but in September he returned to Columbus, and changed circumstances never brought him back to Lick.[64]

By the fall of 1886 word came to Mount Hamilton that Alvan Clark and Sons had finished the 36-inch lens.[65] Newcomb was appointed as the astronomical expert to test the lens before it was accepted, and he took Hall, his successor in charge of the Naval Observatory 26-inch refractor, with him to the Clarks' optical "factory" near Boston. The tests were difficult to make because the Clarks provided only the minimum temporary telescope mounting required by their contract, without a clock drive or accurate setting circles. Newcomb and Hall spent a week in Cambridgeport, but got only one night of reasonably good "seeing", or atmospheric steadiness, in which the star images could be critically examined. They found the lens acceptable in every way, with good definition and color correction, but because of the awkward mounting they were not able to measure the focal length of the lens, an omission that was to come back to haunt them a year later.[66]

Fraser was sent East, to bring the lens to California safely. He traveled with it in a private railway car, the David Garrick. The two elements of the lens were boxed separately. Each was wrapped in many layers of clean, soft cotton cloth, then a thick layer of cotton batting, and then paper. Each element, wrapped in this way, was packed in a wooden box made to fit and hold it without any nails close to the glass, and these wooden boxes were then packed tightly in curled hair inside steel boxes. These steel boxes were suspended by springs inside larger steel boxes, and these boxes in turn were packed in asbestos in larger chests. The whole arrangement was hung from pivots in strong wooden frames, so that each glass element and its box could be rotated ninety degrees each day on the long trip to California. This was intended "to prevent any molecular disarrangement in the glasses and to avoid the danger of polarization, it being feared that the jarring of the train would disturb the present arrangements of the molecules unless the position of the glass should be changed and all lines of disturbance broken up."[67]

At the San Jose station, the boxes were loaded onto a wagon, which Fraser personally drove up to Mount Hamilton on December 27, 1886. There the lens elements were unpacked, reassembled in the cell which would hold them in use, and stored in a safe to await the completion of the telescope.[68] In January Floyd and Keeler tried to measure the focal length of the lens, first by shining collimated light from a smaller lens into it, and then by attempting to use it to focus an image of the distant Sierras, over 100 miles away across the Central Valley. But without any telescope

Fig.3. Captain Richard S. Floyd with the 36-inch lens, soon after it was delivered at Lick Observatory (1886). (Reproduced by kind permission of the Mary Lea Shane Archives of Lick Observatory.)

mounting, they could not handle the huge, heavy lens precisely, and they did not succeed in their measurements either.[69]

A few days after he brought the lens from San Jose to Mount Hamilton, Fraser drove another important cargo up the mountain. This time it was the body of James Lick. After his death in 1876, Lick's remains had been temporarily placed in a vault in the Masonic cemetery in San Francisco, but now that the base of the pier for the telescope was completed on Mount Hamilton, they were to be moved to the mountain. Lick himself had not left any written statement of what he wanted done with his body after his death, but both Floyd and Davidson claimed credit for having independently suggested to Lick that he be buried under the telescope.[2] According to Davidson, he was surprised when Lick at once accepted his half-serious suggestion. Davidson said he then told Lick it might be simpler to cremate his body and deposit his ashes in the pier, but the old man at once rejected this idea, saying, "No sir! I intend to rot like a gentleman."[5]

In the re-entombment ceremonies on Mount Hamilton on January 9, 1887, which Keeler witnessed, all the Lick trustees, headed by Floyd, were present, along with Holden, Davidson and other officers of the California Academy of Sciences, Mayor C.W. Breyfogle of San Jose, and representatives of the Board of Regents of the University of California and of the Society of California Pioneers.[10] The funeral party had come up to the mountain the previous afternoon, and Keeler and Floyd had shown them Saturn and various stars through the 12-inch and another smaller telescope that night, while the casket lay in state in the rotunda of the observatory. James Lick had been a famous freethinker, and one of his heroes had been Thomas Paine, so there was no minister present, nor were there any religious overtones in the ceremony. The casket was opened, Lick's body was identified ("It was evident his wish had been fulfilled," Davidson remarked[10]), and a statement giving the brief facts of Lick's life and the building of the observatory was put into the casket, which was then sealed and lowered into the foundation of the pier. Floyd delivered a short tribute, and the ceremonies were concluded.[70]

From the moment he was hired, Keeler had been assigned to be in charge of spectroscopic work at the new observatory. He was responsible for the design of the spectroscope, which was by now being built in Brashear's shop in Allegheny.[71] A constant stream of letters passed back and forth between the two friends, giving Keeler's detailed ideas, and Brashear's responses to them.[72] Because the telescope was so large, the spectrograph, which had to match it optically, looked "like an enormous thing to hang on to the eye end of a telescope" when it was built,[73] but Keeler understood the principles involved thoroughly, and it proved to be an excellent instrument in use.[74]

All that year of 1887, Keeler worked away on the time service, improving the design of the spectroscope, helping Floyd supervise the completion of the observatory, and observing with the small telescopes whenever he could find the time.[75] His salary had been retroactively raised to $100 per month from the date he was hired, and he had developed a very close relationship with Floyd. He enjoyed his work and wrote William Thaw, his patron in Pittsburgh, "Of course I could not ask for anything better."[71] Meanwhile, the Union Iron Works, the shipbuilding firm which was within the next few years to construct the U.S.S. *Charleston, Olympia, San Francisco* and *Oregon* in its San Francisco yards, was

Fig. 4. Group picture at Mount Hamilton during the Lick Trust days. Standing, left to right, Captain Richard S. Floyd, an unidentified man, Thomas E. Fraser, James E. Keeler, George E. Comstock, Cora Matthews (later Keeler's wife), and at far right, Harry Floyd. Cora L. Floyd is seated between two of the Lick Trustees, and Edward S. Holden is seated, wearing glasses, between Cora Matthews and Harry Floyd (1886). (Reproduced by kind permission of the Mary Lea Shane Archives of Lick Observatory.)

completing installation of the moving parts of the great observatory dome on Mount Hamilton, while in Cleveland the Warner and Swasey Company was building the telescope mounting. By October the observatory was nearly completed, except for the telescope,[76] and in early December the disassembled mounting arrived in San Jose in two railroad cars.[77] It was hauled up the Mount Hamilton road in wagons, and Ambrose Swasey, one of the partners in the firm which had constructed the mounting, arrived to superintend its assembly and erection. Everyone was impatient, for the observatory was far behind schedule, but on Christmas Eve Floyd could report that "[w]e have this moment landed aloft and safely in place the last section which completes the big tube. It looks very majestic. We have nothing but small work to do now to complete the mounting."[78] He was anxious to get a first look through the telescope, and hoped "to launch this Observatory into the Ocean of Science about in February."[79]

The weather turned bad, with alternating snow, rain, sleet and hail, all accompanied by gale-force winds and freezing cold.[80] Old Alvan Clark had died in Cambridge in the summer of 1887, not long after finishing the 36-inch lens, but his son Alvan G. Clark, then 55 years old, had come to Mount Hamilton to help get the telescope into operation. On New Year's Eve the lens was mounted in the telescope, but a severe storm prevented any observations.[81] The snow was several inches deep, the dome and its shutter were both frozen tight, and the high humidity and resultant fog made observations impossible.[82] Finally, on January 2, 1888 the sky cleared, and Floyd, Keeler, Clark and Swasey got their first glimpse of a star through the telescope. To their horror, they found that they could not bring the telescope into focus. The focal length of the lens was not 56 feet 6 inches as Alvan Clark and Sons had stated, but 56 feet, so the telescope tube, made to fit the quoted dimension, was too long. Operations had to be suspended for a few days to allow Swasey to cut a six-inch section off the end of the tube and remount the lens.[83] Floyd was furious and dashed off "Some Notes on the height of the center of Motion of the Great Lick Telescope and on the true focal length of the 36″ O.G. [object glass or lens] found by looking at the first star January 2d 1888 showing an *inexcusable* mistake by Warner and Swasey in making the pier 13 ½ inches too high and an *inexcusable* mistake by Alvan Clark and Sons in giving the focal length of the O.G. 6-inches too long."[84]

As they waited on the mountain for another clear night, Clark, a "terrible old blow and grumbler," was getting on everyone's nerves. He complained constantly about the cold, the observatory, the dome, the shutter, the eyepieces, and in fact about everything except the lens which he and his father had made.[83] Finally, on January 7 the skies cleared again. The dome was frozen solid, and could not be turned, but the shutter could be opened and stars could thus be observed for a few minutes each as the rotation of the earth moved them across the open slit of the dome. Again Floyd, Keeler, Swasey and Clark were present, along with "one of the ladies whose astronomical enthusiasm rose superior to the attractions of sleep," undoubtedly Cora Matthews. They first saw the brilliant double star Rigel, and found to their relief that the telescope now could be focused clearly. They checked the motions and operation of the telescope. At nine the Orion nebula, one of the most striking sights in the sky, came into view, and they saw a wealth of detail in it. Then, just after midnight the planet Saturn became observable. With the 36-inch telescope it was, in Keeler's words, "beyond doubt the greatest telescopic spectacle ever beheld by man. The giant planet, with its wonderful rings, its belts, its satellites, shone with a splendor and distinctness of detail never before equalled."[85] After everyone had had a chance to look at the planet, Keeler had time to study it carefully. The "seeing," or atmospheric steadiness, was excellent, and with the large telescope he could use high magnification and still see the outlines of the rings sharply defined. He could see the innermost faint "gauze" or "crepe" ring very distinctly, and the sharp division between it and the second, or "B" ring. Close to the outer edge of the outer "A" ring Keeler discovered a previously unknown narrow dark gap or division.[86] This new feature was the first astronomical discovery made at Lick Observatory. Because this division is so narrow, it can only be seen under conditions of excellent seeing, but over a year later, on March 2, 1889, another excellent night, it was observed again by Keeler and several other astronomers at Lick.[87] In analogy to the better known Cassini's division and Encke's division, it should be called Keeler's division (or Keeler's gap), but few astronomers are aware of his discovery.[88]

Clark was at Lick Observatory chiefly to complete the polishing of the photographic lens for this telescope. This lens, 33 inches in diameter, when mounted in front of the primary lens would convert the telescope from a visual refractor, corrected to bring all the yellow light to which the human eye is most sensitive to its best focus, to a photographic refractor, corrected to bring all the blue light to a focus. The lens could only be completed by repeatedly testing it in the telescope, then touching it up slightly and testing it again. The observatory was so cold that ice formed on pitchers of water left in the bedrooms overnight, and Clark could not continue his work because of the danger of water freezing on the lens.[89] Keeler helped him with the polishing when it was warm enough to work,

Fig.5. Keeler's drawing of Saturn, made at the 36-inch refractor the first night it was used to observe (1888). (Reproduced by kind permission of Yerkes Observatory.)

and finally, in early February, Clark completed the lens and left the mountain.[90] In the Bay area he visited Charles Burckhalter, an astronomer who was in charge of Chabot Observatory, a branch of the Oakland school system, which had a small Clark refractor for showing children the stars.[91] Clark, a compulsive complainer, poured out a long story of all that had gone wrong on Mount Hamilton, and Burckhalter, who had close ties with the newspapers, passed it on to a reporter. Thus on Saturday, February 18, readers of the *San Francisco Chronicle* were treated to an exposé headlined "A Lick Trust Toy – Blundering Work on the Observatory – Some of the Errors in Construction – A Shaky Pier and Useless Floor."[92] According to this article, the telescope was a disaster. The blame for its alleged faulty design was placed on Floyd, who was depicted as a scientific dilettante who had delayed construction while he junketed around Europe at the expense of the Lick Trust, trying unsuccessfully to educate himself in telescope design. This and a subsequent article, as well as a follow-up editorial, made clear that no blame attached to Clark nor to Holden, who was described as a practical man who should now be given the chance to correct the "numerous blunders" that Floyd and Fraser, his construction superintendent, had made.[93]

The leak was easily traceable to Clark and Burckhalter, and Floyd was furious. He attributed Burckhalter's animus to the fact that he had not received a copy of the first volume of the *Lick Observatory Publications*, and was cutting about his pretensions as an astronomer. As for Clark, Floyd wrote,

> I don't think any living being ever heard Alvan G. Clark praise or unqualifiedly acknowledge the merit of any man's work *but his own* – not even of his own Father or of his brother, George. He is inordinately envious and the Russiun gold medal has turned his head with vanity. He is not a bad old fellow in many ways and is not disposed to unkindly feeling except where his own supreme stinginess or envy is touched. . . And so we have – Mr. Burkhalter the '10 ½ inch' astronomer of Oakland mad because we did not send him a book, in recognition of his fame – old Clark's jealousy unconsciously furnishing a bonanza for Burkhalter's cowardly weapon – and the devlish vicious Chronicle that would attempt the murder of *any man's* reputation for the sake of one sensational heading."[94]

Floyd may have suspected that Holden was also involved in the news-paper attack, for an amplified version of the story soon appeared in his home-town newspaper, the *St. Louis Globe-Democrat*.[95] This article repeated the *Chronicle's* charges, contrasted Floyd and Fraser's "smat-tering of astronomical knowledge" with Holden's expertise, and ended, "The astronomer who points out these defects declares that it will take two years of the best work to repair the blunders committed by these ignorant men, and that if Prof. Holden gets things into working order by the end of 1889 he will be doing well."[96] Some of Floyd's advisers were certain that Holden was the source of the story, but he dashed off a letter denying any connection with the "unjust attacks" in the newspapers, and the captain accepted his explanation and concentrated his ire on Clark.[97] The Lick trustees decided that they would not reply to the attack them-selves, but that one of the astronomers should do it for them, and this duty fell to Keeler.[98]

His letter, published first in San Jose and then reprinted widely in the Bay area newspapers, was a spirited defense of the telescope and of Captain Floyd. Keeler emphasized that every step in the design and construction of the telescope had been made with the advice of the most eminent astronomers in the world, and claimed that the buildings and instruments of the observatory were "the most perfect appliances yet devised for the observation of heavenly bodies." He maintained that the observations they had already made of Saturn and the Orion nebula refuted the charge that the mounting was not stable, and, while admitting that the moving floor was too slow, said that a stronger motor would soon cure that problem. All the instruments had been tested and found satisfactory. The final adjustments of the telescope would take time, and would be made by the astronomers themselves. Keeler ended up his published letter with the words "History will be the judge of how well the trustees of the Lick Trust built the telescope."[99] The trustees felt it was an admirable defense of their work, and appreciated it greatly.[100]

By the end of February the completed spectroscope had arrived from Brashear's shop in Pittsburgh, and Keeler tried it on the telescope. The spectroscope was so large he needed a ladder to get up to the top end of it when it was mounted. He had a wheeled carriage made for the spectroscope, so he could use it as a laboratory instrument when it was not in place on the telescope.[101]

Brashear lost money on the spectroscope, as he did on nearly every large astronomical project he undertook. Andrew Carnegie, the enormously successful Pittsburgh iron and steel magnate, heard of this. Although he had "taken a deep interest in Mr. Brassier [Brashear]" and was resolved that "he is not going to be allowed to fail for want of aid," he did not choose to get monetarily involved himself. Instead he urged Leland Stanford, then a United States senator from California, to secure a "complimentary present" for Brashear to make up his $500 loss. Stanford, another multimillionaire like Carnegie, forwarded this suggestion without comment (and without cash) to the Lick trustees. They declined to take action.[102]

Floyd was able to observe Mars with Keeler in mid-March, but the captain, though only forty-five years old, was in failing health. He had developed severe heart problems, and had worn himself out on the mountain. Digitalis, good claret and rest were the only remedies his doctor could prescribe, and early in April he finally had to leave Mount Hamilton for his Lake County estate.[103] He was seriously ill, and his doctor would not let him return to the mountain.[104] Lick Observatory was his monument, but he never saw it again.

With Floyd out of commission, the responsibility for arranging for a delegation of University of California regents to visit Mount Hamilton to inspect the observatory they were about to take over fell on Lick Trust member Mastick.[105] The newly appointed chairman of the regents' Lick

Observatory Committee, who retained the post until his death eleven years later, was Timothy Guy Phelps, one of the best-known politicians in the state. Born in New York, he had come to California in the Gold Rush days and had soon made a fortune in San Francisco. A lifelong Republican, he had been elected to the assembly and then to the California senate, and had served a term in Congress during the Civil War, where he was President Abraham Lincoln's adviser on Pacific Coast political affairs. Phelps was twice appointed collector of the Port of San Francisco, and had run for a second term in Congress and later for governor, but by then California had gone Democratic, and he was never elected again. He was a down-to-earth, practical man, who knew how to get things done, and he had been a regent since 1880.[106] The other two members of the committee were Judge Hager, the regent who had arranged the appointment of Holden as president of the University three years before, and Andrew S. Hallidie, the inventor of the San Francisco cable-car system.

Accompanied by Mastick, Holden and Swasey, the committee members went up to Mount Hamilton on April 21, and after a good lunch were shown all over the observatory by Keeler. They saw the houses, offices, shops, library and of course, the telescopes. Phelps, a hardbitten old character, chewed tobacco constantly, and whenever he felt the inclination to spit, did so, even when he was standing on the magnificent marble floor of the hall of the Main Building. After a hearty dinner, in which the guests consumed several bottles of claret that the captain had left for them, Keeler showed them the moon, Saturn and Mars through the 36-inch refractor. They all were suitably impressed. After a good night's sleep, they looked over the rest of the mountain, including the water supply, and then returned to San Jose. Keeler had made a very good impression on the regents, particularly on Hager.[107] They were pleased and impressed with all they saw, and were only sorry that Captain Floyd himself had not been able to be there with them.[108]

For a few days at the end of April, Keeler was left in charge of the observatory. The planet Mars was in opposition, near its closest approach to the earth, and he was using the telescope to measure the positions of its two satellites, Deimos and Phobos, each night, while in the daytime he supervised the workmen who were putting the finishing touches to the observatory.[109] Keeler searched carefully to see if he could discover an even fainter satellite of Mars with the new, most powerful telescope in the world, but found none. He continued the measurements every night until April 28, when work on the elevating floor temporarily stopped the observing, and they were used to correct the published ephemerides (predicted positions) of Deimos and Phobos, which had been discovered only ten years before.[110]

On May 1, Holden came up to Mount Hamilton to take over himself. Almost daily he sent one or more letters to Henry Mathews, the secretary of the Lick Trust in San Francisco, giving lists of last-minute needs and jobs that should be finished.[111] Since the observatory still had not been transferred to the regents, Holden had only the authority granted to him by the Trust to spend small sums of money for absolute necessities.[112] An inventory had been taken of all the property of the observatory, beginning with "The Main Observatory Building," ranging through "1 36″ equatorial [telescope] complete and in position," "Large Dome of 36 Inch Equatorial," "1 Brashear star spectroscope complete, on carrier," down to "One single chicken house near Cottage 2," "1 3 foot step ladder," and " 1 ladle, 1 skimmer, 1 egg turner." A deed was prepared conveying all this property to the regents of the University of California, and was signed by Floyd at his Lake County estate on May 19, and then by the rest of the Trustees in San Francisco on May 31.[113] Thus on June 1, 1888 control of Lick Observatory passed to the state, and James E. Keeler became a faculty member of the University of California.

4

Steady growth and excellent achievement

By the time Lick Observatory opened as part of the University of California on June 1, 1888, Director Edward S. Holden had assembled a small but excellent staff of astronomers. Holden himself, then forty-one years old, was a well-known scientific figure as director of the Washburn Observatory of the University of Wisconsin from 1881 until 1885, and then president of the University of California until he took over Lick Observatory. He had been recommended, and probably tentatively chosen, as the future Lick director when he was only twenty-seven years old, and had spent much of his career in preparing for the task, advising the trustees, and ordering equipment, books and supplies for the observatory. He had done a fair amount of visual observational work on planets and stars at the Naval Observatory and at Wisconsin, but he was by no means famous as an observer, and his most important contributions to research were compilations and catalogues of the results of others.[1]

Personally, Holden was a brilliant conversationalist. He had an active mind and a wide range of knowledge, based on avid reading, and he enjoyed discussing and defending his opinions.[2] He was slightly above medium height, with a broad face and brown hair, which he parted in the middle and combed back, and light sandy whiskers. The small, wire-rimmed glasses he wore emphasized his stern professorial appearance, but he projected a feeling of quiet, confident power.[3] Holden's clothes, hairstyle and dining habits all seemed relatively advanced to the unsophisticated Californians of the 1880's, and the Berkeley students had referred to him among themselves as "the Dude" or "Champagne Eddy".[4]

The senior member of the Lick staff, eight years older than Holden, was Sherburne W. Burnham, already famous as a double-star observer. The short, slightly built Vermonter had attended the local Academy, more or less equivalent to a modern high school, in his native Thetford,

Fig.6. Edward S. Holden, first director of Lick Observatory (1886). (Reproduced by kind permission of the Mary Lea Shane Archives of Lick Observatory.)

but had no formal training as an astronomer. He had worked as a shorthand reporter, first in New York and then with the Union Army in occupied New Orleans during the Civil War. There he had first become interested in astronomy, and had bought a small, cheap telescope. After the war ended, he moved to Chicago, where he worked as a court reporter for twenty years. His home in the South Side was near Dearborn Observatory (afterward moved to the Northwestern University campus in Evanston) with its 18 ½-inch telescope, the largest refractor in the world at the time Burnham first settled in Chicago. Seeing it convinced him that he needed a good telescope of his own, and he arranged to meet Alvan Clark, and ordered a 6-inch from him, specified as the best of its size the firm could make, for $800. Observing with this telescope Burnham discovered over 400 double stars, close pairs of stars that are in most cases actual physical companions moving in slow orbits about one another. He had unusually keen eyes and powers of concentration, so that with his telescope he could detect and accurately measure the separation of pairs so close that many other observers could not resolve them (or see them separately). Meanwhile Burnham continued his daytime work as a court reporter to support his growing family, which eventually numbered six children. At times he received permission to use the large Dearborn refractor for his double-star measurements.[5]

Burnham's reputation as an observer grew and, as we have seen, on the recommendation of Holden, Simon Newcomb and Clark, he was hired by the Lick trustees to test the Mount Hamilton site before construction of the observatory began. He spent two months on the mountain in the early autumn of 1879, where he discovered 42 new double stars with his 6-inch refractor, and measured a long list of known doubles as well. He declared the site excellent.[6] In 1881 he returned to Mount Hamilton to observe the transit of Mercury with Holden and Captain Richard S. Floyd, the president of the Lick Trust. Burnham, as the senior and most skilled observer, observed with the newly installed Clark 12-inch; Holden with a 4-inch Clark telescope, and Floyd with a 2 ½-inch portable telescope.[7] Burnham became a close personal friend of Floyd, and they kept in touch through the years.[8] Burnham's hobby was photography, and on one occasion he sent Floyd an artistic shot he had taken of a scantily clothed female model, with the added intelligence that some other pictures he had made in this line would hardly "do for general circulation, though made for a legitimate purpose."[9] Burnham worked at Madison with Holden for one year, but this arrangement ended when ex-Governor C.C. Washburn, who provided the funds to hire him, fell ill and then died, and Burnham returned to Chicago in 1882.[10] Holden and Floyd both agreed that they wanted to persuade Burnham to turn professional again and join the Lick staff as soon as the observatory was completed.[11] Burnham accepted the offer toward the end of 1886, as the 36-inch lens was being completed, but on Holden's advice did not come out to Mount Hamilton until the observatory was turned over to the state.[12]

In addition to Burnham and James E. Keeler, Holden wanted very much to hire George C. Comstock, his former assistant at the University of Wisconsin. When Comstock spent the summer of 1886 at Lick as a volunteer, in charge of the meridian-circle work, part of the idea was that he could see how he might like it as a permanent situation.[13] Although Comstock did like Lick, he preferred Madison, and especially the independence that went with the job as director, so he asked Holden to recommend him as his successor there.[14] Holden first wrote his friend Samuel P. Langley, Keeler's director at Allegheny Observatory, and then Benjamin A. Gould of Dudley Observatory in Albany, New York, and William A. Rogers of Harvard, to find out if they were interested in the position, but after all three had declined he recommended Comstock to President T.C. Chamberlin at Wisconsin.[15] How influential his recommendation was is uncertain, but Comstock got the appointment at Washburn Observatory in the summer of 1887, and Holden had to give up his idea of having him on the staff at Mount Hamilton.[16]

He was certain he wanted a meridian-circle observer, and since he could not have Comstock, he decided instead to hire John M. Schaeberle, a 35-year-old acting assistant professor at the University of Michigan.[17] Several years before, Holden had suggested that he could come to Lick with Comstock and Schaeberle as his assistants, and begin meridian-circle and small-telescope work even before the 36-inch refractor was completed.[18] Holden based his decision to hire Schaeberle partly on the recommendation of Comstock, who knew him well from the days when they were both students at the University of Michigan.[19] Schaeberle had been born in Germany but brought to America by his parents, who settled in Ann Arbor when he was a young boy. He worked as an apprentice in a machine shop in Chicago for several years, but then, becoming interested in astronomy, returned to Ann Arbor, went to high school, and graduated from the University of Michigan in 1876. There he became an assistant at the observatory under James C. Watson, the expert in the old astronomy of position and motions of the planets, and continued as a junior faculty member after Watson left Michigan to become the first director of Washburn Observatory. Schaeberle was a quiet bachelor, who followed a regular routine in his observing and in all the affairs of his life.[20] Holden tentatively offered Schaeberle a job as soon as he knew that Comstock would probably not come, and after Schaeberle accepted, advised him to wait at Michigan until Lick Observatory was completed.[21]

Edward Emerson Barnard, the fourth member of the original Lick Observatory staff, was only three months younger than Keeler, but otherwise their lives were almost complete antitheses. Barnard had been born in Nashville, Tennessee after the death of his father; his mother was very poor and young Edward and his brother were brought up in poverty and privation. The Civil War swept over Tennessee when Barnard was a little boy and, though in later life he could never be brought to talk about this period, evidently the fatherless family kept alive by scavenging and begging. Barnard only attended school for two months, and his mother, who was selling flowers, put him to work in a photographic studio when he was nine years old. His first job was to keep a solar enlarger, a device used for making photographic prints, pointed at the sun. In simplest terms, he was acting as a human clock drive, to keep a primitive telescope aligned with the sun. Barnard, whose mother had taught him to read from the Bible, bought his first book on astronomy from a thief; almost certainly it was stolen property. His first telescope, a simple lens in a cardboard tube, was made for him by one of the men he worked with in the studio. Barnard was the sole support of his mother from the age of twelve but if he had a few nickels to spare he would spend them to pay to look at the stars through the portable telescope of a traveling sidewalk astronomer. Barnard became increasingly interested in astronomy, and in 1876, when he was nineteen years old, he bought a professionally made 5-inch astronomical telescope. It cost him $380, nearly two-thirds of his annual income at the time. He was a keen-eyed observer who often worked all day at the studio and spent half the night examining the skies with his telescope.

In the summer of 1877 the American Association for the Advancement of Science met in Nashville, and Barnard took the opportunity to be presented to Newcomb, its president, probably the best-known astronomer of the day, and the chief astronomical adviser of the Lick Trust. Barnard showed him his observations, and told him he wanted to become an astronomer, but Newcomb advised him that he would never accomplish anything significant by visual observing alone, without an education, particularly in mathematics. Barnard was reduced to tears, but followed Newcomb's advice, found a tutor, and began studying at night instead of observing. Thus the same year Keeler entered Johns Hopkins University, Barnard began learning all the grade-school and high-school mathematics he had missed in his childhood.[22]

In 1881 Barnard married Rhoda Calvert, an immigrant from England, the sister of two of his fellow employees at the photographic studio. She helped him with his English, spelling, and pronunciation, and encouraged him to continue his studies, but he still was able to find some time for observing, and on May 21, 1881 discovered his first comet, while sweeping the skies with his telescope. Such a discovery is, in a sense, a lucky accident, but an accident that happens more than once only to an observer who has long experience, superb eyesight, confidence in his own judgment, dedication, and excellent technique. In his life Barnard discovered over twenty comets, but this first one was never listed or named, because he did not know how to report it to the astronomical community. His discovery encouraged him to begin a systematic search for comets, sweeping the skies near the sun just after sunset and before sunrise, and four months later he found his first official comet, Comet Barnard 1881 VI, and announced his discovery. In those days there was a prize of $200 for each discovery of a new comet, provided by a wealthy American amateur astronomer, and Barnard used the money to buy a small lot and begin building a small wooden house. He continued working in the studio by day, and studying and observing by night, and managed to find one or two comets a year, enough to meet the mortgage payments on what became known as "The Comet House."

His discoveries made him famous in Nashville, and in 1883 a group of his friends and admirers raised enough money to offer him a $500 per year combined fellowship and instructorship at Vanderbilt University. This sum was enough to enable him to quit his job at the photographic studio and devote full time to his studies and observing, although he was still supporting not only his wife but his invalid mother. The university provided a house for him on the campus, where they all lived, and the opportunity to use its 6-inch refractor, mounted in a small permanent observatory. Vanderbilt was a far different place intellectually from Johns Hopkins, but as a combination special student and instructor, Barnard studied mathematics, physics, chemistry, French and German, and improved his knowledge of English and the social graces.[23] He was several years older than the regular students, and to them seemed shy and awkward.[22]

In 1887, Barnard independently discovered the Gegenschein, or counterglow, a large, low-surface-brightness area in the night sky, always opposite the sun as seen from the earth. It had previously been discovered by observers in Germany and England, but it is so faint that it can only be seen on clear, moonless nights, when it is far from the Milky Way, and Barnard was not aware of its existence before he saw it himself. Today we still are not sure whether it is due to reflection of sunlight preferentially back toward the sun by millions of tiny dust particles orbiting in the plane of the solar system, or to emission and scattering of light by molecules and particles in a tail of the earth, analogous to the tail of a comet. Barnard's independent discovery of the Gegenschein demonstrated his very acute vision and powers of concentration while observing.

In 1884 Barnard went to the meeting of the American Association for the Advancement of Science in Philadelphia, his first long trip away from home. En route he visited the observatories at Cincinnati, Allegheny (where he probably met Keeler, although no record survives), Washington, Harvard, Albany and Princeton, and saw New York and Boston. He had to practice rigid economy, and to avoid hotel bills he traveled at night, sleeping on the hard wooden benches of the cheapest railroad coaches.[24] On this trip he met many astronomers and other scientists, and, measuring himself against them, he realized that he could be a professional student of the stars himself.

After he returned to Nashville, Barnard struck up correspondences with several well-known astronomers, including Holden. In October 1885 Holden wrote him that it was definite that he would become the director of Lick Observatory, and Barnard replied that

> I was very greatly delighted to hear that you were going to California to take charge of that grand Observatory. I am so glad, and I am sure no man in the country is so well suited for that great charge as you are. I'm sure I feel that the University of California deserves the highest congratulations for its success in securing one so eminently fitted for the Position. I am not surprised for I have all along said you would be the man for the Lick Observatory. You have my heartiest wishes for your welfare and I hope your health may always be good so you can enjoy the grand instrument which under your charge I know will reveal wonders that man never dreamed of.[25]

Barnard kept up a drumfire of correspondence, reporting his observations of comets and nebulae, and finally asking Holden as a great favor, if he would send him his photograph.[26] Holden did so, in exchange for the picture Barnard had given him, and then outlined his plans for the future of the observatory, into which it was clear he was beginning to think Barnard might fit.[27] The ecstatic Tennessean thanked him for his kindness in sending the excellent photograph, and then described at length his astronomical interests and studies. He dwelt on his experiments in astronomical photography at Vanderbilt, emphasizing his practical on-the-job experience in the studio, said that he would graduate that June, and concluded "that the Lick Observatory with the proper instruments would be the best place in the world to do wonders in Celestial Photography."[28] This letter was the closest approach to an explicit job application that the customs of those days permitted.

Holden considered it very seriously, and by March 1886, when he had selected Burnham, Keeler and Schaeberle for the staff, he was thinking about Barnard as the next candidate; in July he wrote to offer him a job.[29] In his earlier letter, Holden had emphasized that under the terms of Lick's deed of trust, salaries could not begin until the observatory had been completed and turned over to the university. In offering the position, Holden wrote that the duties would commence about October 1, 1887, a remarkably optimistic prediction, but he no doubt expected Barnard, like Burnham and Schaeberle, or like Keeler the previous year, to wait until he was notified that he was needed and could be paid. In this Holden misjudged his man. On receipt of the job offer, Barnard hurried to Tullahoma, a nearby retreat where Chancellor L.L. Garland, the elderly professor of astronomy at Vanderbilt, was resting for the summer, consulted with him, returned to Nashville, resigned his instructorship effective September 1, and sent off a telegram of acceptance to California.[30] He then started making preparations to move, began selling his household effects, and wrote Holden that he would leave Nashville about the middle of September ("I cannot see how I could get away earlier").[31]

Holden must have been startled by Barnard's naive letter, and wrote back asking if he could not delay his departure, but it was too late. By the time the letter came, Barnard had sold most of his possessions except his telescope and his astronomical books, which he planned to bring with him, had arranged for some of his financial support at Vanderbilt to be shifted to David Spencer, a poor country boy who like himself was interested in astronomy, and was waiting impatiently for detailed instructions from Holden on just where to go.[32] Barnard assured Holden that if he and his wife got there a little early they could stop over in San Francisco, and added that if it was only a matter of when his salary was to begin, Holden should not let that interfere in any way, because he was anxious to begin work. Barnard wrote Holden that he loved Vanderbilt, and was only leaving because of the superior research opportunities at Lick. "Another strong inducement," he continued, "is (for I have always had the highest admiration for you personally) that I shall be immediately under your charge, and shall receive the benefits of your training. It will be my earnest endeavor to aid in the progress of the Observatory to the very best of my ability. I am perfectly temperate, neither smoke, chew nor use intoxicating drinks."[33] Although Barnard followed the admirable prohibitions of this last sentence all his life, within a few years the rest of his statement was to seem very ironic indeed.

When Barnard and his wife arrived in San Francisco near the end of September, they found that the telescope mounting was still being constructed at Warner and Swasey's shop in Cleveland, and that the observatory was months from completion. The Lick Trust could not pay his salary, and he could not even live at Mount Hamilton. He had to work to survive, and got a job he must have hated, as a clerk in the San Francisco law office of Jarboe and Harrison, whose senior partner was a friend of Holden's.[34] Keeler soon met the short, stocky Tennessean in San Francisco, and saw him again a few weeks later when he visited Mount Hamilton.[35] When the observatory finally neared completion in the spring of 1888, the Lick trustees hired Barnard to come to Mount Hamilton to inventory the property before it was turned over to the university.[36] He spent two weeks on the mountain on this job, during which he got his first look through the 36-inch telescope with Keeler, who was observing the satellites of Mars.[37] Barnard was eager to get on with astronomy, and was very glad when at long last the trustees handed over the observatory and serious research work could begin.[38]

In addition to Burnham, Schaeberle, Keeler and Barnard, Holden hired Charles B. Hill as combination secretary, librarian and assistant astronomer. He was a former newspaperman who was interested in astronomy, but did no research.[39] When the regents approved the other appointments in January 1888, Hill's was held up in an economy move, although Holden had been assured they all would be passed.[40] By their next meeting, a month later, Holden had managed to persuade the regents that the observatory really did need a secretary and librarian and this last appointment was approved.

When he began actually functioning as director of the observatory, Holden took a $1,000 a year salary cut from the $6,000 he had been making as president. Burnham's annual salary was $3,000, Schaeberle's, $2,000, and Keeler and Barnard started at $1,400 each.[41] The latter amount was the salary Holden had originally offered Keeler, but he had hoped for more, and had asked Floyd to put in a good word for him with the regents.[42] For comparison, of the whole University of California faculty, only the new president, Horace Davis, made more money than Holden. His salary was $6,000 a year, the same as Holden's had been. One faculty member, George H. Howison, the professor of intellectual and moral philosophy, who was supported by the D.O. Mills Endowment, earned $4,000. Three others, John LeConte, professor of physics, his brother Joseph LeConte, professor of geology, and E.W. Hilgard, professor of agriculture, had salaries of $3,300, as did J.H.C. Bonté, the secretary of the university and superintendent of grounds. Nine other full professors had salaries of $3,000, the same as Burnham, while the other Lick faculty salaries were more or less equivalent to those of assistant professors, although they were all called astronomers.[43]

In addition to the faculty members, there were three blue-collar workers at the Observatory. They were John McDonald, the foreman and machinist, whose salary was $900 per year, just a little less than Hill's, Charles Harkort, the janitor, and Chris H. McGuire, the general handyman, each of whom made $720.[44] Harkort was an interesting character, a German who had studied physics and chemistry at Bonn, and then served in the U.S. Army. He could speak German, French and English, read

Fig.7. Lick Observatory, Mount Hamilton, California. The Main Building is in the background, with the 36-inch refractor in the dome on the left, the 12-inch refractor in the smaller dome on the right. The Astronomers' House is in the foreground (1888). (Reproduced by kind permission of the Mary Lea Shane Archives of Lick Observatory.)

Latin and Greek, and was expected to help out in photography in his spare time.[45]

Everyone on Mount Hamilton lived in quarters belonging to the university. Holden took one side of the three-story, brick Astronomers' House, which was built so that it was divided into two completely separate dwellings. It was well designed, sturdy and comfortable, and was very conveniently located next to the Main Building.[46] Burnham had the other side of the Brick House, as it was called, but his wife and large family lived in San Jose, coming up to the mountain for frequent visits, particularly in summer.[47] Schaeberle had two rooms on Burnham's side of the house, and Keeler a room on the ground floor on Holden's side. Barnard and his wife lived in one of the wooden cottages, originally used by the Lick Trust employees who built the observatory, in the saddle a little below the Brick House. Hill and his wife lived in another of these cottages, and the workmen lived with their families in still others.[48]

Each family cooked and ate as a unit, but Schaeberle and Keeler organized a "mess," in which they and any other bachelors who happened to be on the mountain ate together, sharing expenses.[49] Everyone's supplies had to come up from San Jose, at a cost of 1 cent per pound on the daily stage, or 40 cents per 100 pounds on the observatory wagon, which made two trips a month up and down the mountain. Fruit was available in season from the Kinkaids, a farm family five miles down the road to San Jose, and milk and eggs from the Hubbards, another family in Hall's Valley, eight miles further on and halfway to town.[50]

Lick Observatory's purpose was astronomical research, and the astronomers worked at it every night of the year. The 36-inch telescope was assigned on a regular basis, two nights each week to Burnham for work on double stars, two nights to Keeler, for spectroscopic work, and two nights to Holden, for photographic work, mostly on the moon and planets. The seventh night, Saturday, was reserved for public visitors until about 11 p.m., but after they had left one of the observers, usually Burnham, took over the telescope for the rest of the night. Barnard used the 12-inch most nights, giving it up to Burnham if he needed it, while Schaeberle observed with the meridian circle. When Burnham observed with the 36-inch, his program required moving the telescope frequently, a demanding job, and he was assisted by the observatory secretary, first Hill, and later his own son, Augustus J. Burnham. Holden also needed assistance, and Schaeberle or Keeler often worked with him, but Keeler had to get along without an assistant when he was observing on his own nights.[51]

There is more to running an observatory than research. As a young, energetic and responsible staff member, Keeler got more than his share of the housekeeping and technical jobs that kept the observatory going. One of his first assignments after the university had taken over was to sell the chickens that had belonged to the Lick Trust. He got twenty cents apiece for them, which he regarded as a good price for a motley collection of tough old birds.[52]

On a more technical level, Keeler was responsible for the time service, a job which included not only checking the clocks against one another and against the stars, but also making sure that the storage batteries and electric lines for transmitting the time signals to San Jose were in working order. He was also responsible for the earthquake records, which were taken from a seismometer put into service at Mount Hamilton soon after the observatory went into operation.[53]

With his training in physics and experience in the laboratory, Keeler was more familiar with electricity than anyone else on the staff, and Holden gave him the job of getting the 36-inch telescope into good shape, including modifying its controls and adding electric lights to read the circles giving its position.[54] In the original design, oil lamps had provided this illumination. To drive the telescope at the correct rate to follow the diurnal motion of the stars resulting from the rotation of the earth, Keeler developed a simple but ingenious electrical control system. The basic idea was to use the accurate rate from a pendulum clock, read off through an electric circuit by contacts. The resulting pulses of current went through an electromagnet which kept the telescope drive, powered by gravity, in phase with the clock. Since there was no direct mechanical link between the clock and the telescope, there were no sudden jolts as there had been in the system originally provided by Warner and Swasey, the makers of the telescope. Keeler's system depended on the elasticity of the mounting of the electromagnet and on the fact that the force keeping the telescope drive in phase with the clock was applied gradually between the electromagnet and an iron sector. In Keeler's design this force was approximately proportional to the error in phase, and the overall system was thus a primitive form of what we would today call a "hard-wired" servomechanism.[55]

Another job Holden assigned Keeler was to go to San Jose to lay out a standard meridian line, running accurately north and south, for surveyors to use to check their compasses. Because of the secular variation, or gradual shift of the earth's magnetic field, a compass cannot be relied on without checks of this kind at frequent intervals. Endless litigation had resulted all over the United States from neglect of this precaution. In San Jose the traditional standard line was Meridian Road, which was supposed to run north–south along the meridian through Mount Diablo, a conspicuous peak in the chain that includes Mount Hamilton. Holden sent Keeler to San Jose for a few days in the summer of 1889 to lay out the accurate meridian with Charles Herrmann, the Santa Clara County surveyor, and his brother A.T. Herrmann, a surveyor in private practice. They had set up a stone monument about 2,000 feet south of the board fence of the fairgrounds at the end of Meridian Road. With their transit they sighted Polaris, the pole star, and knowing accurately its elongation, or angular distance from the exact north pole, they sighted down to a lantern shining through a hole bored in the fence, and measured the azimuth of Meridian Road. They found the road was a little less than one quarter of a degree off from an accurate meridian line, amounting to a deviation of just over eight feet at a distance of 2,000 feet, and then set up another permanent stone monument to define the true north–south line accurately.[56] This line was used by surveyors in San Jose for many years afterward, but no longer exists.

Keeler, like other astronomers before and since, was often called on to give public lectures, which are supposed to amuse, inform and edify the listeners and thus subtly build support for astronomy and the observatory. He was a good speaker, who used his slides and a blackboard effectively in, for instance, a talk on "Great Telescopes" to the San Jose Y.W.C.A. in 1889.[57] Keeler was also a fluent writer, and he produced the manuscript for a pamphlet on "The Lick Observatory" intended for visitors. This pamphlet, now a booklet, has gone through many revisions and additions down to the present day, but traces of Keeler's outline and phraseology can still be detected in the latest edition. His description of the physical aspect of Mount Hamilton is a good sample of his nontechnical style:

> The view from the observatory peak is a very beautiful one, particularly in the spring, when the surrounding hills are covered with bright green verdure, and the eye looks down upon acres of wild flowers. To the west lies the lovely Santa Clara valley, shut in from the ocean by mountains somewhat lower than the Mt. Hamilton range. Sometimes the entire valley is filled with clouds, rolling onward under a clear sky and bright sun like a river of snow, and this is one of the finest sights to be had at the observatory. The surrounding mountain tops project out of the fog like black islands.
>
> Nearly in front of the observatory lies San Jose, further to the right the bay of San Francisco, while in clear weather the cities of San Francisco and Oakland can be seen between two low hills in the northwest. Across the valley, in the southwest are the two prominent peaks of the Loma Prieta (3800 feet high), on the slopes of which are the celebrated quicksilver mines of New Almaden.
>
> To the north the eye looks down on an endless succession of rounded hills, which in the spring are bright and green, but in the summer become dry and apparently barren. About 19° to the west of north is Mt. Diablo, through which passes the standard meridian for the government surveys.
>
> Along the eastern horizon may be seen in clear weather the snow-capped peaks of the Sierra Nevada mountains, 13,000 to 14,000 feet high. Those just clear of Mt. Copernicus on the right surround the

Yosemite valley, and those on the left are in the vicinity of Lake Tahoe, 150 miles distant.

The high mountain on the south, separated from Mt. Hamilton by a cañon nearly 2000 feet deep is Mt. Isabel, of the same height as the observatory peak.

The slopes of Mt. Copernicus were once the haunts of the robber Joaquin Murietta, and traces of his occupation are still to be found there. A spring on the east side of the peak bears his name.[58]

Keeler was also called on to write a popular article on the work of the observatory, and the many new results flowing from it, for the California astronomical public. It is a skillful job, which emphasizes the positive but at the same time conveys some of the real excitement of research.[59]

Although Keeler's main research effort was on stars and nebulae, he led the first Lick Observatory eclipse expedition, to observe the solar eclipse on January 1, 1889. In those days astronomers were generalists, rather than specialists as today. Solar eclipses provided the only chance to study the faint outer layers of the sun, the chromosphere and the corona, and the region of space immediately surrounding the sun. Pioneer astrophysicists like Langley and Charles A. Young of Princeton did solar research, lunar research, planetary research, stellar research, and any other kind of astronomical research they could think of. If there was an eclipse of the sun, they tried to observe it. Holden had led the eclipse expedition to Colorado in 1878 that had been Keeler's introduction to professional astronomy, and an expedition to Caroline Island in 1883 while he was still on the University of Wisconsin faculty. Thus it was natural for him to send an expedition to observe the 1889 eclipse, whose path of totality passed through California. Many amateur astronomers and photographers were interested in the eclipse, and Lick Observatory's research on it, well publicized in California, helped public relations immensely. A handbook on observing the eclipse, prepared by Holden, was distributed widely, and his efforts to collect all the observations afterward led to the founding of the Astronomical Society of the Pacific.[60]

Holden picked Bartlett Springs, on the path of totality in Lake County in the Coast Range northwest of Sacramento, as the site for the Lick eclipse observations. Keeler was put in charge of the expedition and scouted the area in October 1888, while on a visit with Cora Matthews and the Floyds at their Kono Tayee estate at Clear Lake, just a few miles from Bartlett Springs.[61] The Lick party consisted of Keeler, Barnard, Hill, and Armin O. Leuschner, the first graduate student, who had arrived at Mount Hamilton only a few months before. Leuschner had been born in Detroit, of German parentage, but his widowed mother had taken him back to Kassel when he was only four years old. He received a solid education in a Gymnasium there and returned to the United States in 1886. He attended the University of Michigan for two years and received his B.S. degree in 1888, and then came to Lick where he was accepted as a student by Holden and put to work.[62] Holden had written to Davis, his successor in Berkeley as president, "I have taken it for granted that the University will not wish to refuse instruction to any student & I have written Mr. Leuschner to come out as soon as may be."[63] He also wrote Davis that it would be "pleasanter for all of us to be entirely free of teaching so as to be able to devote all our time to research. But I feel sure our best use to the University is to help to form a school of Astronomy of the highest grade which will attract students (like Leuschner) away from the European Observatories, as he has been attracted."[64]

This was the beginning of the Lick graduate school, forced into being by Holden over the objections of the Berkeley faculty. The president did not even want the observatory in the University of California, and would have preferred it to be part of Stanford, while most of the professors wanted to have all the students and teaching concentrated on the campus.[65] Leuschner, like later graduate students, was supposed to spend half his time at Mount Hamilton doing research, and the other half at Berkeley taking courses in mathematics and a little physics. In his first year at Lick he worked chiefly for Schaeberle, reading the microscopes on the meridian circle and doing two hours' reduction work on the observations per day, for which he received an unfurnished room. On Mount Hamilton he also tutored Holden's thirteen-year-old son Ned two hours a day, and in exchange received his board from Holden. Leuschner paid no fees and had no fellowship.[66]

For the solar eclipse, the telescopes and other equipment were sent on ahead to Bartlett Springs in early December. The advance party, consisting of Keeler, Barnard and Leuschner, together with Ned Holden, left the observatory on December 15; Hill and his wife were to join them at Bartlett Springs closer to the eclipse date. They traveled by rail from San Jose to San Francisco, but when the train started Keeler and Barnard discovered that Ned, an active young teenager, and Leuschner had been left behind on the platform. Keeler and Barnard were naturally worried, but the resourceful tutor managed to recapture Ned and get on a later train, so the group was reunited in San Francisco that evening. They stayed there for two days, making preparations for the eclipse, and to their great relief managed to persuade Ned to join the Jarboes, family friends of Holden's who were going to see the eclipse from Cloverdale, a less isolated location along the path of totality.

The Lick group then went on to Sites, a tiny station on a narrow-gauge railroad line 35 miles from Bartlett Springs. The freight containing their apparatus had not arrived, so Keeler went off to Colusa, the nearest sizeable town, to try to locate it. Barnard and Leuschner had to wait a day for the next "stage," actually an open wagon, to take them to their final destination. It was raining nearly all the time, and the whole area was a vast sea of mud. They left Sites at 7 a.m., and did not arrive at Bartlett Springs, in the mountains at an elevation a little over 2,000 feet, until 6 p.m., hungry, soaked and exhausted.[67] Keeler found the freight and sent it on by wagons, and also persuaded the telegraph company branch manager in Colusa to get the line to Bartlett Springs in working order, so they could receive time signals from Lick. Keeler himself took the stage from Sites to Bartlett Springs on December 24 with the manager of the resort hotel and several other passengers. He reported that "[t]he long rains which had been falling on the *adobe* soil, and the stage travel made the roads muddy beyond anything I had ever seen before. We had to walk up the steep hills, and at short intervals it was necessary to stop and dig the mud off the wheels, in order that the four horses could drag the empty stage. At that time it did not seem likely that our boxes could reach the summit, but as the whole country was well within the limits of totality we were fairly sure of obtaining observations of some kind, even if it should be necessary to set up the instruments on the scene of a possible accident."[68] But they did get to Bartlett Springs at 10 p.m., after fifteen hours on the road, and found that the freight had indeed arrived just ahead of them.

Bartlett Springs consisted chiefly of a resort hotel and about twenty small cottages. Barnard had chosen a clear area in the middle of the hotel croquet ground for the instruments, and he and Leuschner had personally dug holes and set up piers for the instruments. It rained almost continuously, and Keeler groused about the "villainous fare" and the accommodations at the little resort, where the only comfortable room was the bar.[69] Finally on December 28 the continuous rain stopped, and they had mostly clear skies with occasional showers for the next several days. This gave them a chance to focus and adjust their instruments, and rehearse the observing procedure they intended to use on January 1. A few hours after the final rehearsal on December 31, the sky clouded up again, and by eleven that night it looked like there would be rain or even snow before morning.[70]

However, the next day the sky was beautifully clear, and they were able to make all their observations as planned. Leuschner, working with a visual photometer, measured the total brightness of the corona by comparing it directly with a standard candle. Barnard took photographs of the corona, using three different lenses, all on the same equatorial mounting. The largest was a 3-inch diameter, 49-inch focal-length lens ordinarily used at Mount Hamilton to read the scale showing the water level in a distant reservoir. For the eclipse it was stopped down to 1 ¾ inches to improve the photographic definition. Keeler, assisted by Hill, observed the spectrum of the corona, using a small visual spectroscope on the portable 6-inch Clark refractor. At that time it was still not certain whether the corona was an actual physical gaseous envelope around the sun, or if it was some kind of an optical effect. In fact, Charles S. Hastings, Keeler's former professor at Johns Hopkins, had proposed a theory, based on his observations at the Caroline Island eclipse, that the corona was a diffraction effect. His idea was that the emission lines seen in the corona actually were emitted in the chromosphere close to the sun's surface, but during the eclipse were diffracted or bent by the edge of the moon.[71] Keeler's measurements were intended to test this theory. His spectroscope was arranged so that he could see the spectrum of the corona on both sides of the sun at the same time, and as the moon moved across the sun during totality, he could observe how the brightnesses of the two sides varied. Since the moon moved by only a small angular amount during the whole eclipse, if the corona were a physical entity the brightnesses of the two sides would only change slightly, while if it were a diffraction phenomenon, the brightnesses of the two sides would change greatly because of interference effects.

The instant the eclipse began, Keeler was struck by the great brightness of the continuous spectrum of the corona compared with the brightest spectral line, in the green spectral region, called 1474 by the astronomers of that day, but known to us today as [Fe XIV] $\lambda5303$. To Keeler it seemed that this line did vary more rapidly than the moon's motion would predict, and hence he thought it was at least partly due to diffraction, a conclusion we know to be incorrect. The difficulty with his simple picture is that actually the density in the corona is not constant, but is highest closest to the sun; the emission in the $\lambda5303$ line is therefore relatively strong close to the apparent edge of the sun, and quite small changes in the amount of corona that is covered by the moon do cause large changes in the observed brightness. On the other hand, Keeler correctly concluded that the continuous spectrum is not a diffraction phenomenon, and is probably due to reflection from particles, as in fact it is in the outer corona. The inner corona he thought to be self-luminous, resulting from thermal emission from the same particles, heated to incandescence by the sun; this is not true but the correct interpretation of scattering by electrons at very high temperatures was not understood by astronomers until the 1940's. Keeler's published conclusions immediately after the eclipse were thus that Hastings' diffraction theory was partly confirmed (for the emission line) but partly disproved (for the continuous spectrum, the major contribution to the total light).[68] However further analysis and comparison with other eclipse observers' reports led him to realize that the strength of the line increased inward in the corona right up to the edge of the sun, so a few years later Holden correctly reported that Keeler's observations at the 1889 eclipse disproved Hastings' theory for the emission lines as well as for the continuum.[72]

Immediately after the eclipse the Lick party began dismounting their equipment, and they had it all packed by the evening of January 2. It had begun to rain again, and in fact the actual eclipse date was the only day during their stay in Bartlett Springs that it was clear enough for observations in the early afternoon at the time the eclipse occurred. Because of the bad condition of the roads, they stored their equipment at the hotel until the end of the rainy season, and did not get it back to Lick until two months later. Keeler went on to Clear Lake to spend a few more days with Cora and the Floyds,[73] but Barnard was anxious to return to Mount Hamilton and develop his photographic plates of the corona. He hand-carried the plates, still in their plateholders, in his valise. He was ready to leave early in the morning of January 3, but found that two passengers with a trunk had already reserved all the space on the "stage," and there was no room for him. He could not buy them off, nor get them to leave their trunk behind. Finally, "[a]fter a good deal of persuasion – the roads being one mass of mud and the wagon frail – I secured the high privilege of riding on the trunk and holding the valise in my hands, with the clear understanding that I was to walk up the muddy hills. It was uphill all the way to Sites." A hard, persistent rain fell all day, and Barnard had only his overcoat to protect him in the open wagon. The trunk slid wildly in the wagon as they went up and down hills, and every few hundred feet they all had to get out to scrape the accumulated mud off the wheels. At last he reached Sites just as night fell, tired, drenched and muddy, and then went on to San Jose by rail.

But the results were worth it, for when Barnard developed the plates after his return, he found they had been correctly exposed and showed excellent images of the corona. A wealth of fine detail was visible, and there was no doubt that photography had completely replaced drawing as a method for accurately recording the appearance of the corona.[74] All the Lick people, at any rate, thought Barnard's photograph was the best taken at the eclipse, better even than those obtained by William H.

Pickering of Harvard, with a much larger, more expensive telescope.[75] Holden believed that the most important result from the coronal photographs was that they confirmed that the shape of the corona varied with the solar cycle, an idea first put forward by the English astronomer, Arthur C. Ranyard, in 1879. The 1889 eclipse occurred near sunspot minimum, as had the 1867 and 1878 eclipses, and the corona was noticeably elongated in the equatorial plane of the sun, not at all like the 1883 eclipse, which occurred near sunspot maximum, when the corona was much more nearly circular.

Keeler went back to the eclipse track in February, to measure the exact latitude and longitude of Norman, California, a little town near Colusa where the Washington University party under Henry S. Pritchett had observed. They had returned to St. Louis so quickly after the eclipse that they had no time to measure their position themselves, so Holden authorized Keeler to do it for them, at Pritchett's expense.[76] The whole town of Norman consisted of a few scattered houses plus a railroad station, and Keeler could find no place to buy bricks or mortar to make a pier for his transit instrument, so he had to set it up in the chimney of a fireplace in the farmyard of a local rancher. Although he covered the transit with a tarpaulin when it was not in use, one night a horse broke into the yard and nosed the unfamiliar instrument, but luckily did not damage it. Keeler had to rebuild the chimney before he could continue with his measurements. It was a far cry from astrophysics, but he got the latitude and longitude accurate to the last few tenths of a second of arc, equivalent to approximately forty feet, for Pritchett. At least it meant another visit to Cora at nearby Clear Lake. Keeler's expenses for this expedition, recorded in his observing book, totalled just over $17, including his railroad fare, shipping his baggage, and one night at the Lick House in San Francisco.[77]

The Norman episode was a comic end to the 1889 eclipse. The volume of calculations Keeler had to make to get highly accurate positions for every star he might conceivably need to measure was more than sufficient to discourage anyone from ever going on another eclipse, and in fact Keeler never did. But at this eclipse he demonstrated his competence in heading a scientific group, overcoming unexpected obstacles, and returning with important scientific results. Barnard's photographs were the most important results; Keeler's own visual observations were competently made, and carefully and correctly recorded. His interpretation was weak, but the data were published and in the end the correct conclusions were drawn from them.

Eclipse expeditions were exciting scientific adventures, but the bread and butter of Lick Observatory was research with the 36-inch telescope. With the largest refractor in the world, on a site far superior to that of any other observatory then in existence, the Lick astronomers were exploring the universe. One of their highest priorities was to study the planets, to learn as much as they could about their physical makeup. Their main method was simply to observe the apparent surfaces of the planets visually, and to report what they saw. Although photography would provide a permanent record, even today visual observing is far better for recognizing the finest detail from terrestrial sites. The human eye and brain, together, can ignore the moments of poor seeing, or atmospheric unsteadiness, and remember the image as it appeared in the few brief instants of superb definition. In those days, when astronomical photography was in its infancy, and for planetary work at Lick Observatory in the unsure hands of Holden, there was no question but that visual observing was superior. With the large Lick telescope it was possible to observe with all the magnification the atmospheric steadiness would permit, and particularly in the summer and fall the atmosphere is very steady indeed.

Keeler observed all the planets, but particularly concentrated on Jupiter, which shows a wealth of detail on its disk. Because of its short rotation period, he had to work quickly and complete a drawing before the planet's aspect changed appreciably. In his observing book he would in advance trace a circle "from Professor Holden's ink bottle," and then at the telescope rapidly sketch Jupiter in pencil as he saw it, occasionally making quantitative measurements with a micrometer eyepiece to get the most important dimensions exactly correct, but chiefly depending on his own ability to recognize proportions. In the margin of the page and on the face of the drawing itself he would write in the colors of the various features and areas.[78] Soon afterward he would make an accurate india-ink drawing from the observing book sketch, or, on rare occasions, a color drawing.[79] As Barnard, himself a highly skilled planetary observer, testified, Keeler had unusual artistic ability, and his drawings were outstanding reproductions of the visible features of Jupiter.[80]

In his descriptions of the planet, Keeler emphasized that the features seen were clouds and disturbances in its atmosphere. He was fascinated not only by the great red spot (actually pale pink in color, according to Keeler), but also by the small dark spots, the rounded and the oval white spots, and above all the streamers. These brilliant white features seemed to be formed of irregularly rounded or feathery clouds, projected out from the equatorial zone, and gradually left behind by rotation. He emphasized that the smaller features, as they drifted toward the red spot, were always pushed to one side and flowed around it.[81] Keeler's physical

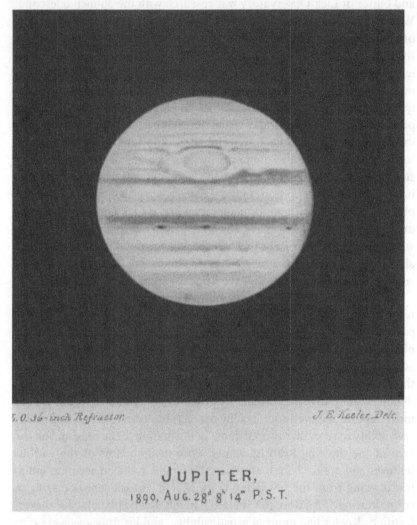

JUPITER,
1890, AUG. 28ᵈ 8ʰ 14ᵐ P.S.T.

Fig.8. One of Keeler's drawings of the planet Jupiter, made with the 36-inch refractor, showing the Great Red Spot (1890). (Reproduced by kind permission of the Mary Lea Shane Archives of Lick Observatory.)

picture of Jupiter with its streams and perturbed flows of "cloud-like matter," was far more in accord with reality, as we understand it today from the Voyager space probes, than were the ideas of Barnard, who thought of the planet as having a plastic or putty-like surface, or of Jupiter specialist George W. Hough, of Northwestern's Dearborn Observatory, who visualized it as having a relatively thin, gaseous atmosphere.[82]

By far the most interesting planet to the public was Mars, on which the

Italian astronomer G.V. Schiaparelli claimed to have discovered "canals", thin, straight, regularly arranged lines, in 1877. Newspapers and popular magazine articles continually asked whether the red planet was inhabited, and Lick Observatory naturally was expected to answer the question. Schiaparelli's purported discovery had been made with a 9-inch telescope, but most of the observers at Lick never could see the canals with the larger telescope at the excellent Mount Hamilton site. Barnard, who probably had the finest eyes of several generations of observational astronomers, never saw them, although the bulk of his work on Mars came several years later, after Keeler had left Lick. Barnard's considered opinion was that the canals as Schiaparelli described and drew them did not exist, but were an illusion. What Barnard could see were much more irregular, broken up details on Mars which he thought observers with small telescopes, anxious to see regularity, linked up subconsciously with the reported straight lines.[83] The subject was so controversial, however, that Barnard never fully published these conclusions.

Keeler, at Lick, tended to keep away from Mars, and there is no evidence in his observing books that he ever saw the canals as Schiaparelli described them. One night Schaeberle, the only true believer (in the canals) on the Lick staff pointed them out to him, but Keeler recorded merely that he could see "some of them" himself, but that they were "very faint and diffuse shadings, not fine lines."[78] Schaeberle on the other hand, a few nights later, believed he had seen two of the canals as double lines, thus confirming a result of Schiaparelli's which, if true, strongly suggested that they were the work of intelligent beings.[84] Modern closeup photographs from space probes have shown that the canals on Mars have no objective reality, and that Barnard, Keeler and the other skeptics were right. Keeler did produce some excellent drawings of Mars, showing the polar caps, "seas" (large grayish green areas), and other large-scale features.[79] These pictures of Jupiter and Mars, together with his exquisite drawing of Saturn, made on one of his first nights with the 36-inch, were very widely distributed and greatly admired by the astronomers of his time.[85]

In addition to drawing the planets, Keeler also continued his measurements of the positions of Deimos and Phobos, the two tiny satellites of Mars, when the planet was near opposition (closest to the earth) and they could be seen.[86] These observations refined the orbital data on the satellites, which had been discovered only a few years before.[87] However, Keeler's real interest in the planets was in their spectra, which could yield clues on their physical character. One of his first spectroscopic papers from Lick was on the spectrum of the rings of Saturn, the subject in which he was to earn his greatest fame six years later. Keeler, in the early research, was trying to check the published work of the English spectroscopist, J. Norman Lockyer, who had reported observing bright emission lines in the spectrum of the ring. This confirmed Lockyer's meteoritic theory, according to which the source of luminosity of many astronomical objects such as nebulae was collisions between small lumps of material, or meteors, with resulting vaporization and conversion of kinetic energy of motion into heat and light. The bright lines Lockyer claimed to have seen would be evidence that there was gaseous, heated material in the ring, and that it did not shine simply by reflected sunlight, as the planets do. Keeler's observations, however, made on several occasions in April and May 1889, showed no trace of bright lines in the spectrum of the ring. Further, he noted that if Lockyer's picture were correct, and the ring were partly self-luminous, then it should be faintly visible in the shadow of Saturn, where the planet itself blocks off all sunlight from striking the ring. However, very careful visual observing, on a night of fine definition, in which Keeler isolated the shadowed region of the ring alone, and found it was completely dark, ruled out this theory altogether.[88]

Keeler also observed the spectra of the outer planets, especially Uranus, in which the great European pioneers of spectroscopy, Angelo Secchi in Rome, William Huggins in England, and Hermann Vogel in Germany, had earlier reported absorption bands in the red or long wavelength spectral region. Lockyer, however, had contradicted their results and reported that the apparent dark bands were actually simply gaps between emission bands. Again, Lockyer's reported observation would tend to agree with his own theory by showing that the planet's atmosphere was self-luminous, and not simply shining by reflected sunlight. Keeler's first observations of Uranus with a small spectroscope, which provided only relatively low dispersion, gave the same impression Lockyer had reported, but as soon as he looked with the large high-power spectroscopes, he saw that there actually were absorption bands on a continuous spectrum. Comparing Uranus with the other major planets, Jupiter, Saturn, and faint Neptune, he could see the same absorption band, at the wavelength $\lambda 6182$, increase regularly in strength from the innermost, Jupiter, to the outermost, Neptune. There was no doubt, even in Lockyer's mind, that it was an absorption band in Jupiter's spectrum; the continuity of its appearance proved it could not be a coincidental gap in the spectra of Uranus and Neptune. Keeler measured several other absorption bands in the yellow and red regions, all of them now known to be due to absorption by the methane molecule, CH_4, and published the most complete description of the spectrum of Uranus up to that time.[88]

He also observed the spectrum of a comet, and reported there were three bright emission bands in it, "identical in position with the carbon fluting given by the blue flame of a spirit lamp" (now known as the Swan bands of the molecule C_2) with a weak continuous spectrum between them.[89] This was not a new result; previous cometary spectra had been observed and described by several astrophysicists in Europe and America, and Huggins had identified the comet bands with the products of hydrocarbon flames.

All Keeler's spectroscopic work was highly precise, and based on accurate measurements of the wavelengths of the features he could see with his visual spectroscope. As a dedicated professional, he was highly skeptical of amateur astronomers and annoyed by the public visitors who, to his mind, cluttered up the observatory and prevented serious work on the first half of each Saturday night. Keeler generally kept these views well concealed, but his attitudes occasionally broke through, as in a report on the research going on at Lick Observatory that he produced for *Scientific American*. It is mostly about the telescope and scientific results, but at the end he wrote, "The public receptions on Saturday evenings interfere greatly with these experiments, as all apparatus must then be removed to fit the telescope for visual observations. Probably few visitors are aware of the hindrance to astronomical work caused by their entertainment, although, as a duty to the public, the sacrifice is always cheerfully made."[90]

He was more scathing in his review of a patent application in which an amateur astronomer proposed to make a mirror for a reflecting telescope from a thin, flat, circular disk of metal, supported around its edge, with a bolt through its center. The inventor's concept was that by turning a nut on this bolt, the disk would be buckled into the shape required for the mirror, a ridiculous idea since the disk would in fact have a form far different from a paraboloid. Keeler sarcastically commented, "The inventor has neglected to mention that by carrying the motion of the nut to a convenient position near the eyepiece a ready adjustment of the focus will be obtained. The method has the great advantages of simplicity and cheapness; its defects will be ascertained by the inventor when he comes to try it."[91] But Keeler loyally overcame his feelings about amateurs and played a prominent part in the Astronomical Society of the Pacific, organized by Holden to channel the enthusiasm and interest generated by the 1889 eclipse.[92] Keeler spoke at several of the Society meetings in San Francisco, including the meeting on January 25, 1890, when Holden was snowbound on Mount Hamilton and Vice President William Pierson, an eminent San Francisco lawyer and an amateur astronomer, had to take the chair. On this occasion Keeler told the members of his work on

Jupiter, and exhibited a selection of his drawings of the planet.[93]

That winter was unusually severe, with heavy snow in the mountains, often rendering the road from Smith Creek to Mount Hamilton impassable for days at a time. The storms began in November, and by December the snow was so deep that the stage could not run for eighteen consecutive days, and even the observatory wagon could not get through for ten days. Supplies and food for the people on the mountain had to be brought up on horseback, through deep snow drifts on the road.[94] Then another series of storms blocked the road for nearly a month in January and February. Food was running short, and observing was out of the question because of the constant fog, so Keeler hiked out to Smith Creek, tying gunny sacks around his legs as improvised snow boots.[95] Leuschner hiked up from Smith Creek a few days later, taking four and a half hours and arriving at the top almost completely exhausted.[96] At one point the snow drifts were as high as the telephone poles and he walked on top of the wires for two hundred feet. Keeler and Barnard were wiser, and stayed down off the mountain. They took advantage of their forced vacation and made a short trip to Santa Cruz, then a sleepy little beach resort, but now the headquarters of Lick Observatory.[97] At Mount Hamilton there were snow drifts as deep as fifteen feet, and some of the cottages were nearly covered except for their chimneys. Through it all Holden and the other astronomers managed to keep the telescopes dry inside the domes, and nobody starved.[98] By the end of March they were back to fairly routine observing conditions.[78]

Keeler devoted much of his observing time in the first year at Lick to stellar spectroscopy. He first looked at the spectra of a good selection of relatively bright stars, to have a firm base of knowledge of the spectra of normal stars, and then began investigating several known to have peculiar or unusual spectra. Keeler and the other Lick observers worked under severe difficulties. They had to stop observing every two hours, to rewind the giant driving clock of the telescope, which kept it moving in synchronism with the diurnal motion of the stars. Rewinding the clock raised the 600-pound weight the height of the pier; all the motive power was supplied by the observer, who had to provide the energy for 320 half turns of the winding wheel to accomplish this task.[99] Furthermore, until December 1889, Lick Observatory was operating without the *Bonner Durchmusterung* star maps, making it impossible for an astronomer to prepare a finding chart of the little area in the sky around each star he wanted to observe.[100] This made setting the telescope on the right stars extremely difficult and time-consuming. Nevertheless Keeler was able to do it, and achieved especially interesting results on several variable stars and emission-line stars.

One object he investigated in detail was β Lyrae, a variable star known to have emission lines from Secchi's early observations. The giant Lick telescope and the highly efficient spectroscope Keeler had designed made it possible for him to see its spectrum in more detail than any previous astronomer. He observed it regularly and could always see the bright Hα and Hβ emission lines of hydrogen, and the emission line at λ5876 in the yellow, called D_3, of helium, an element then known only in the sun and stars, that had not yet been found on the earth. In addition, Keeler could always see the broad Na I D sodium absorption line at λ5893. He could also detect several other fainter emission lines, most of which later also turned out to be helium. As the star varied regularly in light, because of the partial eclipses caused by the orbital motion of the two components of what is actually a close, unresolved double star, Keeler saw that the emission lines were brightest at the same time the star itself was brightest, but that remarkable changes occurred at primary minimum, the deepest part of the eclipse. Then dark absorption lines appeared at the short wavelength side of several of the emission lines, indicating gaseous material coming toward the observer with relative velocities of the order of 500 km/s. He saw an exactly similar absorption line in the same relative position to D_3, and speculated that it might actually be D_3 in absorption. This had never before been previously observed, and Keeler could not quite bring himself to accept it, although his published drawing clearly shows that no other interpretation was possible. However, he did state that the large wavelength difference between the absorption and emission lines at primary minimum could not result from simple orbital motion of the two bodies, no matter what sort of stars they were.[101] This perceptive comment marks the beginning of research on gas streams in binary stars, a subject many astrophysicists have studied over the intervening years, and β Lyrae itself has always proved one of the most interesting as well as complex objects of this type.

Keeler also observed P Cygni, another hot star with a spectrum similar to the spectrum of β Lyrae. He found many of the same lines in P Cygni, but they were relatively brighter and narrower than in β Lyrae, and the continuous spectrum of the underlying star was relatively fainter. Many of the emission lines had absorption components on their short wavelength edges (a form of line profile now called a "P Cygni line"), and Keeler commented that as P Cygni is probably not a binary star, some other interpretation would be required to explain these profiles.[102] This paper was thus also extremely perceptive; today we know that P Cygni

Fig.9. A contemporary etching of the Lick 36-inch refractor (1888).
(Reproduced by kind permission of the Mary Lea Shane Archives of
Lick Observatory.)

and the other stars with similar line profiles have expanding, low-density atmospheres that in many cases are flowing off into space. Keeler also studied other peculiar stars, especially several red stars with spectra far too complicated for the visual method he was using at Lick. He only published short descriptions of them, but returned to work on them far more effectively a few years later at Allegheny Observatory.[103]

At Lick, Keeler measured with unprecedented accuracy the radial velocity, or as the astronomers of his generation called it, the velocity in the line of sight, of a few very bright stars, especially α Bootis (Arcturus) and α Orionis (Betelgeuse). The principle on which these measurements are based is called the Doppler effect. It is that the wavelength of a spectral line is shifted to shorter or longer wavelengths by the relative motion toward or away from the observer of the source that is emitting it. In β Lyrae and P Cygni the Doppler effect is the cause of the wavelength shifts between the absorption and emission components, which arise in different physical regions, as Keeler realized. The wavelength shifts are small, but are exactly proportional to the relative radial velocity, so by measuring the wavelengths of the lines in a star very accurately, Keeler could calculate its motion in the line of sight with considerable precision. All this was known well before Keeler's time, and several astronomers had measured stars' velocities using the Doppler principle before he did. But with the largest working telescope in the world, in the fine Mount Hamilton climate, and with his spectroscope optimized for observations of this type, he had much more light than previous observers, and could measure wavelengths, and therefore velocities, far more precisely than they had done.

Keeler had a very good understanding of optical principles, and the instrument he had designed and Brashear had built was extremely efficient. He clearly realized that not only is the large aperture of the telescope important for collecting as much light as possible, but also that the spectroscope must have as large an effective aperture or beam size as possible, to use the starlight effectively. The Lick spectroscope had an unprecedentedly large effective aperture, 1.1 inches, and was designed and built to be extremely rigid in any position of the telescope, so that false apparent wavelength shifts would not be introduced as a result of flexure of the spectroscope. Many small details, learned by experience in spectroscopy, from the construction of the slit to the convenience and comfort of the observer, went into Keeler's instrument.[104]

For the measurements of the spectra of bright stars, Keeler used a grating in his spectroscope to disperse or break up the light into its separate wavelengths. Gratings can be made to give very high dispersion, but in Keeler's time they tended to be considerably more wasteful of light than prisms. However, Henry A. Rowland, his former professor at Johns Hopkins, had just perfected a process for producing much brighter gratings than previously available, and it was one of these Rowland gratings that Keeler used.[104] He had also realized the advantages of the newly produced Jena glass, from which small lenses could be made with very little chromatic aberration, and had replaced the original lenses in the spectroscope with new lenses of this type. They made it possible to make measurements and comparisons of wavelengths of spectral lines over a wide range without readjusting or refocusing the spectroscope, thus adding greatly to the accuracy of the final results.[105]

The radial-velocity measurements were made by setting the telescope on a star and adjusting the position until the star's image at the focal plane fell on the slit of the spectroscope. Then the observer would see its spectrum in the eyepiece at the viewing end of the spectroscope, and would measure the exact angular readings when a micrometer in the eyepiece was centered on the sodium D lines at $\lambda5890$ and $\lambda5896$. A comparison, or laboratory light source, in this case the same sodium lines produced by salt as an impurity in an oil lamp, was measured immediately afterward, to get the exact zero-point position of the lines. After the last measurement of the night, the spectroscope adjustments were left unchanged until morning, when as a check the wavelengths of the same lines were measured in the spectrum of the sky, which of course is simply sunlight scattered by the earth's atmosphere. The best check on the accuracy was provided by Keeler's measurements of the moon and especially of the planet Venus, made in exactly the same way as the measurements of the stars. Since all the orbital properties of the planets are known with high accuracy from straightforward geometrical measurements and the simplest concepts of dynamics and gravitation, the velocity of Venus at any time can be calculated with extremely high precision. The difference between its measured and calculated radial velocities thus gave directly an estimate of the error of measurement. Typically it was no larger than ± 1 mi/sec, in the units Keeler used, corresponding to about ± 2 km/s in modern terms, and was often much smaller than this.[78] Thus the individual measurements of the star velocities were immediately known to be accurate to better than ± 2 km/s, and the average of several measurements was of course much more precise.

Although he never allowed himself to be photographed wearing them, Keeler customarily observed, and probably also read, with glasses. He had astigmatism, which according to him had been diagnosed as uncorrectable at the eye clinic in Heidelberg during his student days. But from his knowledge of optics he had calculated what lenses he needed, and had glasses ground in Berlin to his own prescription. They had worked, and he found by comparing what he could see with what his friends could, that with glasses he had normal vision. He guarded his sight by never straining his eyes, and he avoided reading fine print except when absolutely necessary.[106] Occasionally when searching for an emission line in the spectrum of a star, such as Hα in the spectrum of β Lyrae, Keeler would take off his glasses, and hold his head so that the blurring caused by his astigmatism would artificially widen the spectrum, and make the line easier to see, but usually when observing nebulae he left them on.[78]

In making these measurements, Keeler was assisted at the telescope by Leuschner, as part of the work expected of a graduate student, during the periods he was at Mount Hamilton. As assistant he would help point and set the telescope, record the measurements as Keeler called them off, occasionally confirm the measurements while Keeler in turn recorded, and operate the comparison light source.

In the early summer of 1890 William Wallace Campbell, whose career was to be closely intertwined with Keeler's for the rest of Keeler's life, arrived at Mount Hamilton to start work as his second volunteer assistant. Campbell, five years younger than Keeler, had been born and raised on a small farm in northern Ohio. His family background was Scots-Irish, and his father died when Wallace, as he was called in the family, was only four. His widowed mother had six children to raise, and they all had to work hard at their farm chores. Wallace soon showed that he had high mathematical ability, which attracted the attention of the teachers in the little country schools he attended. After finishing school Campbell himself taught for a few years, and then at the age of twenty entered the University of Michigan as a student of civil engineering. But he happened to find a book of astronomy in the library, Newcomb's *Popular Astronomy,* devoured it in two days and two nights, and decided then and there to become an astronomer. He learned to observe and to calculate orbits under the tutelage of Schaeberle, and worked as his assistant at the Michigan observatory the last year before he got his B.S. degree in 1886.[107] Campbell, on a trip to Pittsburgh just before graduation, visited Allegheny Observatory and saw Langley, but missed meeting Keeler, who had departed for Lick less than a month before.[108]

After graduating Campbell got a job at the University of Colorado, where he was a professor of mathematics for two years, but in 1888 when Schaeberle joined the Lick staff, Campbell came back to Ann Arbor in the vacancy thus created in its Astronomy Department.[107] He was anxious to get into research work with a big telescope, and in the fall of

1889 wrote Holden and asked for the chance to come to Lick Observatory the following summer. Holden arranged for him to be accepted as a special student, which meant that he could work as a volunteer research assistant at Mount Hamilton, although he was not a candidate for a degree.[109] Many astronomers spent one or more summers at Lick Observatory under this arrangement in succeeding years, including observatory directors Henry C. Lord of Ohio State, Susan J. Cunningham of Swarthmore, and Pritchett, as well as several other less well known scientists.[110] Campbell wanted to work and Holden was glad to let him do so; in return for his room he spent one hour a day transferring data from Schaeberle's observing books to large computing sheets, and he was allowed to observe six nights a week, two with Holden and two with Keeler on the 36-inch refractor, and two by himself with a small 6-inch photographic telescope. Campbell was a hard-working perfectionist, and especially enjoyed the spectroscopic work with Keeler; they soon became close friends, with Campbell always cast in the role of the admiring disciple.[111]

Keeler's most important observational work at Lick, which he began one cold windy night in January 1890, was his measurements of the wavelengths of the brightest emission lines in the spectra of nebulae. Early visual observers gave the name nebulae (Latin for "clouds") to all extended, or non-stellar, objects they discovered that did not move in the sky and hence were not planets. Many of them were small, roughly round, and with fairly high surface brightness; to an observer with a small, imperfect telescope they might look something like planets and they were called planetary nebulae. Other nebulae, generally larger and with less regular forms, were called diffuse nebulae; the "Great Nebula in Orion" (or Orion nebula), is the brightest and best known of these. The spectra of the planetary nebulae, and of some of the diffuse nebulae, consist of bright emission lines, as Huggins had first discovered in 1864, with only a very weak and to the early observers undetectable continuous spectrum. Two of the lines in the spectra of the nebulae were soon recognized as the hydrogen lines $H\beta$, in the blue-green region at $\lambda4861$, and $H\gamma$, in the violet spectral region at $\lambda4340$, known to the astronomers of Keeler's day as F and G respectively. But the brightest lines in many nebulae, two lines in the green spectral region, were not unambiguously identified.

Lockyer believed that the brighter of these two, often called the "chief nebular line" or the "principal nebular line" had the same wavelength as a "magnesium fluting," or in modern terms a band head of the molecule MgO. Furthermore, he maintained that the nebular line was not sharp, as atomic lines are, but rather was diffuse, as molecular bands are, and therefore identified it as arising from magnesium molecules in the nebulae.[112] This interpretation fitted in well with his meteoritic theory. His rival Huggins, on the other hand, believed that his observations showed that the chief nebular line did not coincide with the magnesium feature, and furthermore was sharp, not diffuse, and therefore that it did not prove the presence of magnesium molecules in the nebulae.[113] Keeler, from his first observations of the Orion nebula with the large Lick spectroscope, realized that with this excellent instrument on the 36-inch refractor, he was in a far better position than Lockyer and Huggins, working with their small telescopes and primitive spectroscopes in a poor climate, to solve the problem. The emission lines in the Orion nebula were bright and easily seen, and Keeler was soon convinced that they were sharp and symmetric. He was using his highest dispersion compound prism in the spectroscope for this observation, but resolved to try the grating, which would give even higher dispersion, the next time he observed.[78]

On February 11 he did observe the Orion nebula for the first time with the grating, and was able to see three lines, the two nebular lines and Hβ, as he had with the prism. In spite of its much higher dispersion, the grating did not appreciably weaken the lines, because they were so sharp that they were not broadened even at this resolution. Keeler made several independent settings on the chief nebular line that night, and saw that he would be able to measure its wavelength to an accuracy of a few hundredths of an ångström unit (Å), or "tenth meter" (10^{-10} m) as the spectroscopists of his day said. However it was necessary to measure a comparison line from a laboratory source, as close as possible to the nebular line, immediately before or after the spectral observations of the nebulae. The best candidate turned out to be a line of lead near λ5005, and Keeler was soon overhauling the storage batteries and coil of his spark apparatus, fitted up with electrodes containing lead to generate its spectrum. He carefully adjusted the position of the spark and the diaphragms inside the spectroscope so that the light rays from the nebula and from the spark followed exactly the same paths inside the instrument. By March 20 he had added a Leyden jar (condenser) to his spark circuit, and found this helped produce a strong, sharp lead comparison line. That night he was able to make a good series of settings on the nebular line in Orion, assisted by Schaeberle, who also verified some of Keeler's measurements for him. They showed very clearly that a single measurement had a probable error of only about 0.05 Å, so by repeated settings he would be able to achieve an accuracy considerably better than this. Furthermore, his result for the wavelength of the chief nebular line in

Orion differed by about 1 Å from Huggins' result, a difference far larger than Keeler thought his own measurements could possibly be in error, but still showed that the line had a different wavelength from the MgO band head.[78]

Keeler wanted to repeat these measurements on another night, for Orion was setting earlier each night, and soon would be unobservable in the daytime sky. But one storm followed another, and on the one remaining night in March that was clear, the water supply to the dome failed, making it impossible to operate the floor, a gigantic hydraulic elevator that enabled the observer to reach the telescope in any position. Keeler used the cloudy nights to measure additional comparison lines of iron and "air" (nitrogen) to try to get his wavelength scale fixed as accurately as possible. He had written Rowland, who was systematically measuring the wavelengths of the spectral lines of all elements on a common system to high accuracy at Johns Hopkins, but found that he had not yet studied lead.[114] Keeler therefore had to use earlier, less accurate measurements of the lead line, approximately corrected to Rowland's scale, together with his own measurements of the iron and air lines, as his reference points. Rowland was deeply interested in the nebular problem, and Keeler, fully conscious of the value of having the acknowledged leader of American spectroscopy as his advocate, wrote as soon as he could to give him his preliminary results. He described his methods to Rowland, including of course the use of one of the gratings ruled in his laboratory, and his reduction procedures, but he said that his result differed so greatly from Huggins' result that he did not want to publish it until he had had a chance either to verify it or find the error in his work. If there was one he said, it must be constant, for he always found the same result, and had taken all the precautions he could in adjusting and testing the spectroscope.[115]

In April, Keeler wrote even more diplomatically to Huggins himself, who had suggested the nebular measurements to Holden in the first place, and was naturally deeply interested in learning the results.[116] Keeler described his measurements in detail, as he had to Rowland, but added that when he read Huggins' paper, he realized there must be some source of constant error in his own work, which he would try to find. However he gave Huggins complete freedom to quote him on one result that he considered certain. It was that with the highest dispersion the nebular lines were sharp and narrow, and they could not be the "remnants of flutings" as Lockyer had suggested.[117] This disproof of his rival gladdened Huggins' heart, and he quickly wrote back that Keeler might in fact be right about the wavelength too, but suggested several tests of the apparatus, and also that he ought to try to measure some of the same lines in the spectra of planetary nebulae.[118] Keeler had already made all the tests Huggins suggested, and had planned himself to observe the bright planetaries, which are concentrated in the summer sky, just as soon as he could.

On May 22, Keeler was able to take his first look at a planetary nebula, NGC 6210 (called Σ5 by Keeler), with the large spectroscope. He found the two green nebular lines considerably brighter than in Orion, so there was no doubt he would be able to measure them, but they were somewhat hazy and diffuse, not sharp, a result we now understand as resulting from the expansion of this nebula. Interspersed with the observations of Mars and its satellites he was making in June, Keeler was able to measure the wavelength of the principal nebular line in NGC 6210, and compare it directly with the MgO band head, which he could now produce in his spark comparison apparatus. The comparison immediately showed that the band head was at longer wavelength than the nebular line, which therefore could not be due to MgO.[78] Keeler cabled this result to Huggins in London, sending a confirming letter at the same time to spell out the result in detail, but the exultant old Englishman, not waiting for the mail from America, dashed off a letter to *The Times,* announcing Keeler's result. Huggins wrote Holden "[t]here was no doubt in the minds of the best scientific men that I was right & on other grounds it was all but inconceivable that the nebular line would be due to MgO. Yet L[ockyer's] paper to [the] R[oyal] S[ociety] was so very positive that it was important *for science* that the matter should be cleared up. The telegram which I sent to the 'Times' has been accepted as *final.*"[119]

All that summer, the good observing season at Lick, Keeler, assisted by the eager, energetic Campbell and the thorough, Teutonic Leuschner, plugged away, measuring one planetary nebula after another. At the same time, using the very same methods, he was measuring the radial velocities of a few bright stars, and these measurements, together with the check measurements of planets, gradually convinced him that his results were indeed correct to within a few hundredths of an ångström unit. One crucial case was the star α Bootis (Arcturus) for which Keeler consistently found a velocity toward the sun of about 6 km/s, while Huggins and the observers at the Royal Greenwich Observatory had found a velocity away from the sun of 80 km/s. This large difference in velocities corresponded to over 2 Å in measured wavelengths, just about the same as the difference between Huggins' and Keeler's measurements of the wavelength of the principal nebular line. Keeler tried everything he could to find a source of error in his spectroscope, but no matter what he changed, he

Fig.10. W.W. Campbell, Keeler's volunteer assistant at Lick
Observatory, standing next to the visual spectroscope with which
Keeler measured the accurate wavelengths of the green nebular lines
(1890). (Reproduced by kind permission of the Mary Lea Shane
Archives of Lick Observatory.)

always got the same result.[78] His tests convinced him that the error was not in his own work.

Then he saw a paper in the *Astronomische Nachrichten* by Vogel, at the Potsdam Observatory, who was pioneering the use of photography to record stellar spectra. The photographic plate has the great advantage over the human eye that it can store up the effect of light, and thus by using long exposures Vogel could get much better spectra than any visual observer could hope to see. His paper gave the velocities of a number of stars, measured by the Doppler effect from the photographic spectrograms he had taken, but Arcturus was not among them. However, several of his published velocities for stars were very different from the Greenwich visual observers' results. Keeler wrote Vogel, explaining his nebular measurements, and that they appeared to have a constant error if compared with Huggins' measurements. He said he had found approximately the same difference between his measurements of the velocity of Arcturus and published measurements, quoted his own result, and asked if Vogel had unpublished results for this star. A few weeks later Keeler received a reply from Vogel containing the welcome news that he did indeed have spectrograms of Arcturus, and that they confirmed almost exactly the Lick measurements.[120] Then Keeler had evidence that all his tests had not lied, that his nebular wavelength measurements were also doubtless correct, and that Huggins was simply wrong, almost certainly because of his inadequate instruments. Keeler was quietly exultant.[121] Even before he received Vogel's letter, he had sent a paper to the *Publications of the Astronomical Society of the Pacific,* to get his preliminary results into print, and had forwarded an advance copy of it to Rowland, who he knew would spread the news to other interested spectroscopists.[122] Keeler had also written to Huggins, clearly but diplomatically stating for the first time that his own measurements could not be much in error; after getting Vogel's letter he wrote with even more confidence and definitely stated that his own measurements were better than Huggins', politely alluding to the superior Lick spectroscope.[123]

At the end of summer Leuschner, who had become an instructor in mathematics, went down to Berkeley, and Campbell returned to his faculty job at Ann Arbor. As Keeler had written in his paper, they had helped him greatly, and he missed them personally as well as scientifically.[124] Now Keeler often had to observe alone, except sometimes when he had the assistance of Burnham's son Augustus, who had become the observatory's secretary after Hill left for a better job in San Francisco, and sometimes when Schaeberle helped Keeler.[78] Interruptions occurred from time to time; a new smaller spectroscope, ordered contrary to Keeler's advice, arrived and he had to test it.[125] Holden gave Keeler a long memorandum of instructions on measuring the radial velocities of stars, which he called the "principal spectroscopic work for which the great telescope was designed; namely the observations of the motions of stars in the line of sight for the determining of the motion of the solar system in space." He thought this program would take years to complete, and discussed in detail the types of stars that should be measured, appending to his four-page letter a forty-four-page list of stars, complete with their positions, their bases for selection, and maps for finding some of them.[126] Keeler very politely but sensibly replied that it would be better to see how the measurements went before planning too far into the future. He said that from his experience to that time, many of the stars in the list might turn out to be too faint to observe spectroscopically, and "it seems probable that the plan may be reduced to the severe simplicity of getting everything within reach of the apparatus," as he had previously told Holden.[127] He concluded his letter by describing the high accuracy with which he could measure the radial velocities of the planetary nebulae, and in fact he continued observing them, more or less ignoring Holden's instructions.

While he was fending off these diversions, Keeler continued, every night he could, the hard, steady productive work of measuring accurately the wavelengths in as many nebulae as possible. His measurements had already clearly showed that different nebulae had different velocities with respect to the sun, for the wavelength of the principal nebular line varied from one object to another, as a result of the Doppler effect. To get the approximate laboratory wavelength of the line, Keeler had assumed that on the average the planetary nebulae were moving neither toward nor away from the sun, so that the average measured wavelength for the whole group would be approximately the same as the "rest" wavelength. He recognized that if he could accurately measure the $H\beta$ emission line in the nebulae, he could use it to determine their individual radial velocities by the Doppler effect, since its rest wavelength was accurately known from laboratory measurements. Then, knowing the velocity of each nebula, he could correct the measured wavelength of the principal nebular line measured in it, to get an accurate value of this line's rest wavelength without any additional assumption. The difficulty was that in most of the planetary nebulae $H\beta$ was much fainter than the principal and even the "second" ($\lambda 4959$) nebular lines, and was too faint for him to measure accurately. However, one planetary nebula, $\Sigma 6$ (NGC 6572), was bright enough for him to get a good measurement of $H\beta$, and in September, when the Orion nebula first appeared again in the morning sky just before dawn, Keeler was able to start measuring it again. It has a reasonably strong Hβ line, so he was able to make an excellent determination of its velocity, and therefore also of the true wavelength of the principal nebular line. These measurements of the Orion nebula completely confirmed his results of the previous season, and he now wrote to Huggins that he was "right sorry to say that the recent observations give the same result as the earlier ones, but I do not think it possible there can be any considerable error in them, as I took the greatest precautions to ensure accuracy."[128] Keeler's result for the velocity of the Orion nebula, measured by the Doppler effect on the wavelength of Hβ, was 14 km/s away from the sun, quite close to the modern value. This corresponded to an average velocity for the ten planetary nebulae in which Keeler had measured the principal nebular line of about 30 km/s, rather than zero as he had first been forced to assume.

Keeler also measured, for every nebula, the wavelength of the second nebular line, and found as he had expected, that it always showed the same Doppler effect as the principal line. He determined its wavelength also, with nearly the same precision as the wavelength of the principal line.

Keeler was always highly conscious of the value of discreet publicity. He knew that his career as an astronomer depended on recognition of his work by the scientists who really counted. Certain now that his results were correct, he actively spread the word of his nebular measurements. He had already been in contact with Rowland, Huggins and Vogel, three of the leading spectroscopists in the world; now in addition he wrote William H. Pickering, Agnes Clerke and W. W. Payne, knowing that they would disseminate the news of his finding. Pickering, Keeler was certain, would share his news with his brother, director of the Harvard College Observatory and a powerful figure in the American astronomical establishment whom Keeler had met years before when he had first come north for his education. Clerke was a prolific writer of popular articles and books in England, a partisan of Huggins and a critic of Lockyer, while Payne was the editor of the *Sidereal Messenger,* the only American astronomical magazine of the day. Keeler skillfully described his observations, particularly the measurement of Hβ in the Orion nebula which fixed the velocity, the Doppler shift, and hence the laboratory wavelength of the nebular lines, in individual terms best calculated for each of these three to understand.[129]

He also sent off a short paper to *The Observatory,* the magazine which chronicled the meetings of the Royal Astronomical Society in London and published brief reports of new findings.[130] His paper appeared in print just two months after an admiring review of his work by Edward W. Maunder, an astronomer at the Royal Greenwich Observatory, had appeared in the same magazine. Maunder's review, based on Keeler's paper of a few months before in the *Publications of the Astronomical Society of the Pacific*, emphasized the importance of his work for Lockyer's meteoritic hypothesis. He wrote that Keeler's description of the principal nebular line as sharp and monochromatic, not resembling a fluting (or molecular band), completely confirmed Huggins and was "[d]ead against Mr. Lockyer's theory." The wavelength difference between the nebular line and the magnesium fluting in all the nebulae observed was crucial, he said. "If the chief nebular line is *not* the remnant of the magnesium fluting, the very keystone is knocked away from the arch [of Lockyer's hypothesis], and the edifice, as such, falls to pieces." Maunder considered it still barely possible that the wavelength difference was due to the Doppler effect, and recommended that Keeler measure Hβ in the nebulae to determine their radial velocities.[131] This is just what Keeler had done in the Orion nebula, after submitting the earlier paper, so in *The Observatory* he could emphasize that the crucial test suggested by Maunder had now been cleared up by his further work. It was undoubtedly extremely convincing evidence to anyone who read it with an open mind.

Lockyer, however, could not admit that his theory was wrong. He had only recently published a series of papers in the *Proceedings of the Royal Society of London*, the most prestigious English scientific journal, insisting that the chief nebular line was indeed the "remnant of the magnesium fluting," and claiming that Huggins' earlier measurements were wrong, and that even if they were right, there were many possible instrumental sources of error large enough to explain the apparent discrepancy of 1.5 Å he had found between the magnesium feature and the chief nebular line.[112] Huggins fired off a paper in reply, much like a lawyer's brief, in which he cited not only his own measurements but also "Professor Copeland, Professor Young and Professor Keeler, working with the highest dispersion," all of whom agreed with him that the line was completely sharp and symmetric. Huggins argued that Lockyer's report that the nebular line was diffuse and asymmetric, like a molecular band, probably meant that his spectroscope was imperfectly aligned, which in fact is probably true, although the effects of wishful thinking should not be underestimated.[132]

Keeler himself now wrote a paper replying to Lockyer's claims, on the advice of Huggins, who submitted it for him to the *Proceedings of the Royal Society*. Although Huggins had wanted him to make the paper an all-out attack, Keeler simply presented his observational results without any controversial or partisan statements.[133] In his final editing, Huggins agreed with Keeler's tactics and even removed the aspect of a reply to Lockyer, leaving the paper as a report of new data that cleared up the problem.[134] In it, Keeler emphasized that he had now measured Hβ in the Orion nebula, and thus had determined the wavelength of the chief nebular line independently of any hypothesis about the average motion of the planetary nebulae. He quoted the precision of about ±0.03 Å, derived from measurements of the moon, planets, Arcturus and Betelgeuse, and gave as the final result a difference between the wavelength of the magnesium feature and the chief nebular line of 0.43 Å, far larger than any possible source of error. Further, he said, the nebular line has no resemblance to a molecular band. and finally, if it were the MgO band, other lines and bands which occur in all laboratory spectra of magnesium should be present in the nebulae but were not. It was a model paper with very firm conclusions, and until Keeler's long paper came out several years later, it was the most complete statement of his results.[135]

Although Keeler's paper convinced most of his contemporaries, Lockyer and his partisans could not accept the result, and at a meeting of the Royal Astronomical Society on May 8, 1891, with neither Huggins nor Lockyer present, an argument welled up. It began with the reading of a paper from K.D. Naegamvala of Poonah, India, who had been observing the Orion nebula with his 16 ½-inch telescope and three-prism spectroscope and found that the chief nebular line was sharp under all circumstances, and therefore not the remnant of a magnesium fluting, as Lockyer had suggested. Captain William Noble, a friend and partisan of Huggins, then rose and smoothly congratulated Naegamvala, through the Secretary who had read the paper, saying he agreed with the conclusion found by "our eminent Fellow, Dr. Huggins and his gifted wife," corroborated by Keeler in the United States, also in Germany, and now again in India. Alexander Herschel, the son and grandson of famous astronomers, tried to smooth things over by saying that even if the line was not magnesium, the nebulae might still be cool, low-temperature objects, which he said was Lockyer's main point. But Noble's statement had inflamed Albert Taylor, a Lockyer supporter. He said he had observed the chief line in the Orion nebula with Sir Henry Thompson's 12-inch telescope at Hurstside, Surrey several times. Sometimes the line appeared "fluffy," sometimes it did not. Furthermore, he said, Keeler's observations were not important. The Lick telescope was too big. Its long focal ratio made it quite unsuited for this work. He claimed that the high dispersion Keeler used prevented him from seeing the faint fluting, and that his grating spectroscope had severe light losses, further weakening the supposed magnesium feature. Finally came the unkindest cut of all,

> When we add to these disadvantageous conditions the fact that Mr. Keeler's eyes are not acute, we can easily understand how it is that he cannot see any resemblance to a fluting in the nebular line. I should have been very surprised indeed if he had, for his telescope, his spectroscope, and his eyes are all against him. His negative evidence has had too high a value placed upon it, for negative evidence from him is of very little value, although positive evidence from him would be of greater value than from anyone else.

Noble could not resist asking Taylor if he really believed the nebular line was magnesium. Taylor hotly replied he would not be cross-examined in this way, but that the facts of observation were that Dr. Huggins, Father Secchi, Dr. Vogel, Dr. Bredichin, Mr. Maunder, Professor Lockyer, Mr. Fowler and he himself had all at times seen the line fluffy on its blue edge. It was a last gasp of the tradition of gentlemanly amateur astronomy, in which the consensus of a mass of poor data was taken as defining the true situation, confronted by far superior measurements, obtained with the best professional instrumentation, with which in fact it could not be compared.[136]

The idea that the long focal ratio of the Lick telescope made it ill-suited for nebular spectroscopy was one shared by nearly all the English astronomers, but was completely wrong, as Keeler well knew from his study of optics. It is true that the image of a nebula at the focal plane of the 36-inch telescope, which has a focal ratio F/17, is fainter than it would be in a telescope with a faster focal ratio, but since the collimator of the spectroscope had the same focal ratio, the brightness of the spectral line Keeler saw at the eyepiece depended only on the focal ratio of the small viewing telescope that completed the spectroscope. Furthermore, the large aperture of the Lick telescope made the image of the nebula larger than a small telescope would, and the spectral line it produced at the eyepiece was therefore longer and hence easier to see – a positive advantage of the large telescope. Keeler, in a letter of reply written for publication, coolly but politely said Taylor's remarks were "evidently founded on a misconception of the optical principles involved in the construction of spectroscopes." As to the remarks on his "bad eyes," Keeler wrote "I have no objection on theoretical principles, but merely state that frequent comparisons with the optical powers of many noted observers have failed to show that they are so very bad." Finally, he emphasized that he had not begun his research with the idea of disproving Lockyer's hypothesis. That would have been unfair, he said. "As a matter of fact, my observations were made for the purpose of ascertaining the truth, whatever it might be."[137] Although this ringing statement was undoubtedly true, it omitted the fact that he had little respect for Lockyer, with whom his relations were distant and formal.[138] Whatever his private opinions, Keeler came off far better in this exchange than Lockyer, Huggins, Taylor, or anyone else concerned. .

Keeler's work on the wavelengths of the nebular lines, although not available in full until several years later, were widely known to astronomers from the three papers he had published in 1890-91, and from his correspondence. He made sure to spread the word beyond England and Germany, writing pleasant, respectful letters to astronomical spectroscopists throughout Europe, in which he described how his accurate measurements proved the nebular lines could not be a magnesium fluting, as Lockyer had claimed.[139] In his own country, Keeler was widely regarded as a coming man from then on.[140]

His own director, Holden, was one of Keeler's greatest supporters.[141] He praised Keeler's research highly in his letters to other astronomers, usually however implying that it was joint work in which he himself was actively participating.[142] As early as 1889 he had written to President Daniel Coit Gilman at Johns Hopkins, Keeler's alma mater, recommending him for the Ph.D. degree that "[c]ircumstances have prevented his taking." Holden assured Gilman that Keeler had used his time better in research at Lick than if he had continued for his degree, and that "his course has been one of steady growth & of excellent achievement."[143] Johns Hopkins had strict residence requirements, however, that Keeler could not meet, so he did not get the degree.[144] Nevertheless, Holden continued to recommend him for it, writing once to Ira Remsen, professor of chemistry and acting president in Gilman's absence, and the next year to Gilman again, saying, "I think his last publication on the motion of the planetary nebulae is about the best thing we have done here."[145] The degree would have meant a lot to Keeler, but the Johns Hopkins faculty would not break their rule that degrees would only be awarded for work done on the campus.[146]

Keeler, meanwhile, kept piling up measurements of the nebular lines at every opportunity. Finally, near the end of May 1891, he was able to measure $H\beta$ in NGC 6572, the only planetary nebula, and the only nebula besides the Orion nebula, bright enough for him to do so.[78] These, however, were among the last observations he made at Lick before leaving on June 1, 1891, to become the new director of Allegheny Observatory.

5

A human being first and an astronomer afterwards

At the very time he was making his epochal spectroscopic observations of the nebulae, James E. Keeler was actively seeking a job that would enable him to leave Lick Observatory. In the spring of 1888, even before the observatory began regular operation as part of the University of California, Keeler had confessed to Edward S. Holden, soon to take over as director, that he was tired of Mount Hamilton, after two years there helping in the completion of the telescope and in the initial observations.[1] Mount Hamilton was a tiny, isolated community, twenty-seven miles and over five hours by stagecoach up a winding mountain road from San Jose. Only the astronomers and a few technicians lived at the site. Keeler could joke about his fellow staff member, John M. Schaeberle, a quiet, retiring perpetual bachelor who went to San Francisco only at six-month intervals, but he enjoyed the comforts of city life and human society too much to stay cooped up on the mountain for such a long time himself.[2] Sidney D. Townley, who was a graduate student at Mount Hamilton four years later, decided after only a few months that although it was an interesting, beautiful place, it was too far from civilization to merit consideration as a lifetime home, a feeling Keeler undoubtedly shared.[3]

Keeler was also dissatisfied with his salary, which remained constant at $1,800 per year from July 1, 1888 until he left the University of California.[4] This was comparable with the salaries of assistant professors on the campus in Berkeley, but well below the $3,000 per year associate professors were making at Johns Hopkins, Keeler's alma mater.[5] He considered himself definitely underpaid.[6]

Most of all, Keeler felt the need of better housing than was available at Mount Hamilton. He lived in a few rooms in the basement of the Brick House, the large building that had been constructed as the astronomers' quarters. It was divided vertically into two halves, one of which was

Holden's apartment, with Schaeberle in the basement rooms, while the other side was S.W. Burnham's apartment, with Keeler in the basement. The only other quarters were the undersized, uncomfortable wooden cottages, intended originally for the workmen who built the observatory; E.E. Barnard lived in one of these with his wife. The cottages were small, close, and roughly built, and became very hot in the long, dry, sunny California summers.[7] To Keeler they were completely inappropriate dwellings to ask a wife to share, and that was really one of his main problems with remaining at Lick Observatory.[8]

He had first met Cora S. Matthews on Mount Hamilton in the summer of 1886, when she lived with her aunt and uncle in the Brick House. Her mother, Isabella Matthews, was the half-sister of Cora Lyons Floyd, whose husband, as president of the Lick Trust, supervised the building of the observatory. Keeler worked closely with Captain Richard S. Floyd, with whom his relationship was much like that of a friendly, warmly respectful son.[9] No doubt he was often in the house with the family, and he was soon attracted to the lovely niece from Louisiana. After the Floyds left the mountain, Keeler visited them often at Kono Tayee, their estate on Clear Lake, and it is not hard to believe that at least part of the attraction was Cora Matthews.[10] We have no record of when or how Keeler popped the question, or indeed any personal letters or diaries from this period, but clearly from sometime in 1888 or 1889, Keeler and Cora considered themselves at least unofficially engaged. Certainly by the summer of 1889 he was writing Samuel P. Langley, his former director at Allegheny Observatory, that he needed a house and home of his own, and that he wanted "to marry if I choose and to live happily until I die."[11] Six months later he informed Timothy Guy Phelps, chairman of the Lick Observatory Committee of the Board of Regents, that he intended to get married and had to have adequate housing if he were to stay on Mount Hamilton. Keeler said that he did not want to pressure the regents, but that he did want to know whether they would build a house for him, because if not he intended to look for a position elsewhere, where he could have a home. The regents wanted him to stay, but did not provide a house.[12]

Keeler's other problem, less well-defined, was probably with Holden, his director, although he never admitted it in writing, and vehemently denied it in print.[13] However, the fact is that Holden, although highly supportive, was in many ways a director of the old school. He issued instructions to the astronomers on "his" staff on the scientific work he expected them to carry out, and in describing their results he tended to talk about what "we" (meaning, for instance, himself and Keeler) had done.[14] As we have seen, Keeler quietly and tactfully went his own way, and made sure that other astronomers and physicists learned about what he had done, but he could not have liked continually fending off Holden. Nearly every other astronomer who worked at Lick under Holden came to dislike him intensely; it is almost impossible to believe that Keeler was the single exception. And, however he felt about Holden, in an age of independence he wanted to be independent himself. Thus when George C. Comstock, his friend and co-worker at Mount Hamilton in the summer of 1886, was appointed director of the Washburn Observatory of the University of Wisconsin, a position with distinctly limited research opportunities, Keeler wrote him that it was clearly a step up from Lick Observatory for him, and that he would have accepted it himself if he had had the chance.[15]

In the spring of 1890, just before he began serious spectroscopic observing of the nebular lines in the Orion nebula, Keeler made a strong attempt to get a job at Stanford. This new university, not scheduled to open its doors to students until the following year, was then under construction at nearby Palo Alto. One of the members of the Stanford Board of Trustees was Irving M. Scott, whose Union Iron Works in San Francisco had built several battleships and cruisers for the United States Navy, as well as the steel dome of the 36-inch refractor at Lick Observatory.[16] Keeler had met him in Floyd's house on Mount Hamilton, and now wrote to urge that the new university start an astronomy department, with an astrophysical observatory, and himself as its director. He said he was determined to leave Lick Observatory, but wanted to remain in California if he could. Although Keeler admitted that he could not rank himself with the senior astronomers of the country, like Simon Newcomb, Asaph Hall and Edward C. Pickering, he said he felt he could hold his own with the men of his own age, a modest forecast that his career was to more than justify. He wrote Scott that it was his dream to establish an astrophysical observatory in California that would accomplish more for science than much more expensive, conventional observatories, and that he had a well-thought-out plan that could make it a reality at Stanford.[17] Unfortunately the Stanford trustees did not hire him or any other astronomer at that time, and never did build a research observatory on the campus.

Keeler kept Daniel Coit Gilman, the president of his own alma mater, and one of the most influential figures in American university life, aware of his accomplishments by sending him regular letters and reprints of his papers. When a Johns Hopkins report referred to Keeler as an assistant astronomer at Lick Observatory, he quickly dashed off a letter to Gilman informing him that he was an astronomer, not an assistant.[18]

Finally, in the summer 1890, at the very same time he was carrying out astronomical spectroscopic measurements far superior to any previously made, Keeler made his move for a job he had a real chance of getting. This was the position as professor of astronomy at the Western University of Pennsylvania and director of its Allegheny Observatory, where he had worked under Langley before coming to Lick. Only a few months after Keeler had departed for California, Langley had been named assistant secretary of the Smithsonian Institution in Washington; less than a year later when the secretary, Spencer Baird, died, Langley succeeded him as the head of this combined museum and research institute. He retained his position at Allegheny Observatory on an unpaid basis, commuting back from Washington to Pittsburgh at leisurely intervals.[19] He was becoming increasingly interested in the experimental studies of aerodynamics which he had begun at Allegheny, but had better facilities to carry out at the Smithsonian. Thus after William Thaw, the wealthy Pennsylvanian who had financed the astronomical work at Allegheny Observatory, died in 1889, Langley's interests became almost completely centered at the Smithsonian.[20]

Keeler's friend in Allegheny, telescope maker and astronomy enthusiast John A. Brashear, was naturally concerned at the lack of activity at the observatory. He knew that Langley would resign the directorship before long, and Keeler had always been his candidate to succeed to the position. In his frequent letters to Holden about the instruments he was building for Lick Observatory, Brashear nearly always included a tidbit of news about developments at Allegheny, asking that it be passed along with his kind regards to Keeler.[21] Now he wrote Keeler asking if he would be interested in returning to Allegheny, if and when Langley resigned. Brashear, in his expansive way, thought Keeler might well become the next director at Lick, if Holden ever left, and wrote that he was not sure that his friend would want to come back to the little Allegheny Observatory. Keeler seized the opportunity, and poured out his thoughts to Brashear, or at least those thoughts that he wanted him to know. He wrote that the Lick directorship was too big a job, that it required different talents from his, and that if he did ever get it, it would be years in the future. But he told Brashear confidentially that he had been thinking of leaving Mount Hamilton for some time, and hinted that he might be considering marriage. He spelled out the unsatisfactory housing situation on the mountain. Finally he indicated that although he would not apply for the Allegheny position, he would accept it if it were offered to him, provided that the university administration and trustees wanted him and promised adequate financial support for his research. Keeler said that he did not wish to stand in the way of either Brashear himself or Frank W. Very, Langley's assistant, but he thought that Brashear would not want the directorship and the Trustees would not appoint Very. Keeler closed by saying Lick was a fine place for research "but I am a human being first and an astronomer afterwards."[22]

The very same day, Keeler wrote to William Thaw, Jr., the son of his old patron, whom he knew well and who was an important member of the Western University Trustees' Observatory Committee. Keeler made no mention of his interest in the directorship, but simply asked Thaw if he did not want to join the Astronomical Society of the Pacific, so that he could receive their publications. He sent along a recent issue, from which Thaw could see that Keeler was one of the editors, had contributed several papers himself, and was a highly respected astronomer.[23] Keeler was laying the groundwork for the future.

One of the conditions of employment at Lick Observatory was that the astronomers were entitled to one-month vacations each year, provided they took them in the winter, the poor observing season. Keeler had not left California for nearly five years since coming to Mount Hamilton, and had a lot of vacation time saved up, so he now decided to take two months and go East to solidify his contacts. He planned to visit several of the most important observatories, and naturally to spend a good deal of time exploring the situation at Allegheny.[24]

Keeler left Mount Hamilton after observing on the night of October 30, 1890, and spent a few days in San Francisco, where he described to a reporter some of the research he had been doing, before heading East by train.[25] On the way to Washington, where his mother and married sister lived, he stopped off at Northfield, Minnesota, to visit William W. Payne at Carleton College, and at Ann Arbor, to visit his former assistant W.W. Campbell at the University of Michigan.[26] Payne was the founder of the *Sidereal Messenger,* the only astronomical magazine then published in America. It was not a research journal, but a general magazine combining short papers by professional astronomers and advanced amateurs with news and views, mostly written by Payne himself. Keeler had furnished him with many items, including several progress reports on Lick Observatory.[27]

Meeting Payne in person, Keeler found a bustling, middle-aged man who had grown up on a farm in Michigan, where he had sporadically attended country schools until at the age of seventeen he had become a teacher himself. When he reached twenty-one, Payne had left the farm to study at tiny Hillsdale College, a few miles from his home. He had graduated after five years with an A.B. degree in 1863, and then had gone to law school in Chicago. He ended up as a county superintendent of schools in southeastern Minnesota, where he soon began publication of the *Minnesota Teacher,* a magazine. Carleton College, a Congregationalist school, was founded in 1868; three years later Payne became the third member of its faculty. He taught all the mathematics courses, but had a driving enthusiasm for astronomy, in which his whole training had been a short descriptive course at Hillsdale, a similar course one summer at Oberlin College in Ohio, and a few summers' experience as a volunteer assistant at the Cincinnati Observatory. To the Carleton students he was known as "Uncle Billy," and until his retirement in 1908 he continued to open every class with a prayer, the last teacher at Carleton to do so. The students thought he was lost in the stars, but he had raised the money to build two observatories on the campus for them.[28] Keeler must have been amused by this lively but unprofessional figure, but he cultivated him assiduously and continued to get a good press. Just a few years later, Keeler was to become one of Payne's associate editors on *Astronomy and Astro-Physics,* the successor to the *Sidereal Messenger.*

Arrived in Washington, Keeler divided his time between visiting his mother, sister Elizabeth, and brother-in-law, David T. Day of the Geological Survey, and taking a busman's holiday devoted to astronomical and scientific calls. He went out several times to the Naval Observatory, then located on a hill overlooking Foggy Bottom and the Potomac River near the end of Constitution Avenue. There he discussed research with Hall and with Stimson J. Brown, an Annapolis graduate, first in his class, who had become a professor at the observatory. They were both old-line astronomers of the pre-astrophysics era. At the Smithsonian Institution, Keeler renewed his personal contact with Langley, and no doubt discussed his former employer's plan to start an astrophysical observatory.[26]

Keeler also went up to Princeton and had some real astrophysical discussions with Charles A. Young. Then fifty-six years old, Young had been a pioneer astrophysicist at Dartmouth College, where his grandfather and father had been faculty members before him, professors of mathematics and natural philosophy, and of natural philosophy and astronomy, respectively. By Young's time science was more specialized and the old name had fallen out of use; he himself had been simply the professor of astronomy. With a small telescope and a prism spectroscope he had carefully studied the spectrum of the chromosphere, the thin outer layer of the sun just inside the corona, and of prominences, extrusions of the chromosphere up into the corona, back in 1869, only a year after J. Norman Lockyer in England and Jules Janssen in France had discovered that they could be observed spectroscopically without an eclipse. Young had discovered the emission-line spectrum of the reversing layer, just inside the chromosphere, at the eclipse of 1870 in Spain. He had actively studied the spectra of the various regions of the sun at several eclipses. In 1877 he had moved from Dartmouth to Princeton, drawn by the promise of a large telescope, the 23-inch refractor, which was completed in 1882. Keeler had first met Young in Colorado in 1878, where he had been in charge of the Princeton eclipse party. By the time of his move to Princeton, Young had lost most of the energy of youth, and was spending his time teaching, writing books and magazine articles on astronomy, preparing presidential addresses which he delivered to various societies, and generally playing the role of an elder statesman of science. He was in poor health and practically never observed at Princeton, leaving that part of the research to his assistant. Nevertheless he was a stimulating person, short, plump, and lively (the Princeton students called him "Twinkle" behind his back); he knew hundreds of stories of the earliest days of astrophysics.[29] Young was full of admiration for Keeler's work, and urged his young visitor to stay at Princeton as long as he could and tell him all about his research.[30]

Keeler also stopped at Columbia University, in New York City, and gave a public lecture, which was attended by many scientifically oriented people who had heard of Lick Observatory and wanted to learn more about it.[31] Undoubtedly his talk was very similar to the lecture Keeler gave in San Francisco to a meeting of the Astronomical Society of the Pacific soon after his return from the East. It was a resumé, intended for popular consumption, of his nebular research, blended with a skillful exposition of the principles of astronomical spectroscopy. Thus the listeners could understand that his measurements showed that the nebular lines did not coincide in wavelength with the lines of any known terrestrial source, and certainly not with those of magnesium, as Lockyer had hypothesized. They grasped that Keeler's measurements of the Doppler effect gave the velocity of the earth with respect to the nebulae, and thus, in some sense or other determined the motion of the solar system with respect to a much larger reference system. Keeler's knowledge and air of quiet confidence impressed the audience in San Francisco, as they no doubt had in New York the month before.[32]

The most important part of Keeler's trip, however, was a visit to Pittsburgh and Allegheny Observatory. Undoubtedly he had long conversations with Brashear, discussing the situation at the observatory and the university in depth. It was clear that a new director would soon be needed, although Langley had still not officially resigned. A complicating factor was that the previous chancellor of the Western University, Milton P. Goff, had recently died of pneumonia.[33] No new faculty appointment was likely to be made until after a new chancellor had been named. Before leaving Pittsburgh, Keeler had a long talk with Thaw, who indicated that he wanted the former assistant back at Allegheny as director, but made no definite promises.[31] Keeler returned to Washington with the knowledge that he had Brashear as his advocate who, although not a member of the inner circle, was a dynamo of energy, and also that he had the more important, but lethargic, Thaw as his benign supporter.

In late December Keeler started back for California, accompanying a melancholy cargo. His friend and mentor, Captain Floyd, had been brought to the Atlantic Coast early in the fall to seek the best medical care for his rapidly worsening heart condition. But even the Eastern doctors could not provide a cure, and in October he had died in Philadelphia. His wife, his daughter Harry, and his niece Cora Matthews were all with him, but Mrs. Floyd did not feel up to traveling until the end of the year, and it was then that they brought his body back to rest in a vault in Laurel Hill Cemetery in San Francisco. Keeler escorted the ladies, ending his vacation somewhat sooner than he had planned. They had a private railroad car and followed the southern route via New Orleans, El Paso and Los Angeles to San Francisco, where Keeler was one of the pall bearers in the impressive funeral services at Grace Church on Nob Hill in the heart of San Francisco.[34]

A few days later, Keeler was back at Mount Hamilton, observing the spectral lines of the Orion nebula again on a windy January night.[35] He also kept up his quiet campaign for recognition, writing Agnes Clerke, the historian of astronomy, to tell her about his spectroscopic results on the variable star β Lyrae. He described the emission and absorption lines he saw in its spectrum at various phases, and told her that it was so interesting and so important that he intended to continue his observations a bit longer before publishing any results on it, even though this meant he would risk being anticipated by other researchers.[36]

And Keeler also became formally engaged at this same time, writing to Cora's father in Louisiana to ask his consent to their marriage. Keeler gave a brief account of his life and ancestry, and confessed that although he was a successful astronomer, his financial prospects in science did not compare with what any successful businessman might expect to earn. In this letter Keeler said that he had "been offered the Directorship of the Allegheny Observatory, where I was formerly assistant, and it is possible that I may accept, in which case I should have a more comfortable home than it is possible to have on this isolated mountain."[37] This was a somewhat exaggerated report of the situation; in fact at this stage the offer must have been more of a promise that he would get the job some day than a firm commitment.

Cora's father, William W. Matthews, had been an officer in the Confederate Army during the Civil War; he had been wounded in battle and his family plantation in Louisiana had been pillaged by the Union occupation forces.[38] All that was twenty years in the past, and now he was glad to give his blessing to the engagement of his daughter to the young Northerner, as his wife had been sure he would.

Keeler prodded Thaw, writing that he was engaged and that the regents had promised him some "fairly comfortable quarters" on the mountain, if he would commit himself to stay. There is no documentary record of this promise; probably it involved rearranging the interior of the Brick House so that Keeler's apartment could be somewhat enlarged. In his letter he asked Thaw if he could not let him know whether he was going to be offered the Allegheny position, and if he was, what the salary would be and when he could start. He was trying to make something happen.[31]

Within less than two weeks, Keeler had a reply from Thaw, which again expressed strong interest in having him come to Allegheny when the chancellorship issue had been settled and Langley had resigned; but that was still not a definite job offer. The very same day Keeler also received a letter from Langley, which did contain a firm job offer, to be his assistant again, this time in the new astrophysical observatory being set up at the Smithsonian. Keeler almost certainly had no intention of accepting this position; he valued his independence too much. He had learned how to do research from Langley, but now the learning stage of his life was over and he wanted to be his own master. But he hurried to inform Thaw of the Smithsonian offer, both to show Langley's confidence in his abilities, and to emphasize the need for decisive action at Allegheny.[39] To Langley he wrote a temporizing letter, indicating his gratification that his old director had succeeded in getting the new observatory established, and that he wanted Keeler back with him again. "[I]t never rains but it pours," Keeler said, telling Langley he had been offered two other positions, but not definitely, and that he had been asked not to tell anyone of them. One of course was the Allegheny position, which Langley actually still held, so it would have been very embarrassing to let him know about it; the other "offer" probably just meant the chance to stay at Lick with better housing. Certainly there are no letters in Keeler's private correspondence book or elsewhere indicating negotiations for any other jobs at this time. Keeler asked Langley for a few weeks' delay until he could find out what definite offers the other places would make to him. Knowing of Langley's respect for the Royal Society, England's most prestigious scientific society, Keeler also told him that he had been asked to prepare a paper on his spectroscopic measurements of the nebular lines for it, and that William Huggins had promised to "communicate" (or vouch for) it.[40]

At nearly the same time, Keeler again wrote Gilman, whom he knew might be consulted by the Allegheny trustees. He sent the Johns Hopkins president the first two volumes of the *Publications of the Astronomical Society of the Pacific,* which contained many of his own research papers, thus proving that he was a productive scientist. He told Gilman he had learned that Holden had recommended him for a Hopkins Ph.D. degree, and that although he much appreciated the thought, if he had known of the idea in advance he would have advised his director against it, as it was contrary to the policy of his alma mater. Although the degree was an honor he would have highly appreciated, Keeler grandly said, it would be more valuable to a young man seeking a position than to an established scientist like himself. He also told Gilman, as he had Thaw and Langley, of his engagement to Cora, and said that he had been offered the Allegheny directorship, but had not yet decided whether or not to accept it, and therefore asked that this information be kept completely confidential.[41] Here Keeler was straining the truth, but by then he had probably reached the conclusion that he would eventually be offered the job if he waited long enough, and decided to take the chance.

Then, just a few days after Keeler had mailed these letters, the situation changed further. Cora Floyd, who had been seriously ill ever since her husband's funeral, died on February 27, with her daughter and her niece at her bedside.[42] Mrs. Floyd had inherited a sizeable fortune from her father, and by her own canny investments had built it up to over a million dollars.[43] In her will she left nearly all of it to Harry, and named Cora Matthews, who received a smaller but comfortable legacy, as her guardian. Harry had just turned eighteen, the legal age of adulthood for women in California in those days, and thus could do as she chose. However, her mother had especially wanted Cora to take care of her, so Harry intended to stay with her, at least initially. Keeler and Cora were determined to marry soon; if it was impossible for him to keep a wife in his inadequate quarters on Mount Hamilton, it was doubly impossible to keep a wife and her young, wealthy heiress charge there, even if the regents made some improvements. All this Keeler explained in a letter to Thaw, begging him to arrange for a speedy decision.[44] The crisis had been brought on by the arrival of a letter from Langley, who gave May 1 as the deadline by which Keeler must decide if he wanted the Smithsonian position. Keeler diplomatically replied that he was determined to leave

Mount Hamilton very soon and that as a residence Washington would be his first choice; he promised that he would let Langley have a definite answer by May 1 at the latest.[45]

The person who actually got things going at Allegheny was not Thaw, who was in New York just then, but Brashear, who was galvanized into action by successive visits of Langley and Keeler's brother-in-law Day. Toward the end of March Langley had arrived on one of his infrequent trips, and had told Brashear that he had offered Keeler a position as his assistant in Washington. He then asked what plans the Allegheny trustees had for the observatory. Brashear told him that now that the question of the chancellorship had been decided, they hoped to get a new full-time director who would reactivate the research program. Langley asked if they had made their choice, and Brashear said that they had not, but that they were thinking very seriously of Keeler. This was unexpected and unwelcome news to Langley, who first became angry that he had not been consulted, but then cooled off and admitted that there was no better man for the job. Brashear brought up Very's name, but Langley quickly said he was a good assistant, but totally unfitted to be director of Allegheny or any other observatory. He told Brashear that he would resign his honorary position at Allegheny, and give up the director's house on the grounds, which he was still using on his visits, whenever the trustees wished him to do so. By the next day he seemed completely reconciled to the idea of Keeler's succeeding him, asking Brashear about another possible candidate for the Smithsonian position.

Then a few days later Day came over to Pittsburgh from his home in Washington, and visited Brashear in his sickroom, where he was suffering from a severe case of the grippe. No doubt sent by Keeler, Day told Brashear again of Langley's job offer to his former assistant but emphasized that his brother-in-law would far prefer to come to Allegheny. Keeler could not wait forever, because of the May 1 deadline on Langley's offer, Day reiterated.[46]

Brashear, determined to have his friend back at Allegheny, roused himself from his sickbed and began a busy round of conversations, meetings, and conferences. The immediate problem was that although William J. Holland, pastor of the Bellefield Presbyterian Church in Pittsburgh and long-term member of the University Trustees, had finally been elected chancellor, several other members of the board, including Thaw, had favored another candidate. Two of the trustees threatened to resign. Thaw, however, was committed to bringing Keeler to Allegheny, and Brashear, acting on his instructions, scurried around Pittsburgh and persuaded all the crucial faculty members and trustees except one, James

B. Scott, to fall into line. Brashear begged Thaw, and his younger brother Benjamin, also a trustee, not to resign but to give Holland a chance, for if they left the board, he said, the observatory would surely suffer.[47]

Brashear had a long talk with the new chancellor, and in his enthusiastic way, completely sold him on Keeler, quoting Langley to the effect that his former assistant was one of the very best men for the position in the United States. Holland wrote a humble letter to Thaw, saying he agreed with the minority of the board that he himself was not the best man for the chancellorship, but that since the majority had chosen him he felt he should accept their vote as a call to duty and do the very best he could in the job. He promised Thaw to consult with him, and begged for his support. He urged Thaw to ease Langley out of the directorship in a way that would not hurt him, and get a firm offer to Keeler before May 1. Prompt action was required.[48]

Thaw wrote Langley to inform him officially of the change that was about to take place. Langley replied that he had decided, on his own, that the time had come to sever his honorary connection with Allegheny. Therefore he would resign as of the end of April, and the trustees should feel free to make any appointment they wished to take effect after that date. He added that although he had known nothing of the plan to bring Keeler in as his successor until he received Thaw's letter (which was certainly not true), he thought his former assistant a very good choice indeed (which was true).[49]

Finally on Monday, April 20, an informal installation ceremony was held for the new chancellor at the university. Scott did resign from the board, but none of the other members followed him, and several of them made speeches lauding Holland. In a moment of comedy, Scott's son, who did not know his father had resigned in protest, was the first to lead three cheers for the newly installed chancellor. Brashear had been asked to attend as Thaw's representative, and after the installation the trustees held a meeting to elect a new director of the observatory. Brashear presented the case for Keeler and in the optician's words "[t]here was a very warm feeling manifested toward him & he was unanimously Elected and a Committee nominated to notify him by telegraph."[50] The Rev. Dr. Holland was positively inspired by this first official action of his administration, reporting to Thaw that "I hope that this step will be a good one, and that in coming days we shall have occasion to congratulate ourselves upon our choice. The new term opened today under sunny skies, and everybody seems to be feeling happy at the University. I realize that a world of work is to be done, but that we shall succeed in doing it with the help of God and the friends of the University I do not have a shadow of a doubt."[51]

The salary was disappointingly low. Keeler had originally hoped for $2,400 a year at Allegheny.[37] Langley had offered him $2,000 to be his assistant at the Smithsonian, and Keeler had reported this figure to Thaw early in the negotiations.[39] Although Holland had recommended giving him a slightly larger amount, to be sure he would come, in the end they knew they had him, so economy won out and his salary was set at $2,000.[51] This was only $200 per year more than Keeler was making at Lick, but he felt he had no choice and accepted at once.[52]

The same day he telegraphed Chancellor Holland that he would accept the Allegheny directorship, Keeler wrote Langley to let him know what he had done. He could now tell his former chief that the competing offer had been to be his successor, as Langley in fact already knew from Brashear and Thaw. Keeler delicately made it clear to Langley that he felt he had grown up scientifically, and could never again be anyone's subordinate. But he did it so diplomatically that he kept Langley's friendship, which he would need to operate Allegheny Observatory effectively.[53] Keeler made sure to write to Thaw, to let him know that he had accepted the Allegheny post, and to pass along some of the praise he had been receiving from European astronomers and astronomical publications for his spectroscopic work on the nebular lines.[54] Before the end of April he sent a formal letter of acceptance to Holland, supplementing his earlier telegram with the information that he planned to arrive at Allegheny in July.[55]

All the while he was carrying on the negotiations and writing these letters, Keeler was continuing his nebular spectroscopic observations. By then the Orion nebula, a winter-sky object, had disappeared behind the sun, but he was measuring the nebular lines every night he could in the planetary nebulae NGC 6210, NGC 6572, and NGC 6543.[35] He was highly conscious of the fact that although Allegheny would be a far more pleasant place for him and his bride-to-be to live, he would have only a small telescope in a poor climate, and his observational capabilities would be drastically reduced.[56]

Keeler probably told Holden that he was actively negotiating for the Allegheny directorship very soon after his return to California at the end of 1890 from his trip to the East. Holden, without the benefit of any search committee, advertisements or recommendations, immediately decided that if Keeler did leave, Campbell would be the best choice to hire as his successor. Certainly, just a few days later, Holden wrote Campbell that in the near future he might possibly be able to offer him a permanent position at the Lick Observatory. At the very least, Holden said, he wanted Campbell to come for another summer.[57] Campbell was anxious to do so, and said that if the permanent position were offered, in all probability he would accept it. As to the summer, he certainly wanted to work at Lick again, but this time he could not afford to do so unless his travel and living expenses were paid.[58] Finally they agreed that Campbell would come for $300, his absolute minimum necessary expenses for the summer.[59] But then, toward the end of March, Keeler evidently told Holden that he felt pretty certain the Allegheny position would come through, and the Lick director immediately wrote Campbell that he had every reason to believe that he would soon be able to offer him a permanent position.[60] Holden still did not reveal that it was Keeler's job he would probably be getting. Campbell quickly replied that he was exceedingly grateful and that "I shall accept the offer when it comes with the feeling that it is the chance of a lifetime ... I would keep the *high standards of the Lick Obsy. work* constantly before me."[61] He was under consideration for a faculty job at Hamilton College in New York, for which Keeler among others had strongly recommended him, but he assured Holden that if the Lick job came through it was the only one in which he was interested.[62]

Thus when the telegram from Holland finally came, Keeler immediately notified Holden, and wrote a formal letter resigning his position effective June 1, 1891, to accept the new post.[63] The very same day Holden forwarded this letter to Martin Kellogg, longtime professor of ancient languages at the University of California and then in his first year as its acting president. Holden recommended that Keeler's resignation be accepted and that the regents adopt a resolution expressing their high appreciation of his astronomical work at Lick Observatory, and wishing him every success in his new position. Holden further recommended that Campbell be appointed in Keeler's place, at the same salary, and added that the new astronomer would accept and could be there to begin work early in June.[64]

The following day, Holden wrote a long letter to Campbell explaining that Keeler was resigning to accept the Allegheny directorship, and that it was his job that his former assistant would now take over.[65] Simultaneously, and no doubt at Holden's suggestion, Keeler wrote Campbell a very friendly letter, confirming that he was leaving. He said that he had put in a good word for Campbell, but that it had not been necessary, for Holden had already wanted to hire him.[66] Campbell admitted privately that he felt that he did not know much about spectroscopy, but Keeler reassured him, writing truthfully, "[y]ou could count the men in this country who are familiar with astronomical spectroscopy on the fingers of one hand." Campbell at least had some experience at Lick Observatory, he said, and need not worry, he said, for whoever took the job would have to work hard at becoming a full-time spectroscopist.[67] At the May regents' meeting, Campbell's appointment was officially confirmed, at a salary of $1800 per year, the same as Keeler's had been.[68] At this meeting, Keeler's resignation was accepted, and the regents issued a handsome statement of appreciation for his work at Lick together with best wishes for his future success at the Western University of Pennsylvania and its Allegheny Observatory, along the lines Holden had suggested.[69]

To Holden, as to nearly everyone else in astronomy, it was clear that Keeler's move to Allegheny was a step up. Although astronomically he would not be nearly so well off, the independence and prestige of the position as director more than made up for it. And although the Allegheny Observatory was small and poorly equipped, Langley had made it famous among scientists. Pittsburgh was one of the great metropolises of the East, a center of culture, and far more attractive for a family man than Mount Hamilton, as Holden well knew. Keeler left with the respect and best wishes of his former director and all the Lick staff.[70] Yet it was not clear to everyone that Keeler had made the right decision. A year later Campbell was offered the job of professor of astronomy and director of the observatory of the University of Michigan, a closely parallel situation to Keeler's Allegheny position. Campbell, however, turned down the Ann Arbor job; Lick Observatory was the center of his world and all he wanted to do was to stay there.[71]

On May 26, 1891, Keeler made his last observation with the 36-inch refractor, examining the spectrum of the large, faint planetary NGC 3587, the Owl nebula. With his low-power spectroscope he could see it had the usual emission-line spectrum indicating a gaseous nebula, but the lines were too faint to measure accurately with the large spectroscope. He must have sighed as he closed up the dome that night, and then wrote several pages of suggestions on improvements that could be made to the spectroscope, and hints on using it, for Campbell's guidance.[35]

A few days later Keeler left the mountain, traveling by train to New Orleans.[67] On June 16, he and Cora Matthews were married at Oakley Plantation, the home of her parents, in West Feliciana Parish, Louisiana.[72] Unfortunately, nearly all the Keeler family's personal papers, scrapbooks, letters and diaries were destroyed in the disastrous 1923 Berkeley fire, and no surviving description of the wedding ceremonies has been discovered.[73] Keeler himself, in a letter to Holden, only reported that he had been married and that they had had a pleasant time in

Fig. 11. Cora Matthews Keeler, James E. Keeler's bride, whom he met on Mount Hamilton (1893). (Reproduced by kind permission of the Mary Lea Shane Archives of Lick Observatory.)

Louisiana, a part of the country he had always wanted to see. He found life in the old plantation homes interesting, but after so many years in California he considered the Louisiana climate oppressively hot and humid. By early July the newly married couple were settled in Allegheny, and Keeler was hard at work at his new, old observatory.[74]

6

The ablest spectroscopist in this country

When James E. Keeler and his wife moved east in the early summer of 1891, Pittsburgh was one of the ten largest cities in the United States. Its population was nearly a quarter of a million people; Allegheny County, which contained Pittsburgh and the entire metropolitan area surrounding it, had a population of just over half a million. Pittsburgh had grown from the little English settlement around Fort Pitt at the Forks of the Ohio, where the Allegheny and Monongahela Rivers join, to become the industrial hearth of the nation. Rich seams of coal from the hills north of the Allegheny provided the fuel to convert iron ore from the Great Lakes region into steel in the mills built by Andrew Carnegie, Henry Clay Frick, Henry Phipps, and a host of industrial workers. The country was expanding as the railroads obliterated the frontier, and the steel that built the railroads came from Pittsburgh. It was a dirty, smoky city, but there were theaters, concerts, opera, and a Carnegie Library in Allegheny. The early settlers of Pittsburgh had been largely English and Scots-Irish in origin, but the factories drew immigrants, at first chiefly from Ireland and Germany, then in increasing numbers from Italy, Poland, Austro-Hungary and Russia in the years just after the Keelers arrived.[1]

Allegheny Observatory was located on a hill in the separate city of Allegheny, across the river north of Pittsburgh. The observatory had been built in 1861, by a group of amateur astronomers, who had incorporated themselves for this single purpose, but after a few years they had lost interest and the Western University of Pennsylvania had taken it over in 1867, with Samuel P. Langley as its first director.[2]

The Western University of Pennsylvania was a small but old institution, founded originally in 1787 as the Pittsburgh Academy. Although not directly connected with any church, it was strong in the tradition of Scots-Irish Presbyterianism. The university had originally been located in

122

Fig.12. Allegheny Observatory, Allegheny, Pennsylvania (1898). (Reproduced by kind permission of Allegheny Observatory.)

downtown Pittsburgh, but had moved several times. In 1882 the university had been shifted to temporary quarters in the Presbyterian Seminary in Allegheny, but by 1890 it was in a new campus on the top of Observatory Hill, above most of the smoke and grime of the city. In addition to the observatory itself, there were only two university buildings, the Main Building and Science Hall, for a faculty of fourteen and a student body numbering just over a hundred undergraduates. The year before Keeler came, the Western University of Pennsylvania had granted three bachelor's degrees in liberal arts, four engineering degrees, and two honorary degrees.[3]

The campus was located at the end of the Perrysville Avenue streetcar line, making access simple for students and visiting astronomers alike. Although the university had previously offered only a traditional classical academic program, at the time of the move to Observatory Hill the trustees made a conscious decision to build up the scientific and technical educational facilities so badly needed in industrial Pittsburgh.[4] During Keeler's years at Allegheny, he was to see the engineering faculty and course offerings grow, existing local medical, pharmacy and dentistry schools merge with the university, and new mining and law schools open within it.[5]

The new chancellor, William J. Holland, was nine years older than Keeler. Born in Jamaica of American missionary parents, he had been educated at Amherst and at Princeton Theological Seminary, and had been ordained in 1874. He had been pastor of the Bellefield Presbyterian Church in Pittsburgh from that year until he accepted the chancellorship, and had served as a trustee of the Western University since 1889. Holland was a man of many scholarly interests – natural history, languages, art, literature and education as well as theology.[5] His office was hung with pictures he himself had painted, illustrating his travels in South America, and his collection of mounted butterflies was supposed to be the largest in the United States. The chancellor was an urbane, successful professional, wealthy in his own right, and Keeler got along famously with him.[6]

When he came to Allegheny, Keeler, as director, had the right to live rent free in the Observatory House, immediately adjacent to his office. Langley, a bachelor, had used it, and Keeler first intended to do so with his wife, but he soon decided that with Harry Floyd, Cora's wealthy cousin, living with them, the house would not do.[7] It was in poor repair and Langley, an inveterate hoarder, had taken all the movable furnishings, including even the window shades, with him to Washington.[8] The new director stayed in the Observatory House only until fall, and then moved into a new house that he rented at 200 Perrysville Avenue, just in front of the observatory.[9]

Allegheny Observatory had been nearly deserted for several years, after Langley had become secretary of the Smithsonian Institution but continued as "honorary" (unpaid) director of the observatory. Only his assistant, Frank W. Very, was working at the observatory, most of his energy taken up by the time service on which it depended for income, and by a little experimental work for Langley.[10] One of his more fascinating duties was to keep fireflies alive for Langley's research on the nature of the light they emit.[11] At one time Langley had considered hiring a second, part-time assistant, and the outstanding applicant for the post had been George W. Ritchey, a young student at the Cincinnati Observatory.[12] However, Langley decided that Allegheny's finances could not stand the expense of another employee so Ritchey stayed at Cincinnati, later to go on to a career as a highly creative but eccentric astronomical optician.

Very, who had undoubtedly hoped to inherit the directorship himself, was not at all happy to have Keeler, formerly Langley's junior assistant, return as his superior. He had sent a stiff note to William Thaw, Jr. when Keeler was appointed director, offering his resignation as a matter of form, but making it clear that he wanted to stay on, with more independence than he had enjoyed under Langley.[13] Keeler advised Very not to act hastily. He said that he planned to work on stars and nebulae with the refractor himself, and suggested that Very, in addition to the time service, could do as much independent research as he wanted, on the sun and the earth's atmosphere, along the lines initiated by Langley.[14] But, Keeler insisted, Very must work under his general supervision and report to him; the observatory could not have two masters. Keeler promised that as long as it was clearly understood that he was in charge, he would give Very as much scientific freedom as he could use.[7] Very's title was upgraded to adjunct professor of astronomy,[15] but his salary remained unchanged at $1,000 per year. Keeler worked hard to raise support for Very's research on "radiant heat," arranging for the Weather Bureau to provide some of the equipment he needed for his laboratory measurements of the absorption and emission of infrared radiation by air.[16] As director, Keeler used his contacts with other physicists to borrow an air pump that Very needed for this project.[17] There was essentially no scientific interaction between the two of them, but since there was no money for new positions at Allegheny, as long as Very stayed Keeler could not hire an assistant to help with his own projects. Very's research project on atmospheric radiation dragged on forever, and he only completed it and wrote up his report to the Weather Bureau years after he had finally left Allegheny in 1895, for a better paying but temporary job at Brown University.[18]

Keeler well knew what he was getting into at Allegheny. It was a one-man observatory with a small telescope in a poor site, by California standards. Not only is there much less clear weather in Pennsylvania, but the "seeing," or steadiness of the atmosphere, is far worse, as Keeler had almost forgotten.[19] Furthermore, Allegheny Observatory was very poorly equipped with auxiliary instrumentation to use on the telescope. There had not been much to begin with, and Langley had taken almost every prism, grating, mirror, lens and telescope he could possibly use to Washington with him.[8] He regarded these instruments and apparatus, bought originally for his use by William Thaw, his wealthy patron, now dead, as the property of the Thaw estate, to be kept in his custody. Before he left Washington for his annual summer trip to Europe in 1891, Langley gave orders that the instruments not be returned to Allegheny while he was away. At the very same time, he was trying to get Keeler to send him a galvanometer he had left at Allegheny.[20] His tactful young former assistant did send the galvanometer to Washington, and, at the same time, urged his old chief to return the instruments and apparatus which Keeler insisted were the property of Allegheny Observatory. He particularly wanted back the 3-inch Clark refractor, in order to use it as a finder telescope on the main Allegheny Observatory 13-inch refractor.[21] Langley dragged his heels and Keeler had to apply more pressure through William Thaw, Jr., now the chairman of the Western University of Pennsylvania Board of Trustees' Observatory Committee, and the undoubted heir of the astronomical part of the Thaw estate.[22] In the end Langley did return the small telescope and most of the instruments and apparatus, except for two large mirrors he needed in his solar work, one of them a telescope objective, which Keeler could not use. Langley bought the mirrors from Allegheny Observatory for the Smithsonian for $350, providing the funds Keeler needed to buy a new rigid steel tube for his 13-inch refractor, in place of the flimsy old wooden one.[23]

Keeler wanted to use the telescope to do frontier astrophysical research. Because of its small aperture, he could not compete on equal terms with larger telescopes, in particular the Lick 36-inch in the hands of W.W. Campbell, on stellar spectroscopy. But for nebulae, whose images are large compared with the slit of the spectroscope, such as the Orion nebula, the size of the telescope is less important. Hence Keeler planned to concentrate his efforts on nebular studies, beginning by repeating and checking the measurements of the principal nebular lines he had made at Lick Observatory.[24]

First he needed a spectroscope. The only instrument at Allegheny suitable for use on the telescope was a small, low-dispersion visual spectroscope, fine for preliminary survey work, but not at all capable of producing accurate wavelength measurements.[6] Keeler knew the days of visual spectroscopy were over. The earliest pioneers, Josef Fraunhofer and Angelo Secchi, Gustav Kirchhoff and Charles A. Young, had worked visually; they had no other choice. But photography had come into astronomy, beginning with crude, very slow wet plates. Henry Draper, the wealthy New York amateur, had obtained some good plates of the moon in the 1870's, and then of the Orion nebula in 1880. William Huggins, the English astronomer, had taken the first spectrogram, or photographic record of the spectrum, of a star in 1863. It had been only a faint, blurred poor effort, but by 1891 photography had advanced considerably. The introduction of dry plates in 1878 was a major step forward. For scientific purposes photography has the great advantage of providing a permanent, objective record, independent of the sensitivity of the observer's eye or his skill in drawing or describing what he sees. Even more important, the photographic plate can integrate, or add up, faint light in a way the human eye cannot, so that the spectra of faint stars and nebulae can be photographed with long exposures even though they cannot be seen. In checking his Lick nebular measurements, Keeler had come to realize that Hermann Vogel, with his little 9-inch telescope at Potsdam, using a photographic spectroscope (or spectrograph) had been able to measure much more accurate stellar radial velocities than anyone had done visually, except Keeler himself for one or two very bright stars with the giant 36-inch refractor on Mount Hamilton. Now Keeler decided to do the same with his 13-inch telescope at Allegheny for nebulae.[25]

Keeler had been promised financial support for his research before he accepted the job at Allegheny, and in this, his first year as director, he got it. Mrs. William Thaw, the widow of Langley's patron, announced that she would provide the funds to pay for a new photographic spectrograph for the observatory. She gave Keeler carte blanche to order the instrument he needed. His friend John A. Brashear, whose optical and instrument shop was only a few steps from the observatory, was to do the work.[26] Keeler, with his unparalleled knowledge of optics and practical spectroscopy, designed an instrument that could be used with either a grating, a single prism, or a train of three prisms. It was highly rigid, so that long photographic exposures could be taken without any blurring of the spectrum by flexure. Like all Keeler's instruments, this spectrograph had a large aperture, making it efficient for recording the spectra of faint stars and nebulae. It could be used either visually or photographically, but was designed especially for the latter method of observing.[27]

Although he was convinced of the great advantages of photography for astronomical spectroscopy, Keeler had never actually taken a photographic spectrogram.[28] At Lick the large spectroscope had been designed for visual use, and he had done all his planetary, stellar and nebular research there with his own eyes. His colleagues S.W. Burnham and E.E. Barnard had both been excellent photographers, although only Barnard had done much astronomical photography, using a small, wide-field camera to record star clouds in the Milky Way. Now Keeler busied himself learning photography. He bought a photographic outfit, including a camera and darkroom equipment, from Burnham's friend Gayton Douglass, a Chicago photographer and amateur astronomer.[29] Keeler began taking pictures, experimented with developing and processing the plates (all photography was done on glass plates at that time), and bombarded Douglass and Barnard with questions on practical technique.[30] He quizzed Henry A. Rowland's assistant, Lewis Jewell, to find out all the photographic methods used in the spectroscopic laboratory at Johns Hopkins.[31] On a visit to Baltimore, Keeler talked with Rowland himself, and learned that the most sensitive, fast plates have only poor resolution, or ability to record fine detail, like close spectral lines. On the other hand, Rowland explained to him, slower plates have better resolution, because they are less grainy. This was a point Keeler's former director at Lick Observatory had never grasped in three years' experience of planetary and lunar photography with the 36-inch refractor.[32] The young Allegheny astrophysicist, in contrast, was applying himself to learn all he could from the research workers in this new and rapidly expanding field, so that he would be an expert by the time his photographic spectrograph was completed.

In the meantime, he was getting his telescope and observatory ready. The first time he tried to observe the spectrum of a comet, Keeler found the shutter of the dome was so badly bent and corroded that he could not force it open.[6] One of the university trustees who was on the Observatory Committee, C.G. Hussey, a wealthy physician, gave Keeler the money for a new shutter.[21] Mrs. William Thaw, Jr. contributed the funds to buy a new drive clock for the telescope, to make it follow accurately the diurnal motion of the stars across the sky.[33] Brashear checked over the telescope, cleaned it up, and installed the new steel tube, bought with the proceeds from the mirrors sold to the Smithsonian. The members of the Junta Club, a group of Pittsburgh businessmen to whom Keeler gave a talk on astronomy, provided matching funds to help buy a new dome for the observatory.[34] The whole building was even given a new coat of paint, for the first time in almost ten years.[35] Everywhere Keeler could see progress in his little research kingdom.

While he was waiting for Brashear to finish his new spectrograph, Keeler busied himself with several projects that he could carry through without it. The first came up when he, his wife, and her ward, Harry Floyd, went down to Washington for a Christmas visit to Keeler's mother. Besides stopping at the Naval Academy at Annapolis, attending the President's reception at the White House on New Year's Day, and other sightseeing with the ladies, he called on Langley, Rowland, and several other prominent scientists, including Mark W. Harrington, head of the U.S. Weather Bureau.[36] Harrington, who had just resigned his post as professor of astronomy at the University of Michigan to take the Weather Bureau job, planted an interesting idea in Keeler's mind. Harrington had been planning to translate the recently published book on astronomical spectroscopy, *Die Spectralanalyse der Gestirne,* by Julius Scheiner, of Potsdam Observatory, into English. It was the only modern treatise on the subject. But now, in his new position, Harrington found he had no time to do translations, and wanted someone else to take over the task. Young George E. Hale, in Chicago, had expressed some interest in translating the book, but besides knowing little German, he also found himself too busy. Hale suggested Keeler, as Edward S. Holden, the Lick Observatory director, had earlier, and Harrington liked the idea.[37] Keeler was not too busy, and was quite interested in doing the translation. Scheiner's book was already a classic, and Keeler intended to revise it extensively, introducing Rowland's system of wavelengths in place of the slightly different system the Germans used, and to add a chapter on diffraction gratings. By February he had decided he would definitely go ahead with the task.[38] Holden cautioned Keeler that Scheiner would never accept all the proposed revisions, and advised him to translate the book as written, but to add all his corrections and new material as appendices.[39]

Now, just as Keeler was about to write Scheiner that he was willing to substitute for Harrington and work on the book, a new problem arose. Young, the benign old Princeton spectroscopist, wrote Keeler, saying he had heard that he was thinking of translating Scheiner's book. However, Young continued, "a young friend of mine" had similar plans, and it would be a great pity if there were two competing editions, especially as the young friend had the cooperation of Scheiner.[40] Keeler instantly wrote back that he had indeed intended to translate the book, but not knowing anyone else was also considering it, had put off consulting Scheiner until his own ideas on just what he wanted to put into the new version had solidified. Keeler added that if Young's friend had already reached an agreement with Scheiner he would instantly withdraw in his favor. On the same day he received Young's letter, Keeler wrote to Scheiner to explain the situation. He said that if Scheiner had already reached an agreement with a prospective translator, he did not want to interfere, but that if Young were mistaken and no final decision had yet been made, he would like to be considered on the same basis that Harrington had been.[41]

Keeler soon learned that Scheiner had indeed decided on another translator. The young friend turned out to be Edwin B. Frost, an instructor at Dartmouth, who had studied briefly with Young at Princeton. He had just returned to America after two years in Germany, where he had worked under Vogel at the Potsdam Observatory and seen Scheiner daily. Frost not only had Scheiner's approval, but he had a signed contract with Ginn and Co. to publish the book.[42] He invited Keeler to collaborate with him on the translation, but the Allegheny astronomer declined and instead offered to send Frost a list of the corrections that he had already compiled. The younger man agreed that trying to do the translation as a cooperative effort would hardly be worthwhile, but that he was grateful for the corrections. They were chiefly Keeler's observational results on nebulae from Lick Observatory, which had not yet been published. Frost was glad to incorporate them into the book.[43] He found Scheiner difficult to work with, because the German astronomer did not believe any of the newer American results if they had not been confirmed at Potsdam, preferably by himself. Still, Frost discovered that English is a more compact language than German, and he could therefore translate everything Scheiner had written and add about forty pages of new material without increasing the total number of pages over the original edition.[44] When the book came out in 1894, it was a great success, and it remained the standard English-language treatise on astronomical spectroscopy for many years. Keeler had waited too long to write Scheiner; if he had acted promptly after his talk with Harrington he probably could have done the translation, and reaped the royalties and praise that came to Frost.

A strong contrast to Keeler's hesitancy was provided by Campbell, his former assistant at Lick Observatory. When Keeler left for Allegheny, Campbell was hired in his place, and he brought with him to Mount Hamilton the half-completed manuscript of a book on practical astronomy. It was based on a course he had been teaching at the University of Michigan. At Lick, Campbell stole time from his research to push the book through to completion, and as soon as it was published, he began actively soliciting favorable reviews and statements for use in advertising it.[45] Campbell's book filled a real need, as it contained all the astronomical methods used by surveyors and civil engineers, and soon it was in a second edition.[46] Meanwhile the best Keeler could do was to write a short, popular article for the *Century* magazine on his drawings of the planets, which, although it was accepted, was not published until several years later, when he had become briefly famous.[47]

In the late summer of 1892, Mars had one of its unusually close approaches to the earth, and Keeler, along with nearly every astronomer in the world, observed it. These close approaches or oppositions, as astronomers call them, occur when Mars is near the perihelion of its orbit, closest to the sun, at the same time that the earth, moving faster along its smaller orbit, passes by it. They occur only twice every fifteen years. In the days before space vehicles and interplanetary probes, these favorable oppositions provided the closest glimpses of Mars that earthbound astronomers could get. On every clear night that summer, Keeler observed Mars visually with the 13-inch Allegheny refractor. Although a large telescope is superior if the atmosphere is unusually steady, Mars and the other planets are bright enough so that a small telescope such as Keeler's is adequate for this kind of work under most conditions. Each time he observed Mars, working rapidly he would draw on a previously prepared outline the face of the planet as he saw it. Mars rotates on its axis once every 24 hours and 37 minutes, a little longer than an earth day, so Keeler would see it from a slightly different angle, or as astronomers say, a slightly different phase, if he waited a few hours. He carefully noted the time and calculated the phase of the planet each time he observed it. For this work on Mars, Keeler had prepared a small globe on which he had carefully drawn all the surface features shown on the charts published by G.V. Schiaparelli, the Italian planetary expert, who had observed it in previous years. Each time Keeler observed Mars he finished his sketch before he looked at the globe but then, as soon as it was completed, he compared his drawing with the globe as seen from the same phase. He would then immediately write down any differences he saw between his own conception of Mars and Schiaparelli's.[48]

In general Keeler saw the same large surface features on Mars that Schiaparelli had seen, the so-called continents, seas, bays and the like. There were some differences, but they were only differences in detail. In addition, however, Keeler emphasized the diffuseness of the Martian features, and said there were no sharp edges or lines. In this he contradicted Schiaparelli, who had reported the existence of narrow dark lines on Mars, which soon became known in English as canals. This report had gripped the public imagination. The French astronomer Camille

Flammarion became the leading spirit in propagating the thesis that Mars was inhabited by a race of intelligent beings, perhaps older and wiser than mankind here on earth. Soon the ideas of canals, oases, growing seasons, a network of waterways, a Martian civilization were current everywhere.[49] For a month before opposition Mars was bright and easily visible in the sky each night, and visitors flocked to observatories from Allegheny to Lick, eager to catch a glimpse of the mysterious red planet. Keeler, as a hard-boiled professional astronomer, considered the public excitement ridiculous, but tried to channel it into financial support for his observatory, by inviting wealthy potential donors to view Mars with his telescope.[50]

Some professional observers thought they saw canals on Mars, but most, including the keen-eyed Barnard, Keeler's former colleague at Lick Observatory, did not.[51] When a selection of Keeler's Mars drawings were published, Barnard, a convinced admirer of his friend's artistic skills, praised them extravagantly. He said they were superb, and looked more like Mars had appeared to him in the telescope than any other drawings he had seen.[52] Several other experts, including William H. Pickering, who had been observing at the Harvard southern-hemisphere station in Peru, had similar high opinions of the accuracy of Keeler's drawings. Langley, who knew the limitations of the Allegheny telescope all too well, was equally impressed. Even Percival Lowell, the wealthy Boston orientalist, who was racing to complete the construction of an observatory at his own expense at Flagstaff, Arizona Territory, so that he could begin studying Mars at the next favorable opposition, liked Keeler's drawings.[53] The praise from Pickering was somewhat embarrassing, for Keeler considered the Harvard observer's published drawings of Mars to be nothing but crude diagrams, not at all realistic representations of the planet's appearance. Furthermore, he thought Pickering's written descriptions of Martian features were almost unintelligible, as in fact they seem today.[54]

In addition to his visual planetary work, Keeler occupied himself with writing up his Lick observational results for publication. First, however, he got into print an interesting lecture he had given at the Academy of Science and Art in Pittsburgh, in the autumn of his first year as director at Allegheny. The Academy was essentially an upper-middle class lecture society, which gathered once a month to hear popular presentations by local or visiting professionals. Keeler, as an important new scientific figure in the area, was asked to join the Academy and was elected to its council soon after his arrival, and was often called on for advice as to whether purportedly scientific contributions from members should be accepted for its publications.[55]

In his lecture to the solid burghers of Pittsburgh, Keeler gave a perceptive review and synthesis of the current astronomical ideas on nebulae and their connection with stars. He projected a slide of an excellent photograph of the Andromeda nebula, M 31, taken by Barnard, and pointed out its spiral structure. He showed that the stars which could be seen in the picture followed the same spiral arrangement as the nebula, and therefore must be part of it. He explained how a spectroscope can be used on a telescope to study the composition of celestial subjects. But, in the case of the Andromeda nebula, he told the audience, the result is disappointing. All that could be seen was a very faint continuous spectrum, apparently showing the nebula to be made up of solid or liquid objects. But, Keeler advised the audience, we must be cautious for the spectrum is so faint this testimony is inconclusive. Next he showed a slide of the Orion nebula, in which Huggins had seen a bright-line spectrum back in 1864, thus proving it to be a gaseous nebula. He showed several pictures of planetary nebulae, drawings made at Lick by Holden and John M. Schaeberle, as other examples of gaseous nebulae. Every planetary has a central star, he said, and that fact cannot be a coincidence. The spectrum of the nebula shows the hydrogen lines plus the two other bright lines that cannot be accounted for by any known terrestrial substance, he told them, referring to his own work. "The nebulae then, appear to be immense masses of glowing vapor, containing hydrogen and unknown substances. From the character of the lines, the gases seem to be very hot and extremely tenuous."

He contrasted this description with the meteoritic hypothesis of the English spectroscopist J. Norman Lockyer, that the bright nebular lines came from a magnesium band at very low temperature, but showed clearly how his own research had demolished this hypothesis. Keeler then went on to outline the idea that all stars are formed by condensation of nebular material. He cited the Pleiades and the stars in the central part of the Orion nebula as examples of stars clearly associated with nebulosity, but whose spectra are no different from the spectra of other stars seen outside of nebulae. Our sun is an ordinary star, he said; it too must have been formed by the same process. Thus from modern observational data, we are led, he said, to the same conclusion that Immanuel Kant and Pierre Laplace had reached by theoretical speculation years before, that the sun and solar system had formed from a rotating nebula.

Keeler described the regularities in the solar system, that the planets all move in orbits in the same direction and in more or less the same plane, and explained how these regularities would be expected to occur from the contraction of a rotating gas cloud. He stated the idea that the distribution of mass in the nebula would be nonuniform, and sketched how this would lead to the growth of dense condensations as a result of gravitation. He, in common with all the scientists of his time, considered gravitational contraction to be the energy source of the sun, and therefore gave an upper limit of its lifetime as ten million years. But, he concluded, "[i]t is, of course, possible that something unseen by our imperfect vision may intervene to avert this dismal end."[56]

Although some of the ideas in it are wrong – we now know that thermonuclear reactions fuel the stars and that the sun's lifetime will be ten billion years, not ten million – this lecture in its published form is strikingly perceptive and modern in tone. All the ideas in it were current among astronomers, but Keeler put them together in coherent form, and linked them clearly to observational data. Some of the ideas and problems in this paper are ones he worked on during his whole career, and some of them are problems we still work on today.

Most of Keeler's writing effort in his first year at Allegheny went into his magnum opus on the measurements of the wavelengths of the nebular lines. He had explored briefly the possibility of publishing this work in England, because at the time there was no professional astrophysical journal in the United States. However, Holden wanted to have it appear in a *Lick Observatory Publication,* and Keeler was glad to agree to this, provided he could get a supply of reprints to distribute from Allegheny.[57]

Before writing the paper, he wanted to redetermine in the laboratory the exact wavelength of the lead comparison line he had used for the reference point of his nebular measurements, but he was unable to find time for this experiment. Nevertheless, he completed the manuscript and sent it off to California, in the first winter after his return to Allegheny.[58] But his paper did not go straight to the printer, because no other contributions were available to be published with it, and although it was an important paper, it was not long enough to make up a complete volume by itself. During this delay, however, the wavelength of the lead comparison line was measured by Jewell in the laboratory at Johns Hopkins, so Keeler got his manuscript back from Holden and changed all the wavelengths to the Rowland system.[59] There were further agonizing delays, for Burnham, the double-star expert, did not want his measurements to be published in the same volume with any other observer's work, and insisted that Volume 2 of the *Lick Observatory Publications* should contain his own results alone. As month followed month, and Keeler's work remained unpublished, he finally wrote up a short paper for the journal *Astronomy and Astro-Physics,* placing on record the results of his nebular wavelength measurements.[60] He considered getting his long paper back from Holden again, and publishing it elsewhere.[61] Then at last, just a few days less than three years from the date Keeler had left Mount Hamilton for Allegheny, his complete nebular paper went to the printer with a few last-minute corrections and additions included, and a few others, that arrived a little too late, omitted. It appeared in Volume 3 of the *Lick Publications,* together with a paper consisting of drawings made from photographs of the moon, taken with the 36-inch refractor, and an accompanying text by Ladislaus Weinek, a Czech astronomer.[62]

In his seventy-page paper, Keeler spelled out in detail the methods he had used to measure the nebular spectra. He described the Lick spectroscope and the Rowland grating he had used in it. He described his spectroscopic measurements of the Orion nebula and the bright planetary nebula NGC 6572, in which he had been able to see and measure the Hβ emission line of hydrogen, and thus determine the radial velocities of these two objects. Then from his measurements of the two green nebular lines in these nebulae, knowing their velocities, he could determine highly accurate values of the rest wavelengths of the two lines. In addition, from the wavelength of the stronger green line, λ5007.05, measured in this way, he could determine accurate radial velocities for five other planetaries in which Hβ was too faint to see. Keeler then used his measurements of the fainter green line in these five nebulae to determine better its wavelength, which he found to be λ4959.02. He described all the checks he had made by similar spectroscopic measurements of Venus and of bright stars, cases in which the radial velocities are accurately known from solar-system orbital data. The wavelengths he determined were all based on the wavelength of the lead comparison line supplied by Jewell, λ5005.63. Modern measurements place it at λ5005.43, showing Rowland's wavelength scale was in error by only 0.20 Å; with this correction applied, Keeler's wavelengths of the nebular lines agree very nearly perfectly with the best present-day determinations. His wavelength of the brighter nebular line was correct to 0.01 Å, while his measurement of the fainter one was in error by only 0.10 Å. From these measurements it was certain, he concluded, that no known terrestrial substance has a strong line at the portion of either of the two nebular lines. He also emphasized that λ5007.05 is always sharp, not shaded as Lockyer had described it, and definitely not part of the magnesium spectrum. Keeler recognized that the relative intensities of the two nebular lines always had the same ratio of brightnesses, and thus it was probable that they were due to the same substance, but that their ratios to Hβ differed from one nebula to another.

Finally, Keeler also reported his measurement of a fainter nebular emission line at λ5876 in the yellow spectral region, which showed conclusively that it had the same wavelength as the line known to astronomers as D_3, an unidentified line observed in the chromosphere of the sun.[63]

Keeler's paper was instantly recognized as a classic, and widely acclaimed. A glowing review by the Swedish spectroscopist Bernhard Hasselberg, solicited by Holden who had sent him the proofs of Keeler's paper before it came out, appeared in the *Publications of the Astronomical Society of the Pacific*. Hasselberg focused on Lockyer's contention that the chief nebular line was due to the magnesium fluting. Although Huggins' measurements had seemed to show it was not, Lockyer had persisted in his opinion, but now Keeler had conclusively proved he was wrong. Hasselberg concluded "[t]he reader will find a rich reward and assuredly agree with the present writer in congratulating Professor Keeler on having made, in this work, the first really important progress in a domain of spectroscopy of precision in which the most eminent spectroscopists have hitherto in vain exerted themselves."[64]

Even before Hasselberg's review appeared in print, congratulatory messages were pouring in to Keeler, many of them from eminent astronomers, such as Huggins, and important opinion molders, such as Agnes Clerke. Frost, in complimenting Keeler on his paper, could not resist joking that the combination of it with Weinek's paper gave Volume 3 "a distinct Kangaroo character".[65]

After he had finished the long nebular paper, but before it came out, Keeler busied himself in writing up some of his other Lick observational results for publication. One was an interesting use of the chromatic aberration of the 36-inch refractor lens, usually a hindrance in observational work, as a positive asset. Chromatic aberration results from the fact that the index of refraction of glass, or its bending power, varies with the color of the light, that is, with its wavelength. For this reason, with a single one-element lens only one color can be brought to a focus at a time. All refracting telescopes are therefore made with two lenses of different types of glass, combined to compensate approximately for this defect, in a so-called "achromat". In fact the compensation is by no means perfect, but an achromat can be made to bring the light of one broad range of wavelengths nearly to the same common focal point, while the departure from this focus of light of other wavelengths is much smaller than it would be in a single-lens telescope. All telescopes intended for visual use are achromatized, or "corrected" as astronomers say, to bring all the yellow light to which the eye is most sensitive to the same focus, while the blue and red rays increasingly diverge from it. Keeler was very familiar with this property, for he had accurately measured the color curve, or variation of focal point with wavelength, of the Lick 36-inch lens soon after it was put into use.[66] He realized that, in effect, the lens serves as a crude, low-dispersion, visual spectroscope, for by measuring accurately the point at which light from a star comes to a focus, he could get a quantitative idea of its color. When he had observed NGC 6720, the well-known Ring nebula in Lyra, a moderately large, low surface-brightness nebula, with the 36-inch refractor, he had found that the focus for its faint central star corresponded to the focal point for yellow light, different from the nebula itself, which is green. Thus his observations showed that the central star does not have the same spectrum as the nebula. The same was true for several other planetary nebulae with central stars bright enough to be observed spectroscopically, which Keeler had seen all had continuous spectra. Clearly the central star of the Ring nebula in Lyra also had a continuous spectrum.[67] This conclusion contradicted a published report by Scheiner, who from a photographic spectrum he had taken thought this star had the same spectrum as the nebula, which would have meant it was not as "condensed" as typical stars. He had used only a very small telescope, and probably actually had not recorded the spectrum of the star, which is quite faint, at all. No doubt Keeler felt a little guilty but real pleasure at having corrected the German who had given his book to Frost rather than him to translate, and who did not like to believe the Americans' results.

During the cold Allegheny winter of 1893, Keeler wrote up his Lick spectroscopic results on the variable star β Lyrae, which had been observed photographically by Edward C. Pickering at Harvard and by Aristarch A. Belopolsky at Pulkova in Russia. Keeler had been able to observe emission and absorption lines in the green and yellow spectral regions that their spectrograms could not record, and, as we have seen, had concluded that the system is a complicated combination of stars and gas streams.[68] Thus, within a few short years, he had made important astrophysical contributions to the problems of stars, nebulae, and gas streams in double-star systems.

Long before Keeler's nebular paper finally went to the press, Brashear finished his spectroscope for the Allegheny telescope. Keeler tested it, first on the daytime sky, which shows the solar spectrum resulting from scattered sunlight, and then on the planet Jupiter. These were the first photographic spectrograms Keeler had ever taken, in late October 1892. He found the new instrument worked very well. The only problem with it was that its sensitivity dropped off very rapidly at wavelengths shorter than λ4340, the position of the Hγ line of hydrogen. This resulted largely from the absorption of violet and ultraviolet light by the glass prisms that spread out the light to form a spectrum. Keeler had wanted to get high dispersion, and had therefore had to use a "train" or series of prisms made of dense flint glass; they spread out the spectrum more than lighter glass, but unfortunately also absorbed more strongly at shorter wavelengths. Another possibility would have been to use lighter glass prisms, and a camera on the spectrograph with a longer focal length, to make up for the lower angular dispersion by magnifying the image of the spectrum more.[69] The longer focal length, however, makes the image fainter, so the compromise Keeler had made was nearly as good as any. Within a year, Keeler replaced his dense prisms with somewhat lighter ones, at the expense of some loss in dispersion.[70]

As he analyzed his first attempts at photographic spectroscopy, Keeler realized that not only the glass of the prisms in his spectroscope but also the yellowish flint glass in the lens of the Allegheny refractor strongly absorbed the blue and violet light from stars and nebulae. The smoky Pittsburgh atmosphere was an additional source of absorption, strongest at the shortest wavelengths. Everything conspired against photographic spectroscopy, because the ordinary plates then in use, though sensitive enough in the blue and violet spectral regions, were not at all sensitive to the green, yellow and red light which the atmosphere, telescope lens and prisms transmitted most strongly. Therefore Keeler decided not to use only these plates, as other astronomers were doing, but to push out as far as he could to longer wavelengths.[71] He began a ceaseless correspondence with makers of photographic plates, trying to find the fastest of the new orthochromatic plates, sensitive in the yellow and green spectral regions, that were just coming into use.[72] When he contrasted his observational situation with that of Campbell at Lick, who had a far superior telescope and a clearer, steadier atmosphere, Keeler sometimes feared that there would be no discoveries left for him to make at Allegheny.[73]

But Keeler had some advantages on his side. Most important of all was his keen, analytical mind, honed by his training and practical research experience at Johns Hopkins and in Germany. He followed closely every development in laboratory spectroscopy, constantly bombarding the physicists with requests for their most recent data, especially measured wavelengths and charts of the spectra of individual elements.[74] Keeler seized every opportunity to see and inspect spectrograms taken by other astronomers, and adopted any improvements in technique he could learn from them.[75] He pumped his old friend Henry Crew for information about the new photographic developer metol that he was using in his spectroscopic laboratory, and Harvard Director Edward C. Pickering, on the secrets of hypersensitizing plates with erythrocin.[76] And, if it was often cloudy or smoky at Allegheny Observatory, whenever it was clear Keeler could use the telescope himself, and did not have to share it with anyone. Thus he made himself an expert in stellar spectroscopy.

Astronomers were trying to understand the spectra of the stars in those early days of photographic spectroscopy. Earlier visual observers like Secchi, Huggins, and Lockyer had described the spectra of some of the brightest stars, and tried to arrange them into a coherent classification scheme. Photography made it possible to get much better data, showing many more spectral features than the human eye could recognize and describe accurately. At Potsdam, Scheiner was obtaining spectrograms of stars in the ordinary "photographic" or blue spectral region. At Harvard, under Pickering's direction, a mass-production program of stellar spectroscopy was just getting under way. The objective was to obtain spectra of as many stars as possible and, from their appearance, arrange them into a classification scheme, in the hope that the results would reveal the relationships between the stars, and the ways in which they evolved. Keeler's plan was to take high-quality spectrograms, which necessarily would mean restricting himself to fewer objects, and by comparing their spectra with laboratory data, deduce which elements were present and what the temperatures and other physical conditions in the stars were. Working with his orthochromatic plates, he had the "visual" (green and yellow) spectral regions almost completely to himself.[77]

Initially Keeler took spectrograms of every bright star he could. His interest was in studying the similarities and differences of their spectra and learning whether one type of star evolved into another.[78] He and the other astrophysicists of his day believed that they might solve this problem by finding groups of stars with similar chemical (or elemental) compositions, but with, for instance, different temperatures. Members of such a group, if it existed, would then be more likely to be more closely related to one another than stars with different compositions. The general picture astronomers had was that stars condensed from "nebular material", contracted and became hotter, then in some way that was contrary to the known laws of physics (this was usually glossed over), cooled off, became fainter and eventually disappeared. Keeler, however, tended to distrust far-ranging conceptual schemes, and to emphasize the complexities shown by the observations themselves. He clearly realized that good, trustworthy data were all-important, and that even the most grand and beautiful theory had to yield to hard facts. Thus in reviewing a long paper in which Lockyer claimed that new photographic spectra confirmed his earlier visual spectral spectroscopic studies, Keeler concentrated on the

English astrophysicist's observational findings. He pointed out several errors in the identifications of spectral features, which he clearly implied were based chiefly on Lockyer's desire to make the observations fit his meteoritic theory. Keeler emphasized the wavelength discrepancies in Lockyer's observations, particularly in his purported identifications of "carbon flutings" (bands of carbon molecules) in emission in nebulae and emission-line stars. No theory could be confirmed by such inaccurate measurements.[79] Although he was kinder to Vogel and Scheiner's spectral classification scheme, which was supposed to be closely related to stellar evolution, Keeler reiterated that no general picture of this type fitted all the observed cases.[80]

In his own research, Keeler began by taking spectra of as many bright stars as he could. Following up on a suggestion of Scheiner, he studied the relative strengths of various magnesium lines in stellar spectra. From laboratory work of Kayser and Carl Runge in Germany, the way in which these lines strengthened or weakened with temperature was known. Keeler was thus able to arrange the stars in a temperature sequence, ranging from Betelgeuse and Antares at the coolest, through Arcturus and Capella at approximately the same temperature as the sun, on through Vega and Sirius, to the hottest stars, Deneb and Rigel. Since Rigel was believed to be a young star, because of its location in Orion and the similarity of its spectrum to those of other Orion stars which are clearly involved in nebulosity, this last result was quite new and unexpected, but it was also quite correct.[81] Keeler's study of stellar spectra convinced him "that all stars cannot be fitted into a single scheme of development and there are groups of stars (like those in Orion, for instance) that follow a course of their own. And I doubt whether a star like γ Cassiopeiae . . . will ever pass through the stage represented by α Orionis [Betelgeuse]."[82] Today we know that the most important parameter in determining the evolution of a star is its mass, and that γ Cassiopeiae is not massive enough to become a high-luminosity red supergiant as Betelgeuse did, and as several of the other stars in Orion that he studied will do. From his careful comparison of his high-quality spectrograms, Keeler had an understanding, unrivalled in his time except by Antonia C. Maury at Harvard, of the features in the spectra of different types of stars. She was the most creative of Pickering's assistants.[83] But atomic physics was not advanced enough for Keeler or Maury or anyone else to integrate this understanding into a coherent physical picture until thirty years later.

Keeler never published a complete paper on this research on the spectra of the cool stars. He gave a report on it at the dedication of Yerkes

Observatory in 1897, and showed slides of his spectrograms. He described them to the assembled astronomers, and showed how the various spectral features changed from one star to another. By that time he realized that the lines and bands that are strengthened in these "third type" stars are those that are strong in the laboratory at relatively low temperatures, and that these stars must therefore be relatively cool. Only a brief abstract of this talk was published.[84] Keeler planned to write up his results after he left Allegheny for Lick Observatory, but he never found time to do so.[85]

Keeler was recognized as the leading expert in astronomical spectroscopy of his time. He wrote a series of four articles for the magazine *Popular Astronomy* on the spectroscope and its applications that were very widely read and consulted. They contain a masterful blend of descriptions of simple experiments, explanations of the principles of spectroscopy, practical details on designing, adjusting, and using prism instruments, and examples of how laboratory data on the spectra of the elements can be used to interpret the physical properties of the stars.[86] Keeler visited his friend Brashear's instrument shop on Perrysville Avenue almost daily, and was frequently called on to test and evaluate spectroscopes being built there for other observatories.[87] He was often asked for advice by traditional astronomers, like David Gill of the Royal Observatory at the Cape of Good Hope, who were beginning to get into spectroscopy. Keeler was always willing to take the time to give them carefully thought out suggestions on the design and use of their new instruments.[88] At times he even rendered opinions on telescope lenses, as when he advised William H. Pickering that an 18-inch objective Brashear had just made was very good, and that he should buy it if he could. Pickering borrowed it, and within a few months it was in use in the largest telescope of the new observatory that Lowell was building in Arizona.[89]

In addition to his research and service activities, Keeler was deeply interested in teaching. Soon after he arrived at Allegheny, he drew up a plan for a course of graduate studies in astronomy and astrophysics at the Western University of Pennsylvania.[90] He confided to Barnard that he planned to limit the enrollment "as I don't want too much teaching, although a little will suit me very well."[91] In the first year he had only two applicants for the course, "one of them ... a preacher", and although the description was printed in the Western University catalogue every year, only two graduate students were registered in all the time Keeler was there. One was Breading Speer, the physics teacher in Pittsburgh's Central High School, who spent the summer of 1892 studying with Keeler, and the other was Armin O. Leuschner, the former Lick Observatory graduate student, who was at Allegheny part of the summer of 1893, learning spectroscopy.[92] A few other prospective students, such as Professor J.N. Hart, of Maine State College, Orono, investigated the ambitious little graduate program, but ultimately decided not to come. The quoted tuition fee was $100 per year, with shorter times "in proportion", meaning $25 for the summer, but Leuschner, as a professional colleague, did not even have to pay this. Keeler estimated the cost of room and board in one of the many boarding houses near the university as only about $6 per week, but still Hart decided to stay in Maine.[93]

Like many other professionals, Keeler frequently counseled young would-be students on what subjects they should study as preparation for careers in astronomy. He always recommended mathematics and physics, plus chemistry as well. Biology and geology were not essential, but French and German were, to the extent of reading knowledges at least. Latin and Greek were useless. It was all well and good to be attracted by the "imaginative side" or romantic view of astronomy, "but it is well to point out that the path to professional knowledge is the unromantic xyz of algebra, and anyone having a distaste for mathematics is not likely to achieve success in astronomy as a vocation, although he may cultivate it with pleasure and profit as a recreation." According to Keeler "[w]ith few exceptions, all astronomers have gone through this kind of training. Those that have not are either handicapped, or they have made up their deficiencies by hard study in their few leisure moments". No doubt he was thinking of his former Lick colleagues Burnham and Barnard. As a career, Keeler concluded, astronomy "is agreeable work to those who love it, but there is no money in it."[94]

Although he had only two graduate students, Keeler never gave up trying to attract them, but he always insisted they would have to have at least regular college training in mathematics and physics before he could accept them.[95] His livelihood did not depend on having students, and he could afford to joke about it, as when he told Campbell he intended to adopt his new book for his course in astrophysics "but I am afraid the demand will not be very greatly increased on that account".[96]

There was more demand for his services as an Extension lecturer, where the only requirement to sign up for his course was to have an interest in astronomy. In those days before movies, radio, or television, such courses were very popular and well attended. Keeler gave an Extension course nearly every year, usually in Pittsburgh, but sometimes as far away as Franklin, Pennsylvania, requiring an 80-mile train ride each way, and nearly two days of his time each week. The train to Franklin was a local that made every stop and took five hours to make the journey. Yet it was worth it, for over three hundred people attended his lectures in the Opera House and their interest in the course held up to the end.[97] Usually Keeler gave eight to twelve lectures in such a course, always scheduled for one evening per week. He covered all of astronomy as it was then known, but with special emphasis on spectroscopy and the latest astrophysical results on the sun, planets, stars, and nebulae. A key feature of his lectures was the many slides he showed, including his own drawings of the planets, photographs of star fields, comets, nebulae, and his own spectrograms.[98] He received $10 per lecture plus traveling expenses, a not inconsiderable supplement to his salary of $2,000 per year, but more important was the enjoyment he found in teaching, and the intellectual rewards the students received from his lectures.[99]

Like astronomers today, Keeler had many other dealings with the public beside the controlled, well-organized Extension courses. Most frequently these came in the form of letters, frequently from people with strong opinions on particular subjects, usually cosmic in their nature, and hence in their view deserving consideration by an astronomer. To Keeler such people were all "cranks", but he usually answered their letters as politely and considerately as he could.[100] One request, from an agnostic crank, caused Keeler to reveal himself more fully than he usually allowed himself to do. This man, P.S. van Mierop, invited Keeler to deliver a lecture before his Pittsburgh Secular Society. Keeler declined, on the grounds that the lectures were scheduled for Sunday afternoon, a day calculated to offend the sensibilities of the religious members of the community in the 1890's. Van Mierop, in a long reply, showed by his choice of words that he was not only an agnostic but a socialist, dedicated to "enlighten[ing the] people on that day about things of which we know something" and thus to changing society.[101] Keeler, like most astronomers of his time, was profoundly conservative and dedicated to the ideals of property and the Republican party. A few years later he was to refer to the 1896 Democratic Convention in Chicago at which William Jennings Bryan delivered his "Cross of Gold" speech and won the nomination for President as "a menagerie of cranks", and to hope that his correspondent was watching it "from a place of comparative safety in the gallery".[102] Keeler, Allegheny Observatory, and astronomy in general depended on the largesse of wealthy men like James Lick, William Thaw, and the members of the Junta Club. He could not afford to offend them, and at a deeper level, his feelings and general point of view were identical with theirs. Thus he quickly replied to van Mierop that although he agreed with much of what he said in favor of Sunday lectures, if he gave one himself and thus pleased some people, he would surely at the same time offend others, among them some of his good friends. "Why should I do it?", Keeler asked. "Remember that I have no principles to support in doing so, as you have. You have your 'principles and rules' to live up to, and in many of them I have no faith. With me it is a question of policy, and no policy could approve of a course in which there is something to lose and nothing to gain."[103] He seldom expressed his inner thoughts so directly, but that maxim was one of the guiding principles of Keeler's life and career.

Most of his other jousts with cranks were more amusing than revealing. To a physician who sent him a copy of a book he had written "disproving" the theory that the source of the sun's luminosity was gravitational contraction, Keeler replied thanking him for his gift, but saying "[s]ince you ask my opinion I am compelled to say that I do not think the scientific value of the book is great. The only objection to Helmholtz' solar contraction theory is that it seems to set a limit to the existence of life in our system, and this (in our opinion) ought not to be; but our opinions as to what ought, or ought not to be have very little to do with the matter". He added that in a "few thousand years" the contraction theory would be settled one way or the other by direct observation, but that other methods of proving or disproving it might be discovered sooner. Less than half a century later, Richard Atkinson and Fritz Houtermans were able to show, by theoretical reasoning based on experimental data from nuclear physics, that thermonuclear reactions, not gravitational contraction, are the energy source of the sun and stars.[104] And when another crank sent the Allegheny director a long and complicated scheme for imposing a new system of time keeping on the world, Keeler wrote him that even if it were excellent "its practical adoption is an absolute impossibility. It would be easier to make a republic of China and an absolute monarchy of the United States."[105] He always believed in paying attention to practicalities, and not in wasting time tilting at windmills.

When a distant relative sent him a clipping from the *San Francisco Daily Report* about a new theory of the universe by one Solomon J. Silberstein, which purported to explain "the causation of its origin and orderly development; also the primitive cause of force and matter" and to prove "that Newton's law of gravity is a fallacy", Keeler wrote him that it showed a very insufficient knowledge of elementary mechanical principles. "Discoveries are not made by digging into the barren rock, but by following out the vein of ore, the direction of which is indicated by the work of previous investigators". Anyone who claims that Kepler's third law does not follow from gravitation does not deserve a hearing, Keeler concluded, but added politely that he trusted that nothing in his reply would offend his relative.[106]

Keeler's main interest at Allegheny was always in his research. A constant threat was the possible loss of the time service, for the income it provided was the main source of funds for operating expenses. For instance, in the fiscal year ending May 31, 1893, Allegheny received over $3,000 from the Pennsylvania Railroad and the cities of Pittsburgh and Allegheny for the time signals it furnished, $300 from the Weather Bureau to finance Very's research, and $385 in miscellaneous contributions; while the total expenses were just under $2,700.[107] The time service was in danger, however, because the Naval Observatory, as a government agency, had begun to furnish the time free in Washington, and Western Union could supply it anywhere that it had a telegraph line for considerably less than Allegheny charged. A vigorous conflict was waged in the years at the end of the last century between the professional astronomers and the U.S. Navy Department, over the issue of getting the Naval Observatory out of the hands of the admirals and under civilian control.[108] This battle was fought largely on scientific grounds, but was perhaps more realistically seen by Keeler, who recognized the problem as soon as he took hold at Allegheny. He reported at that time to William Thaw, Jr. that he was worried that the income was precarious, but that "[i]f only the Naval Observatory passes into civilian hands the time service would probably be safe again."[9]

The blow fell in 1893, a year of "bad times" that we today would call a depression. The Pennsylvania Railroad gave notice that it intended to switch to the Western Union time service. Keeler instantly alerted Chancellor Holland and Benjamin Thaw, brother of William Thaw, Jr. and a member of the Observatory Committee, and provided them with arguments to use to convince the railroad officials to continue to pay for the Allegheny time service. Thaw went straight to the Duquesne Club and had a "little talk" with his friend Robert Pitcairn, the general agent of the Pennsylvania Railroad in Pittsburgh. In the next few days Thaw lobbied with three vice presidents of the railroad as well as with the general manager, all of whom he knew personally. As a result, the railroad continued to subscribe to the time service, but at a reduced rate. The observatory income dropped by $1,400 a year, but Thaw managed to get several friends, including some of the railroad executives, to subscribe personal contributions of $100, which somewhat softened the blow.[109]

In the fiscal year ending May 31, 1894, the income from the time service was $2,733, nearly the same as the previous year, for the loss was only beginning to be felt. But the following year the income had dropped to

$1,766, supplemented however by $800 in contributions from Thaw's family and friends, including $200 from Chancellor Holland himself. The time-service income remained approximately stable near $1,700 per year for the rest of Keeler's stay at Allegheny, while the sum received in contributions gradually declined to only $200 in his last full year. The university itself never provided any money except Keeler's salary.

The largest item of expense for the observatory each year was $1,000 for the salary of Keeler's assistant, at first Very, who did not in fact assist him, and later Henry Harrer, after Very left late in 1895. The next largest item was generally about $500 for books for the Observatory library, and smaller amounts for minor auxiliary instruments, coal to warm the building, and maintenance.[110] Funds were so short that on one occasion, when Keeler found he could get a complete set of the *Philosophical Transactions of the Royal Society of London* for the years 1867 through 1883 at the unusual bargain price of £16, or about $5 per year, he claimed the observatory budget could not afford it. However, when the Carnegie Library in Pittsburgh turned down his request to buy the set, Keeler did manage to scrape up the necessary funds and order it.[111]

In addition to his spectroscopic work on stars, Keeler continued his nebular research at Allegheny, as he had planned when he came. As soon as his photographic spectrograph was completed, he began taking long exposures on the Orion nebula.[112] He could see numerous emission lines on his plates, some of them reported earlier by Huggins, others new. To measure the wavelengths accurately, he soon developed the procedure of using a "decker" or mask for covering or uncovering various parts of the slit of the spectrograph without disturbing the plate. Thus he would take a long exposure of the nebula through the central part of the slit, and then, by moving the decker, expose a comparison spectrum on either side of the nebular spectrum, immediately adjacent to it. Usually for this comparison he took the spectrum of the sky, illuminated by the moon at night or, if the moon were down, by the sun the next morning. Thus he could measure the wavelengths of the nebular lines with respect to the wavelengths of the solar absorption lines, accurately known from Rowland's work. Occasionally, working in the longer wavelength spectral region, he used a laboratory source – salt in a spirit flame – to produce the sodium lines for his comparison spectrum. Keeler took long exposures, but found that even after he had added stiffening rods to his spectroscope, flexure of the instrument made it unprofitable to extend the exposure beyond two hours on either side of the meridian, or four hours per night. He experimented with even longer exposures, for which at the end of the night he left the plate in the spectrograph all the next day, and exposed it again the following night, but they were not successes; generally temperature changes caused expansion or contraction of the instrument and resulted in blurred images of the spectral lines.

Keeler took exposures in various parts of the Orion nebula, all however, within the central bright region near its center. He did this to check a result of Huggins, who had found on one spectrogram he had taken in 1889 an apparent change in the spectrum, in which although the hydrogen line Hγ at λ4340 was visible, the shorter wavelength lines Hδ, Hε and the nebular line λ3727, which he had previously photographed, had disappeared. On the next spectrum of the Orion nebula he took, in 1890 (there are long cloudy spells in England!), Huggins found the spectrum returned to its earlier appearance, with the missing lines restored. As Keeler stated, Huggins' result "seems very remarkable", but deserved to be tested observationally. The Allegheny astronomer found that his spectra of the Orion nebula were all essentially the same; he never found that any lines had disappeared. He very delicately suggested that the abnormal spectrum Huggins obtained in 1889 "was an idiosyncracy of the particular plate on which it was obtained" – that is, that it had not been completely developed or that part of the light beam had been blocked inside the spectrograph. To soften the blow to the aging pioneer astrophysicist, Keeler wrote that this interpretation was suggested by various puzzling things that he had seen in his own experience, but that he would have been more ready to accept this hypothesis if he had taken the plate himself. Diplomatically, Keeler sent a copy of his paper to Huggins with a note assuring him that he meant nothing personal, and that their results differed only on details. The old man replied handsomely that as scientists they were only interested in the truth, and that he personally could "not but wish the results of all good work, such as yours, should be fully published." However, he insisted, he had kept the 1889 spectrogram and it still looked the same to him.[113]

In the Orion nebula, the emission line that could be photographed furthest in the ultraviolet was λ3727. On one of his best spectrograms, Keeler said, this line looked distinctly double, split into two components, the one of shorter wavelength being the stronger. All the hydrogen lines of the nebula that showed on this plate were single and sharp, as were the solar lines in the comparison spectrum. Nevertheless, Keeler said, he had learned from his experience that faint details seen on only a single plate must be regarded with suspicion. In spite of this disclaimer, he reported his measurements of the individual wavelengths of the two components, λ3725.8 and λ3728.0. In fact he had seen for the first time the two components of what we now know as the [O II] λ3727 doublet, and the splitting he had measured, 2.2 Å, is very close to the accurate modern value, 2.75 Å. He did not see the line double on his other spectrograms because the focal length of a spectrograph camera such as Keeler used varies with wavelength, so that ordinarily the extreme ultraviolet is slightly out of focus. Probably this one plate was better focused for λ3727, and therefore resolved the close doublet. It was an observation that was far ahead of its time.

Keeler also took a spectrum of the brightest star in the central bright part of the Orion nebula, θ₁ Orionis A, also called Bond 628. The star's spectrum showed absorption lines of hydrogen, coincident with the hydrogen emission lines of the nebula. There were also several other absorption lines, including λ4471, a strong line characteristic of the "Orion stars", or stars we now call O and B stars. Several of these stars are in the constellation Orion, and both θ₁ Orionis A and θ₁ Orionis B, the two brightest stars within the central bright part of the nebula, as well as θ₂ Orionis, just outside it, belong to this class. By his accurate wavelength measurements, Keeler proved that this absorption line in the Orion stars' spectra is identical with the emission line λ4471 that is seen in the Orion nebula, thus confirming a suggestion made earlier by Scheiner. Furthermore, Keeler showed that λ5876, another emission line seen in the nebular spectrum, is also identical in wavelength with an absorption line he measured in the spectrum of Rigel (β Orionis), one of the bright Orion stars. Thus his spectroscopic measurements had demonstrated an "intimate relation" between the Orion nebula and the stars near it; namely that they both contain the element that gives rise to the spectral lines λ4471 and λ5876. Furthermore, as Keeler said, since Edward C. Pickering had shown from his spectroscopic studies at Harvard that nearly all stars in the sky have spectra that can be regarded as made up of four sets of lines, one of them being the lines of the Orion stars, these measurements firmly connected the Orion nebula, or nebulae in general, not only with the Orion stars, but with essentially all the stars. In other words, Keeler had shown that the stars contain the same elements as the nebulae – at least the element hydrogen, as was known from Huggins' early work, and the element responsible for the λ4471 and λ5876 lines – and thus could in principle be regarded as having formed from nebular material. A few years later, as we shall see, astronomers and physicists recognized that these two lines come from helium, now known to be the second most abundant element in the universe. All stars and nebulae contain helium, but it is observable in O and B stars because of their high temperatures, and in nebulae like the Orion nebula because of their high degree of ionization.

Keeler also measured carefully the wavelengths of several other absorption lines in the spectrum of θ_1 Orionis A, but still found that the two green nebular lines did not coincide with any known stellar absorption lines. He remarked that Lockyer had found some coincidences with absorption lines seen in the spectrum of α Andromedae, but that its spectrum contains so many lines, all of roughly equal intensity, and also that the wavelength tolerances accepted by the English spectroscopist were so large, that this result was not really significant at all. Keeler's paper containing these results on the spectrum of the Orion nebula and the Orion stars is an excellent piece of research, drawing important conclusions from accurate observational data.[114]

Keeler published this paper, based on spectra he had taken at Allegheny in the fall and winter of 1893-94, the following summer. One reason that he got it into print so quickly was that he realized that Campbell, his former assistant, was working along very similar lines at Lick and was fast on his heels, if not actually overtaking him. When Keeler had left Mount Hamilton, Campbell had been hired in his place, and had soon modified the spectroscope, adding a camera so that he could use it photographically, instead of only visually, as Keeler had done.[115] Campbell, in contrast to Keeler, had been trained only in the old mathematical astronomy at the University of Michigan. But he was a demon of energy and plunged into spectroscopy after only one summer's experience working with Keeler.[116] He intended from the first to concentrate on measuring the radial velocities of stars by the Doppler effect, just as earlier astronomers had concentrated on measuring the proper motions, or tangential motions of stars, before the days of spectroscopy.[117] It was a very important fundamental program, as all the classical astronomers agreed,[118] and Campbell devoted his life to organizing and carrying it out.

However, in his early days, Campbell was a scientific prospector, and when Nova Aurigae, a "new star", flared up, in 1892, he began observing it spectroscopically. He immediately saw that its spectrum contained many emission lines, some of them the same lines Keeler had seen in planetary nebulae and had shown to his young assistant.[119] To measure the wavelengths of these lines and study them himself, Campbell began taking spectra of planetary nebulae, then of the Orion nebula, and almost before he realized it, he was working along very similar lines to his former chief's. He wrote and told him what he was doing, and said he would not go on with this line of work if Keeler intended to do it himself.[46] Actually Campbell could no more have stopped his own research than he could have stopped the flow of Niagara Falls. Keeler of course congratulated him on his successes in spectroscopy, and urged him to continue, especially on the radial-velocity work.[50] At Lick, Campbell was soon taking spectrograms of the Wolf-Rayet stars, a small group of unusual objects with broad emission lines.[120] Keeler had observed them earlier while he was at Lick, but had published nothing about them. Their spectra contain some of the same lines as the nebular spectra, λ5876 and λ4471 in particular, so Campbell was again forced toward Keeler's field of specialization. He reported that he was interested in nebulae, and said that he wished that Keeler's long paper on the Orion nebula had come out. He promised to give full references to it in any paper he published on the subject, a promise he kept handsomely.[121]

At last in February 1894, in a long letter about spectroscopy, Campbell wrote Keeler that he was working on the Orion nebula and was especially interested in checking some of the strange results reported earlier by Huggins. He explained that "on the spur of the moment" he had borrowed a copy of Keeler's best spectrogram of the nebula from Leuschner, who was now back in California after his summer stay at Allegheny, but that "[s]ince thinking it over I have come to the conclusion that I was not doing the right thing" and apologized.[122] Keeler quickly replied that he could see that they both were working along very similar lines. He proposed that they write up their results for publication independently, and arrange to have both their papers published in the same issue of the journal, *Astronomy and Astro-Physics,* which he was then editing.[123] Campbell, somewhat disingenuously, wrote that he was surprised to learn that they were both working on the same topics. He explained that his own results depended not only on his spectra of the Orion nebula, but also on the spectra of several planetary nebulae he had observed earlier. He had nearly finished the paper, a long one, two or three weeks before, but he would now touch it up and send it off. Campbell told Keeler that he had taken nearly all his spectrograms of Orion in October, when the nebula can be observed early in the morning, just before dawn. "I shall want to put a few dates in my paper, just in the usual manner, for the reason that the English & Europeans are probably working on the same line", he added. In fact he wanted to make sure he received credit for whatever priority he had over Keeler, his former mentor. He made it clear that he did not want to wait for publication until Keeler's paper was finished.

To this same letter Campbell added in a postscript that his salary at Lick had been raised from $1,800 to $2,400, and that the University of California regents had promised to build a new house for him on Mount Hamilton. That would make him "contented", he purred.[124] Keeler, whose salary had never been raised above $1,800 at Lick, which he had left because the regents had told him they could not build him a house, now found himself with a $2,000 job at Allegheny. From his salary he had to pay rent for a house for his family because the observatory house was so dilapidated. His former assistant, with little training in astrophysics but with a large telescope in a fine climate, was effortlessly skimming the cream of astronomical discoveries that he himself, with his small tele-scope and smoky, cloudy atmosphere, had labored mightily to make. Keeler must have been more than a little discontented, and he must have wondered if he had really made the right decision when he left California.

Whatever he felt, Keeler could do little but agree with Campbell's plan. He promised to print Campbell's paper as soon as he received it, and said he would try to have his own paper finished and in the same issue, but if he could not complete it in time, it would just have to wait. In the closest thing to an admission that perhaps he wished he had stayed at Lick that Keeler would permit himself, he closed his letter to Campbell by writing:

> No one has followed your success in the line of astro-physical work with more pleasure than I. Do you not find it more interesting than the old astronomy? But the old astronomy is the best training for the new. How about the new houses on Ptolemy [the ridge on Mount Hamilton where Campbell and Barnard's houses were being built]? Are they actually in course of construction? You must be very busy, but with such employ-ment work is happiness. I congratulate you heartily on your well-deserved advancement.[125]

Campbell completed the manuscript of his paper on March 20, 1894, dated it, and mailed it to Keeler to be published. He said, however, that if Keeler could finish his paper in time to get it in the same issue of *Astronomy and Astro-Physics,* he should change the date on the manu-script (so that the two papers would appear to be simultaneous), but that if Keeler's paper had to go in a later issue, he should leave the original date on the manuscript (to protect Campbell's priority against "the English and Europeans"). And, in expressing his thanks to Keeler, he stated his own credo:

> I am greatly obliged for the kind words written about my astrophysical work. No, I regret having to give some time to the *old* work, & want to get out of it entirely. In spectroscopy one can cut loose from traditions and roam as free as he likes. And there is so much waiting to be done, and the nights are few & short.[126]

As soon as Keeler received Campbell's manuscript he sent it, un-opened, to the publisher, with instructions that it must go into the May issue. He specified the articles, previously accepted, that could be left out to make room for it.[127] And he worked hard to complete his own paper, so that he finally was able to send it off to the publisher on May 1, just over a month later than his young former assistant's paper.[128] Then at last he read it, for it was now in print, but only the first half of it, for Campbell's paper had proved too long to squeeze into the May issue *in toto*. All the Orion results were there though, and Keeler at once saw that although Campbell's data were more complete than his own, their essential results were in very good agreement.[129]

Campbell's paper was indeed a very complete work. Like Keeler, he recognized that the lines at λ5876 and λ4471, seen in emission in the Orion nebula, were also visible as absorption lines in several of the Orion stars. With the large Lick telescope he had been able to get good spectrograms of several stars in Orion, including the two fainter stars in the Trapezium, which showed that all of them had similar spectra. He also recognized several fainter lines that Keeler had not been able to record on his spectra, including λ4922 and λ4713, that are emission lines in the nebula and absorption lines in the stars; these, like λ5876 and λ4471, we now know are helium lines. The most important result he found, that Keeler had not seen, was that the relative strengths of the two principal nebular lines, λ5007 and λ4959, varied from one spot in the nebula to another with respect to Hβ, the nearby hydrogen line. The ratio of strengths of λ5007 and λ4959 was always the same, about 4 to 1 he estimated, but both varied together with respect to Hβ. In the central brightest parts of the nebula, Campbell found λ5007 the brightest line, about four times as strong as Hβ, while in regions of medium brightness, it was almost equal to Hβ, and in the outermost, faint parts of the nebula, Hβ is considerably stronger than λ5007. At Lick, Keeler had earlier seen differences of this same type between different planetary nebulae, but at Allegheny the small scale of his telescope had the result that his spectrograms averaged over the whole central and medium-brightness parts of the Orion nebula, while its outer, faint parts were too faint for him to photograph, so he never observed this effect there. Like Keeler, Campbell saw and reported that several other strong nebular emission lines, including λ5007, λ4959, λ3869, and λ3727 do not appear as absorption lines in any stellar spectra. Although he was unfailingly respectful in his references to Keeler's work, the young Campbell severely criticized Huggins for the errors in his earlier work on the Orion nebula spectrum. In the remainder of his paper Campbell gave his results on the planetary nebulae, recording their differences from the spectrum of the Orion nebula, as well as their similarities to it. His paper, which contained much more data than Keeler's, was excellent.[130]

The same issue of the journal also contained Campbell's long paper on the Wolf-Rayet stars, comparing and contrasting their spectra with the spectra of nebulae, the Orion stars, Nova Aurigae and the solar chromosphere.[131] Campbell had submitted the manuscript only a few weeks after the nebular paper, and it was another excellent piece of work, as Keeler instantly recognized.[132] He must now have clearly realized that he would only be able to make important new discoveries if he concentrated on areas that Campbell, with his large telescope, transparent atmosphere, and incredible personal energy, had not yet begun to explore.

Yet, all the while, Keeler was acting as an unpaid consultant for Campbell and Holden on the new spectrograph that Brashear was making for Lick Observatory. Holden had persuaded D.O. Mills, the wealthy San Franciscan, to provide the money for this photographic instrument, to be built to Campbell's specifications for radial velocity work.[133] Keeler advised Campbell on the overall principles that should be followed in designing it, checked the specifications, tested the optics, and constantly made the little decisions that were necessary so that Brashear could push on with the work. On one occasion Holden even complained to Keeler that Brashear was wasting too much time on other jobs, when he should have been going ahead full speed on the Lick instrument. Keeler diplomatically explained the circumstances that had delayed the job in the shop, and politely assured Holden that the work had been proceeding at a good rate since Brashear had hired a new foreman.[134] When the completed spectrograph arrived at Mount Hamilton, Campbell exulted "it is a dandy!!" and hastened to thank Brashear for his painstaking work in building it.[135] However, the pressure of other tasks kept Campbell from even trying to use it in earnest for nearly a year. At first the new Mills spectrograph did not perform up to expectations, and Campbell was constantly in touch with Keeler, seeking advice on further tests to diagnose the problem, and reporting his results.[136] After many trials, which included returning the prisms and camera lenses to Brashear, Campbell himself finally located the problem. It turned out that the lenses in the camera, although giving perfect images on axis (at the center of the spectrum), suffered badly from chromatic aberration in the outer part of the field. Once they were replaced, the spectrograph proved to be an excellent instrument, and Campbell and his assistants used it for many years in his major long-term research program, the measurement of the radial velocity of the stars.[116,137]

In his early years at Lick, while he was first making his name as an astronomical spectroscopist, Campbell did nothing to make himself well liked by the English and European high priests of the science. He was a poor farm boy who had made his way upward against great odds, by his own hard work. He believed in the American dream, and had little but contempt for anyone, no matter how famous, who did not work as hard as he did, and did not find the same results that he did, which in his own mind at least were always completely and utterly correct. When these pioneers, men like Huggins, Vogel, and Heinrich Kayser in Germany, saw their work criticized in print by this young American upstart, they tended to reply in measured tones, based on the authority of their observations, often made years before with small telescopes, in poor climates. They soon found that Campbell could strike back hard.[116]

When he published his paper on nebulae, Campbell told Keeler that he "regretted" Huggins' observations were wrong "but it is not my fault".[122,126] When Huggins then published a quiet, restrained "note" on this paper, arguing that the young American had not quite understood what he meant in places, Campbell fired back into print with a harsh reply. In it he "corrected" Huggins' "misstatements", quoting the exact words from the Englishman's original papers, and insisting that he himself was right.[138] And in the main, he was. Likewise when Kayser, who had been one of Keeler's teachers in Berlin, criticized a paper of Campbell's on the spectrum of a comet, the Lick observer instantly got into print with a strong defense. His personal letters to Keeler indicate that he thought of Kayser as a dictator very similar to the German Kaiser![124,126] Campbell believed that the quality of research was the only thing that counted, and he was convinced that his could measure up to anyone's:

> I agree that "age and long experience" are entitled to a considerable degree of respect. Youth, of course, has no such claim to offer, but *conscientious work*, by whomsoever done, is entitled to consideration, even from those of age and experience. Now the truth is Kayser and Huggins rode rough-shod over some of my work. They did not treat it *on its merits* ... In my opinion *age* is no excuse for injustice: it only adds to the culpability...
>
> I deeply regret the spirit of controversy which seems to be creeping into astronomical literature. On the other hand it is not right for any man, however young, who happens to collide with the "authorities" to lie down and let them pass over him.[139]

Years later, when Campbell, full of age and experience, was the long-time, highly authoritarian director of Lick Observatory, these words would have made interesting reading indeed for nearly everyone who worked under him.

Even Keeler felt a bit of Campbell's lash, although not to nearly the same extent as Huggins, Kayser, and several other famous astronomers.

The question had to do with the most efficient type of camera to use on a spectrograph, which had come up in Campbell's reply to Huggins' note. In it the Lick spectroscopist had stated that a short focal-length camera was best for extended objects, like nebulae, just as it was for stars, as everyone already knew. Keeler, as the acknowledged American expert in astronomical spectroscopy, felt he had to correct the second half of this statement, which he believed was wrong. In a very polite, diplomatic note, he wrote that the advantage of the short-focus camera for extended objects was well known to laboratory spectroscopists, and for instance was even mentioned in the *Encyclopedia Britannica* article on spectroscopy. However, for a star, he argued, the focal length of the camera does not matter, because for a given linear dispersion of the spectrum on the plate, whether achieved by using a single prism with low angular dispersion and a long-focus camera, or a train of prisms giving higher angular dispersion together with a shorter focal-length camera, all the light is spread out over the same area and hence has the same photographic effect. In fact this is not quite correct, but Keeler went on to say that the case of a star is quite different from an extended object, and that a large telescope has an advantage over a smaller one in the number of cases to which a short focal-length camera is applicable.[140] This is correct, and indicates that he probably nearly fully understood the situation.

Campbell would not admit that he was wrong, and after a friendly correspondence, published a reply, which appeared along with Keeler's brief reply to it, and ended the printed discussion. Campbell, in his note, said that what he meant was that the shorter focal-length camera, used with the *same* angular dispersion, of course did reduce the linear dispersion, but did at least enable the observer to get some kind of spectrum recorded on the plate. This is true, but is not at all what he had meant in his earlier note, as Keeler politely pointed out in his final rebuttal, at the same time emphasizing that he and Campbell were in agreement on most of the points in question.[141] Neither of them appeared, from the printed record, to realize that in a real telescope a star image is not a geometrical point, but actually has a small but finite size as a result of "seeing" (atmospheric unsteadiness) and, in the refractors of their day, of chromatic aberration. Thus star images are actually spread out into images like little nebulae, usually larger than the slit width, but shorter than its length, so a short focal-length camera is advantageous, because it allows the slit to be opened wider, letting a larger fraction of the starlight reach the plate. Campbell certainly did not understand this at the time, or he would have stated it in his reply; Keeler may have understood it, for he said in his private letter:

I am not sure that I see wherein you suppose me to be mistaken, since by your own argument, entirely in line with my own views, the case is different for a star and for an extended surface. In fact the exposure in the former case is proportional to the length of the camera; in the latter case to the square of the focal length.[142]

This statement is correct (for a given telescope and given effective aperture of the spectrograph, which is no doubt what Keeler meant), but he did not include it in his published note. One reason it was harder for astronomers of that day to grasp is that they did not see the star image on the slit of the spectrograph while guiding during an exposure as we do today, and thus did not realize how much of the starlight is lost. They guided using the light reflected from the front face of the prism, behind the slit, and only saw the light that had entered the spectrograph. A few years later, Campbell did realize that the real advantage of the short-focus camera, even for stars, was that the spectrograph slit could be opened wider, letting more light onto the plate.[137]

Keeler's objections to using a short focal-length camera were not merely theoretical; he was ever the practical research worker. The real problem with a short camera, as he wrote Campbell, was that it would give only a very small length of spectrum that was sharp, or in astronomers' terms, that would have good definition. He got better spectrograms with a long camera and lower-dispersion prisms than he did with a short camera and stronger prisms.[143] Keeler's shortest camera, with a focal length of eight inches, had a focal ratio of F/7, relatively "slow" in modern terms, yet although it was based on an advanced lens design for its day, it had a "frightfully small field". Even his 16-inch camera had a field of good definition only about half an inch long, so that it recorded less than 500 Å of the spectrum in good definition at one time. This was the real problem with using a short focal-length camera in Keeler's and Campbell's time.[144]

Keeler made this same point in reviewing a paper on the design of astronomical spectroscopy by Frank Wadsworth, who, although an expert in theoretical optics, tended to overlook practical details. The observer, Keeler said, usually needs a particular resolving power in the spectrum to solve a particular problem; this requires a certain minimum dispersion, which then sets the other parameters of the instrument. The coarse grain of photographic plates determines the apparent widths of spectral lines in nebulae and most stars, not the intrinsic widths of the lines, and this can never be ignored in the design of a spectrograph. The most important consideration is how the system works in practice. Several of the conclusions he drew from these principles were the same as those in the paper he was reviewing, but Keeler added that he "ventured to believe that the best spectroscopes of modern construction are more correctly designed than Professor Wadsworth seems to think." Wadsworth's idea of using a large number of small prisms in series, rather than a single large one, Keeler criticized on the practical ground that it would be impossible to mount them so that the optical system would be stable and give consistently measurable results. Wadsworth's scheme of using very high angular dispersion would require a short-focus camera which would necessarily have only a limited field, which would be "decidedly objectionable ... [e]xcept for a few special purposes." In conclusion, Keeler wrote, "[i]n making the above remarks I would not at all be understood as underestimating the valuable discussion of Professor Wadsworth, which contains many novel and interesting features; but I wish to point out that experience cautions us not to accept the conclusions too hastily, although they are founded in theoretical principles which are undoubtedly correct."[145] He knew it was all too easy, in framing a theory, to overlook some of the relevant considerations, such as the mechanical stability of a prism train, and to oversimplify some of the actual complications, such as the off-axis aberrations of a camera lens system.

As Keeler's scientific career flourished, his family life was continuing and expanding alongside it. His wife's cousin, Harry Floyd, at first lived with them at Allegheny, although in the icy, smoky winters she usually escaped on visits to her relatives and friends in Louisiana.[146] The Keelers' first child, a boy, was born on a very cold morning on January 10, 1893. Holden, in a typical astronomer's message that Keeler must have enjoyed, wrote "the light of *Nova Keelerii* reached me only today. I congratulate you sincerely. Bring the boy up on bright lines – make him Class III (Vogel). With the best wishes for the whole group of three – A, B, a, Sincerely ..."[147] The son was named Henry Bowman Keeler, the first name for Keeler's grandfather, uncle and brother, all of them dead, and the middle name for Cora Keeler's mother's family name. Henry was a bright, lively boy, and Keeler was intensely proud of him.[148] As a baby, he was known to the family as "Kidlums", but he soon outgrew this nickname.[149] According to Keeler, Henry was so lively that "most of the pictures I have taken resemble windmills, from the multiplicity of arms and legs in them", and it is a fact that every existing photograph of him as a boy does show some part of his anatomy blurred by motion.[134]

The Keelers' second child, a daughter, was born on July 16, 1894.[150] She was named Cora Floyd Keeler for her mother, and for the Captain and his wife who had brought her parents together on Mount Hamilton eight years before. After little Cora, the Keelers had no more children, but Keeler enjoyed domesticity and always recommended it to his friends, especially if they were already engaged and about to be married.[151] As Henry and Cora grew older, they accompanied their mother on long trips to Louisiana during the cold Pittsburgh winters.[152] Harry moved back to California, spending much of her time at Kono Tayee, the Clear Lake estate she had inherited.[153] Keeler was particularly fond of the children, and often entertained them with his drawing skills, producing rapid pencil sketches and caricatures of them, their playmates, their toys and their pets. He and his wife were particularly close friends with the Brashears, and with James McDowell, Brashear's son-in-law and chief assistant, and his wife.[154] By this time Keeler had joined the Junta Club, and also got together regularly on Monday nights with a group of cronies he called "the boys", including Brashear, W. Lucien Scaife, one of the Trustees of the Western University of Pennsylvania and a member of its Observatory Committee, and Reginald J. Fessenden, the professor of electrical engineering.[155] It was a pleasant life that revolved about science.

Keeler was always turning over in his mind ideas on improving the efficiency of his observing techniques, and in the fall of 1894 he came up with a scheme which was speedily adopted by nearly every major observatory. The problem was that the refracting telescopes then in common use did not bring the blue and violet light to a sharp focus. Their lenses were corrected, as Keeler knew well from his measurements of chromatic aberration, for a good "visual focus" of the yellow rays, necessarily leaving the shorter-wavelength light badly out of focus. Therefore, much of the blue light did not enter the slit of the spectrograph and was lost. It was possible to refocus the telescope for any one wavelength in the blue spectral region, but the focal length varied so rapidly with color that the other wavelengths would still be poorly focused, including the yellow light, so the observer would not be able to accurately "guide" the telescope and keep the star image in the slit. For direct photographic work the Lick refractor had been provided with a large "photographic corrector", 33 inches in diameter, to overcome this fault. This extra lens had been designed to be put into the telescope tube in front of the main 36-inch two-element lens, and to convert it from a visually corrected objective to a blue, or "photographically corrected" objective. This corrector was the lens that Alvan G. Clark had been finishing in the cold Mount Hamilton winter of 1888. It had never performed satisfactorily, probably partly because of flexure in the telescope tube and partly because Clark had not really polished it to the exact final form. In any case, because it was so large and heavy, it was not easy to put into the telescope or take out, and it was not used except for direct photographs. Keeler realized, however, that for spectroscopy only the star image exactly on the geometrical axis of the telescope had to be corrected; in technical terms the field required was extremely small. Hence instead of a large ungainly lens near the main objective, it was possible to use a small lens near the focus to change the correction from visual to photographic, and thus get all the blue and violet light through the slit and onto the photographic plate in the spectrograph. Keeler studied the optical requirements, and soon had designed a two-element lens system, which brought nearly all the photographically active light to the same focus as the visual light, for his own Allegheny refractor. He was vaguely aware that such lenses had been previously used, but could not find any published descriptions of them. He wrote a short paper on the subject, which was published first in the *Transactions of the Astronomical and Physical Society of Toronto,* which had just elected him a Corresponding (foreign) Member as an "honorable though modest recognition" of his services to astronomy, and then in the *Astrophysical Journal.*[156] When Huggins saw the paper, he wrote Keeler that it was he who had used a similar lens years before, and supplied the reference. His note had actually contained little information and no general description. Nevertheless Keeler politely thanked him, and published the historical correction in the next edition of the journal.[157]

In the spring of 1895, Keeler made his outstanding discovery at Allegheny Observatory, the one that briefly carried his name around the world. Like his earliest spectroscopic work at Lick, it involved a planet in the solar system, always a fascinating subject to the public. Soon after he had put his new photographic spectrograph into operation, two years before, Keeler had begun experimenting with measuring the rotational velocities of the major planets. Jupiter and Saturn were both known to rotate rapidly on their axes, from observations of spots and other features that can be seen on their apparent surfaces. Timing these motions gives directly the speeds of rotation of the planets. Keeler realized that their rotational velocities could also be measured spectroscopically, by the Doppler effect, as could the rotational velocities of planets without apparent surface features, like Venus and Uranus, and even the velocities of the particles in Saturn's rings. These rings, which appear to be solid disks in the equatorial plane of the planet, were known from theoretical work, published thirty years before by the English physicist, James Clerk Maxwell, to be actually composed of many small particles, tiny "moon-lets", each in its own individual circular orbit. Maxwell had shown that the rings could not be solid objects, like sheets of stone, nor fluids, because either of these situations would be unstable under the strong gravitational force of the planet, and would quickly break up. The only possibility was that the rings were composed of individual particles, orbiting independently under gravity. All serious physicists and astronomers accepted this interpretation, but it was based completely on theoretical reasoning, and there was no observational evidence to confirm it.

Keeler recognized that the Doppler effect provided a method for measuring the rotational velocity of the rings or indeed of any object, as long as it was going around fast enough. For a rotating globe, such as a planet, a point on the equator on one edge, or "limb", as astronomers say, would be moving away from the observer, while a point on the equator on the other limb would be moving toward the observer with an equal and opposite velocity. These velocities would show up by the equal and opposite Doppler shifts in the wavelengths in the spectra of the two limbs. The sun's rotational velocity had been measured in this way many years before; most American astronomers believed Langley had originated it, but Keeler seemed to take special delight in pointing out that actually the Swedish spectroscopist Nils C. Duner had been the first to apply the method successfully.[158] For a planet, the slit of the spectrograph could be placed along the image of its equator, and because of the projection effect the lines in the resulting spectrum would be straight but inclined, redshifted by the maximum effect of the rotational velocity at one limb, and blueshifted by the same amount at the other. Furthermore, because the planets shine by reflected sunlight, the effect of the rotational velocity is approximately double the effect it would have if the planet were self-luminous like the sun; this results from the fact that one limb is rotating away from both the light source (the sun) and the observer (on the earth), while the other limb is rotating toward them both.

As for the rings of Saturn, Keeler realized that if they were indeed composed of particles, the orbital velocities of these particles would be determined by the gravitational force of the planet. This force decreases with increasing distance from the planet, so that the velocities of orbiting particles also decrease outward in a perfectly definite way, the velocity at any point being calculable from the mass of the planet. This result had been discovered empirically by Kepler nearly three centuries before, for the case of the planets orbiting about the sun. It was very well understood from Newton's laws of motion and of gravitation, and had been verified observationally for the satellites of Jupiter and Mars. Keeler intended to take a good spectrum of the rings of Saturn and measure their rotational velocity by the Doppler effect, and thus confirm that they were indeed composed of particles moving in orbits under gravitational forces.

He had first tried to observe the spectrum of the rings soon after getting his new spectrograph into operation, two years after his return from Lick Observatory to Allegheny. At that time he was working in the ordinary spectral region, but the combination of the strong absorption in the violet of the 13-inch lens, the poor transparency of the atmosphere, and the yellow color of Saturn and its rings combined to make the spectrum too faint to measure.[159]

In the spring of 1895 he tried again, now using the orthochromatic plates that better suited his telescope and sky. From a short paper he saw in the *Comptes Rendus,* Keeler realized that the French astronomer Henri A. Deslandres was probably working on very similar lines. In that paper, Deslandres had described his measurement of the rotational velocity of Jupiter on spectrograms he had taken the previous November. Deslandres also emphasized the fact that for rotating objects like planets that shine by reflected light, the Doppler effect is twice as large as it would be if they were self-luminous.[160] This point had in fact been well known for many years, as Keeler stated in a short note in which he described his own spectra of Jupiter. They showed very clearly the expected slope of the lines from one edge of the planet to the other, just as Deslandres' spectra did.[161] But, as Keeler wrote his note, he was surely thinking that if Deslandres had already taken spectra of Jupiter to check its rotation, it would not be long before he also took one of Saturn, now coming into opposition where it is best placed for observations. And if the Frenchman took a spectrum of Saturn, he would certainly try to measure the rotation of the rings, as well as of the planet itself.

Keeler wanted to be first. He quickly readied his spectrograph, and built a special guiding device to hold Saturn's image steady, centered on the slit. Finally, on the night of April 9, 1895, he got a good 2-hour-long exposure on the planet and its rings. While he was guiding the telescope, he went over in his mind what the spectrogram might show, and realized for the first time that it might even be possible to see the predicted decrease of orbital velocity outward, from the inner to the outer edge of the rings.[162] For they had an appreciable width, and the orbital velocity at the outer edge would therefore be expected to be significantly smaller than the velocity at the inner edge. On the other hand, if Maxwell's reasoning were somehow wrong, and the rings really were solid, as they looked to the eye or on photographs, their velocity would increase outward, just as the velocity increased outward on the apparent disk of the planet. The spectrograph would clearly distinguish between these two possibilities.

When Keeler had finished the exposure on Saturn, he moved the decker, or mask on the slit of the spectrograph, and took a comparison spectrum of the moon on either side of the planetary spectrum. Then he developed the plate. As soon as he took it out of the photographic fixer and looked at it with a magnifying glass, he could see that the plate was well exposed. In the center was the strong spectrum of Saturn itself, with all its lines inclined, showing the essentially rigid body of the planet. On either side was the strong spectrum of the moon, with its lines straight up and down. In between was the much weaker spectrum of Saturn's rings. Although faint, the spectrum was exposed well enough so that Keeler could see that in it the lines were not inclined as in the spectrum of the planet, showing that the rings did not rotate as a solid body. Studying the lines carefully, he could see that they were slightly inclined in the opposite direction to their slope in the main body of the planet itself. His spectrogram contained the first direct, observational confirmation that the rings were indeed composed of separate individual particles, orbiting the planet exactly as predicted by Newtonian dynamics.

Keeler at once sat down and wrote a letter to his good friend George Ellery Hale, co-editor with him of the *Astrophysical Journal*. Keeler began his letter "I made an interesting spectroscopic find last night, that will make a good article for the Journal. It is a *spectroscopic proof of the meteoritic constitution of Saturn's rings.*" With a clear little sketch and a few well chosen words he explained the whole idea. He was proud of what he had done, and called it "the prettiest application of Doppler's principle that I have seen".[163]

That night the Allegheny skies miraculously stayed clear again, and Keeler was able to take a second excellent spectrogram of the rings, which confirmed the first in every way. He worked away at measuring the plates, reducing his measurements, and writing up the results as quickly as he could, for he was anxious to get his paper into press before Deslandres stumbled onto the same result.[164] Hale fired back his "[c]ongratulations on your most interesting and valuable find on Saturn! It is by all odds the prettiest application of Doppler's principle I have seen, and is certainly the most important discovery made in a long time. What a lot of fine fish there are in the sea – for the right kind of fisherman!" He urged Keeler to send him the completed manuscript of the paper as soon as he could, and promised to do his best to get it into the May issue of the journal.[165]

Eight days after he had taken his first spectrogram of the rings, Keeler had finished his paper and was able to mail it off to Chicago. He correctly predicted that the drawing it contained, modeled on his sketch in the earlier letter, would be reproduced in many textbooks. Hale personally shepherded the manuscript through the press, had an excellent "cut"

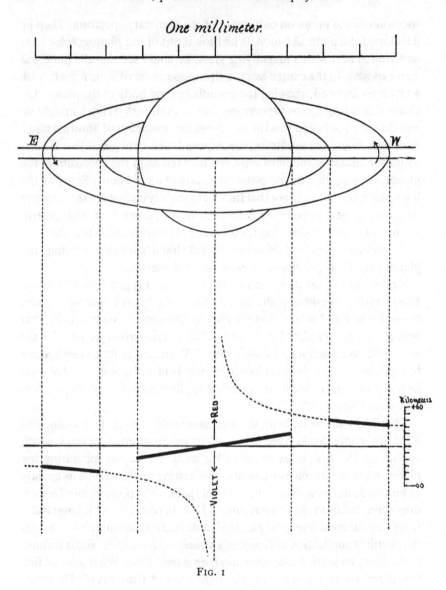

FIG. I

Fig.13. Keeler's drawing showing the spectroscopic demonstration of the fact that Saturn's rings are composed of myriads of tiny particles, each in orbit about the planet, which itself rotates as an essentially solid body.(Reproduced by kind permission of the Mary Lea Shane Archives of Lick Observatory.)

made of the figure, and got the paper out in the May issue as he had promised.[166]

In his manuscript, Keeler described in physical terms Maxwell's "classical paper" which "firmly established" the "hypothesis that the rings of Saturn are composed of an immense multitude of comparatively small bodies, revolving around Saturn in circular orbits", and then went on to say that he himself had recently obtained direct spectroscopic proof of this conclusion. He gave his measurements of five absorption lines in the spectral region $\lambda\lambda 5324$–5429 on his plates, taken at a dispersion of approximately 30 Å/mm. He showed they led to a velocity of rotation of Saturn of 10.3 ± 0.4 km/ s, in excellent agreement with the velocity 10.3 km/s deduced from the rotation period of the planet. Furthermore, he showed his measurement gave 18.0 ± 0.3 km/s for the mean orbital velocity of the particles in the ring, in very good agreement with the value 18.8 km/s computed from gravitational dynamics. Finally he said that although he could not accurately measure the inclination of the lines in the spectrum of the ring, he could see that they had very slightly the opposite inclination to the lines in the spectrum of the planet, as predicted by Kepler's "third law".[159]

Keeler's results were speedily confirmed by Campbell, using the new Mills spectrograph that Brashear had built for Lick Observatory. Because of the larger image size of Saturn at the 36-inch refractor, Campbell's spectrogram of the rings was correspondingly wider than Keeler's, and the Lick astronomer could therefore actually measure the velocity difference between the inner and outer edges. He found it to be 3.1 km/s, in good agreement with the computed 3.9 km/s, and this completely confirmed Keeler's results, obtained with his much smaller telescope.[167]

Keeler was as conscious as most astronomers of the value of a little newspaper exposure for himself and his observatory, so he not only sent word of his discovery to his professional colleagues, but also filled a local reporter in on its significance. Scientists, even more so in his day than now, were not supposed to be publicity-seekers, however, and Keeler insisted that when a "persistent reporter" had gotten "a few words" from him on the subject, he had no idea he would make more than a "local item" out of it.[166] The news was picked up by the Associated Press even before Keeler's scientific paper was completed, and of course was sensationalized by the reporter. He treated Keeler's result on the rings of Saturn as a completely new discovery, that revealed for the first time their true nature. Undoubtedly Keeler had carefully explained that it was an observational confirmation of an unusually well founded theoretical result, but most news writers want a more exciting story than this, and tend to miss the nuances of a complicated historical explanation. A follow-up story equally garbled, soon appeared, quoting unnamed English astronomers as saying Keeler's "theory" on "the nature of Saturn's rings" was "merely confirmatory of the idea laid down twenty years ago by Prof. Trouvelot of Harvard University". Then the Pittsburgh reporter countered with a more nearly complete story saying that Keeler had finished the measurements of his spectrograms and confirmed that the velocities he measured confirmed "the principle known as Kepler's third law", and thus furnished "the first direct proof of a view which has long been held, and demonstrate[d] the power of the telescope."[168] Although the item was hardly earth-shaking in the judgment of newspaper editors – the *Philadelphia Public Ledger* for instance carried it on page 6, in the "Pennsylvania News", devoting less space to it than the stories on the Annual Session of the Eastern Board of the Reformed Church in America held in Lancaster, on alleged favoritism on tax assessments in Reading, or on Easter brides – it was widely published.[169]

Century magazine's editor now decided to capitalize on Keeler's name and publish the article on "Picturing the Planets" it had bought from him three years before. His drawings of Mars, Jupiter, and Saturn were excellent, but they were not particularly well reproduced in the magazine. Keeler's prose, although correct and admirably professional, was disappointingly dull. Worst of all, he pooh-poohed the idea of life on Mars, writing "taking into account all that has been learned of this planet by years of study, the nearest approach that can be made to an answer is that Mars may possibly be habitable. This is certainly unsatisfactory, but no statement more definite is justified by the evidence."[47] At the very same time, in the *Atlantic Monthly*, Percival Lowell was describing vividly his theories of vegetation, water, canals, intelligent life and a dying civilization on Mars.[49] Keeler was not asked for any more articles by the *Century* magazine.

Keeler was bombarded with requests for information from professional astronomers, who had read the newspaper stories and wanted to know just what he had done. He answered them all, describing the main points of his work and how it confirmed Maxwell's theoretical work, and referred them to his paper, which would soon appear in the *Astrophysical Journal*.[170] He drew up a short semitechnical paper giving the main facts about his work, denying that he had made any "claims", and referring to the more complete paper in the *Astrophysical Journal*. One version or another of this short paper was published in many journals, including not only *Popular Astronomy* and the *Publications of the Astronomical Society of the Pacific*, but also *Science*, then as now a magazine with a very wide circulation among scientists in all fields.[171] Keeler also wrote a letter to Edward W. Maunder, the secretary of the Royal Astronomical Society in London, which soon published it in its journal, along with a reproduction of the famous spectrogram.[172] In his tactful, diplomatic way, Keeler had inserted a reference to Maunder's earlier work on the rotation of Jupiter, which the secretary naturally appreciated.[173] Keeler wrote a follow-up paper for the *Astrophysical Journal,* working out some of the more abstruse details of the Doppler-effect method applied to the rings.[174] By this time Deslandres had also confirmed Keeler's result; like Campbell, the French astronomer had been able to measure the actual velocity gradient in the rings. Deslandres, however, maintained that these spectra did not prove the rings were necessarily composed of myriads of particles, but only that they were not solid. In principle, at least, they might consist of many narrow, wire-like rings, next to but not connected to one another, he wrote.[175] Keeler replied that this possibility was far too artificial to be considered seriously.[176]

Astronomers everywhere were impressed by Keeler's work on Saturn's rings. It was so simple, they all realized, that they could have done it themselves if they had only thought of it. But they had not thought of it, and he had, and had pushed it through to completion with his little telescope at the edge of a smoky city. His result was a simple, straightforward and immensely convincing demonstration of the correctness of the theoretical reasoning of a great mathematical physicist. Congratulations rained in on Keeler from his scientific colleagues, and from friends like Charles H. Rockwell, the wealthy amateur astronomer who had helped pay for his education at Johns Hopkins years before.[177] The Astronomical and Physical Society of Toronto raised his status from Corresponding Member to Honorary Member, citing his work which "has made your name immortal, in that you were the first to detect the motion of nebulae in the line of sight and to prove, experimentally, the composition of the rings of Saturn; these, in addition to other signal original work".[178] These are the two outstanding spectroscopic research results that astronomers of Keeler's day thought of when they heard his name. A few years later, when he was awarded the Rumford Medal of the American Academy of Arts and Sciences, for "an important discovery ... in light or heat", the citation referred to Keeler's "application of the spectroscope to astronomical problems, and especially for his investigations of the proper motions of the nebulae, and the physical constitution of the rings of the planet Saturn, by the use of that instrument".[179]

Even before 1895, Hale who, as we shall see, wanted desperately to hire him for the staff of the new Yerkes Observatory, described Keeler as "the ablest stellar spectroscopist in this country".[180] After the Allegheny director had published his rings-of-Saturn work, astronomers and astrophysicists universally held this opinion. A young student at Princeton, taking Young's course in astronomy, wrote in his notes on the lecture on Saturn the important names to remember – Galileo, Huyghens, Cassini, Herschel, Bond and Keeler.[181] Small wonder that a would-be astronomer, trapped in a business career in Marengo, Illinois, only a few miles from the new and famous Yerkes Observatory with its largest refracting telescope in the world, should write not to its Director, but to Keeler in far-away Allegheny for advice on how to become a spectroscopist, nor that Keeler could find time to answer him with practical, helpful suggestions.[182]

Along with the kudos from his fellow scientists, Keeler also received a large number of crank letters, brought forth by the newspaper publicity on Saturn. One of the most amusing was from a man who was the "proprietor" of a medium in Mount Clemens, Michigan. According to him this medium, a young girl named Elfa, had returned the previous summer from a trip to Saturn, which she had made as "her disembodied self", and had told him that the ring of Saturn was composed of innumerable small bodies so close together that no one had yet noticed from Earth that the ring was not actually solid. "How much do I reproach myself for not making Elfa's description of Saturn's rings public at the time! What an incontestable verification your discovery would have been of her psychic power", he concluded. Even this letter Keeler answered politely, saying that while he was not an "impregnable skeptic" he could not accept as facts the various phenomena of spiritualism without testing them – "this is regarded by believers (possibly with justice) as unreasonable". In the case of Saturn's rings, he concluded, their "meteoritic constitution" was an old idea, and his observations had simply provided a new and fresh test of its correctness.[183]

Keeler was not content to sit back, rest on his laurels, and answer letters. He plunged at once into preparations for applying the same Doppler method to determining the rotational velocity of Venus. This planet, closer to the sun than the earth is, has only vague and indefinite shadings that can be seen at times on its disk by practiced observers. Its period of rotation was generally believed to be short. But according to Schiaparelli, who had studied it closely, its rotation period was 225 days, indicating it rotated on its axis with exactly its orbital period, keeping the same face always toward the sun. However, the eccentric Austrian astronomer, Leo Brenner, who had made a career out of claimed discoveries with a very small telescope, believed he had measured the rotation period as 23 hours, 57 minutes, 36.2396 seconds![184] In any case, because of its small radius, Venus' rotational velocity would be much smaller than those of Jupiter or Saturn, and Keeler realized he would need higher-dispersion spectrograms to measure it by the Doppler effect. Hence only two weeks after getting the spectrograms of Saturn, he was writing Young at Princeton, to ask if he could borrow his high-dispersion spectrograph, originally intended for work on solar prominences. Keeler had already tried his own spectrograph on Venus, and found that with the most dispersion he could muster, a train of three prisms, he could get a beautifully sharp spectrum with only two minutes exposure time. He estimated that on it he could have detected the equatorial velocity of Venus if it were as large as 10 km/s, but he had seen no sign of it. Instead of the train of prisms, with his spectrograph he could alternatively use a grating to give a very high-dispersion spectrum, but the light loss was too great, and even bright Venus would not produce a measurable spectrogram. Keeler thought that with Young's solar-prominence spectrograph, which also had a train of prisms, he would have about five times higher dispersion than with his own instrument using prisms, and still be able to get a good plate in about half an hour. He planned to use orthochromatic plates, as he had on Saturn, and work in the yellow-green spectral region, to minimize absorption of light in the glass of the prisms.[185]

The generous Young was only too glad to lend Keeler the spectrograph, but had to admit that more than ten years before it had met with a "bad accident". It had been dropped off the end of the telescope and had smashed on the floor, leaving one of the prisms broken, another chipped at the corners, and "the metal work ... a good deal demoralized." Keeler then recalled that Haverford College had a similar spectrograph, that was not used at all, and thanking Young, made arrangements to borrow it instead.[186] As W.H. Collins, the Haverford astronomer, had warned Keeler, this spectrograph also was in rather poor optical shape, but he and Brashear soon eliminated all the sources of stray light within it and had it lined up and ready for use. It was a five-prism instrument, arranged so that the light went through the train twice, giving the equivalent dispersion of ten prisms. However, when he tested it, Keeler found that the prisms were quite light, so their dispersions individually were small, and overall the total dispersion was not much higher than that provided by his own Allegheny spectrograph. This was not powerful enough, so Brashear quickly constructed a special very high-dispersion spectrograph, with a train of eight dense (strongly refracting) prisms, and unusually long-focus collimator and camera lenses. It was a one-shot instrument, built in a week, largely of wood instead of steel and brass, since it was not intended for sustained use. With it Keeler could get a well exposed spectrogram of Venus in 15 minutes, at a dispersion of 7.5 Å/mm, about four times as high as he had used on Saturn.[187] Even this was not enough, and Keeler and Brashear made further changes in the spectrograph, which improved the dispersion by another factor of two, to 4 Å/mm. The exposure time required was still only about half an hour, which presented no problems. But the trouble was that with Keeler's little telescope, the image of Venus was so small at superior conjunction that the resulting spectrogram was too narrow to show the slope of the lines. Superior conjunction occurs when the planet is on the opposite side of the sun from the earth, and hence as far as it ever gets from the observer. Unfortunately this is the only time that the entire disk of an inner planet, like Venus, is visible, and when the rotational velocity may therefore be measured accurately. By November Keeler had further improved his technique so that he now could use the grating, giving even higher dispersion than Brashear's jerry-built wooden spectrograph, but even this was not enough. He definitely needed a larger telescope, and would not be able to measure Venus' rotational velocity, or set a meaningful upper limit to it, until he somehow got the opportunity to use one.[188]

One of the fringe benefits of an astronomer's life is the chance for foreign travel. Holden, when Keeler had first gone to Allegheny as director, had warned him against it, advising him to stay at home and improve his contacts with the wealthy people who could finance the observatory. "With your influence always *there* in Pittsburg[h] you will get a lot that would be missed if you were like the Sun God [Langley] – Sometimes gilding the top of Aetna, sometimes that of Whitney!"[37] Keeler had followed this advice for five years, but when the opportunity arose to visit England after his work on Saturn's rings had made him scientifically famous, he seized it.

The occasion was the week-long meeting of the British Association for the Advancement of Science, held in Liverpool in September 1896. Keeler was invited to attend as a distinguished foreign guest of the B.A.A.S., which would take care of all of his expenses while he was at the meeting. As soon as he had made arrangements for his family and the observatory, he accepted with alacrity.[189] He wanted a traveling companion, and tried in turn to persuade his friends Hale, Henry Lord, and Crew to go with him, but Hale's wife was expecting a baby, so he could not leave home, Lord wanted to go but was "not very well fixed for spare money", and Crew had the same problem: "There is nothing in the world would please me more than to join you on your trip to Liverpool this autumn. Only the absence of those 'depreciated, demonetized, demoralized' silver dollars [a reference to a William Jennings Bryan campaign speech] stands in the way."[190]

Keeler had to pay his own traveling expenses to and from Liverpool, and decided to take a small, slow ship rather than a luxury liner, to save money. He had never been to England before, and wanted to travel around and see some of the country after the meeting. The Extension lectures he had given would help provide for some of his expenses on the trip.[191]

Keeler took passage on the *Circassia,* a small steamer of the Anchor Line, and sailed from New York on August 29. The fare for the crossing to England, with a lower berth, was $60. In Liverpool he stayed as a guest in the home of D.M. Drysdale, an affluent member of the local host committee for the B.A.A.S. meeting.[192] The Association was and is a mass organization, numbering among its members some of the most renowned professors in the United Kingdom, as well as village science teachers and amateur researchers. The Liverpool meeting was attended by almost 3,200 people, including "41 Foreigners and 873 Ladies." The new president, who took office on the first day of the meeting, was Sir Joseph Lister, succeeding Sir Douglas Galton. In his address Lister described recent advances in medicine, including the early uses of the newly discovered Roentgen rays (X-rays), anesthesia, microorganisms, antiseptic methods (his own contribution), and vaccinations. The next day, at the opening session of the Section of Mathematical and Physical Sciences, Keeler heard its new president, the physicist J.J. Thomson, present his address, in which he referred to the work of Maxwell and Lord Kelvin, and spoke at length on Roentgen's discovery.[193]

That same day the astronomer Isaac Roberts presented his paper "On the Evolution of Stellar Systems" in which he described his photographic work with a reflecting telescope. It was a highly speculative paper, in which he gave few facts but showed many slides, which he emphasized were permanent objective data "not subject to the disturbing effect of human or personal imperfections." His photographs, Roberts said, had not been retouched, and showed objects as they undoubtedly exist in the sky. Although he never said anything about the telescope he had used, it was a 20-inch reflector. With it he was able to take long exposures and record the images of much fainter stars than the human eye could see. He described at length lines and groups of stars, all of which he believed were physically related associations, although actually most of them were probably random groupings. Roberts also had faint photographs of a few of the largest spiral nebulae, all of which he linked in an evolutionary sequence which went back to Lockyer's meteoritic dust.[194] Although

Roberts' speculations, based on very little solid evidence, must have seemed laughable to Keeler, the photographs themselves he studied carefully. A few years later, he himself was to make tremendous strides forward in this very field, but at Liverpool his main interests were still in spectroscopy. He was convinced that "a very large reflector, *mounted with as much care as one of our great refractors,* would be the best of all instruments for stellar spectroscopy."[195] Keeler did not live long enough to put this program into effect, but his friend Hale later built Mount Wilson Observatory and started Palomar with just this idea in mind.

At Liverpool, Keeler gave his invited paper on the last day of the meeting. He spoke of his own work on the spectroscopic measurement of the velocities of rotation of the planets, as well as of the work of Deslandres, Campbell, and Belopolsky. Keeler showed slides of his spectrograms of Jupiter, which demonstrated the effect very clearly, and he described how accurately they could be measured. The final result he quoted for its equatorial velocity was 12.94 ± 0.27 km/s, in very close agreement with the value derived from timing the rotation period. He also spoke of his results on Saturn and its rings, and at the end told of his still unsuccessful attempts to measure the rotational velocity of Venus.[196] It was a good survey paper, allowing the listeners to see a recognized expert in his field, who could take them from a clear and obvious example, through his most important discovery, to the very frontier of research.

After the meeting, Keeler traveled in England. He went to London, and met some of the eminent English astronomers with whom he had corresponded, but never seen before. He visited Huggins at his house at Tulse Hill, in south London, and had lunch with Gill at the Royal Societies' Club in St. James' Street.[197] Then he went for a short tour in England, and was back home in Allegheny, in the midst of his domestic duties, by early November.[198]

The two months in England broadened Keeler's horizons tremendously. He had received welcome recognition, and had seen and talked with many of the most famous scientists of the age. Very probably in his own mind he had measured himself against them, and had found himself not wanting. To fulfill his own potentialities, he could see even more clearly than before, he had to have access to a larger telescope at a better site. He needed advanced students to work with him, and scientific colleagues to share in the excitement of his research.[199] That quest was to take him even closer to Hale, and then suddenly and unexpectedly back to Lick Observatory and his early death.

7

I have really counted more on you than on all the others together

James E. Keeler first met George Ellery Hale on July 10, 1890 at Mount Hamilton. The Lick spectroscopist saw before him a short, slightly built young man, bubbling with energy and enthusiasm. He had dark brown hair above an already high forehead, gray eyes, a large mouth, and a small chin.[1] Then 22 years old, Hale had graduated the previous month from the Massachusetts Institute of Technology, and two days later had married his childhood sweetheart at her home in Brooklyn, New York.

According to Hale it was the hottest day of the year, but later a strong cool wind and a little rain had given them some welcome relief. He and his bride spent two nights at a hotel in New York City, and then went on to Niagara Falls, where they mingled with the other honeymooners and saw the traditional sights. After one night at the Falls, they took the train to Chicago, arriving "home" at Hale's parents' house much sooner than expected, surprising everyone there. After a week with his parents they continued their honeymoon trip to Manitou, Colorado. By this time Hale felt a little sick, which he attributed to drinking the water in Denver, so they did not go up to Pike's Peak, but were able to enjoy the Garden of the Gods instead. Next came San Francisco, and then Yosemite, where "we had a most glorious time. There is an immense amount of water in the falls, and everything was perfect. The Yosemite Fall by moonlight was as fine a thing as I ever saw." With a dollar tip, Hale succeeded in persuading the caretaker of the Mariposa Grove to put up a sign designating a 337-foot tall sequoia as the M.I.T. Class of '90 Tree.[2] Hale also found time at Yosemite to write a long letter to Henry A. Rowland, the outstanding laboratory spectroscopist of the day, outlining his own vague plans for graduate study at Johns Hopkins University and in Germany.[3] Then the newlyweds went on to Mount Hamilton.

Lick Observatory was the high point of the honeymoon trip for Hale.

Fig.14. George Ellery Hale, Keeler's young friend and brother astrophysicist (1892). (Reproduced by kind permission of the American Institute of Physics.)

He and his wife rode up on the stage from San Jose and at noon arrived at the summit where they were welcomed by S.W. Burnham. Now the senior staff member at Lick, he had lived in Chicago for many years and had been Hale's boyhood mentor in astronomy. Burnham was a perfect host, and showed young George and Evelina all over the observatory in the afternoon. Hale also had a long talk with John M. Schaeberle, another member of the Lick staff, on the solar corona, their mutual field of interest. That evening Keeler was observing planetary nebulae spectroscopically at the 36-inch refractor, and Hale worked with him, while

Burnham showed Evalina stars, planets and nebulae with the 12-inch refractor at the other end of the Main Building.[4]

Years afterward, Hale recalled that he never forgot his first sight of the 36-inch telescope, then the biggest refractor in the world. As they observed in the dark, relieved only by the light of the stars and the occasional faint spark of the comparison spectrum, he could hardly see what was happening, but he was enchanted by the long tube pointing up in the great round dome, toward the slit that seemed to him to be an opening into heaven.[5] Keeler was working on his measurements of the wavelengths of the spectral lines in planetary nebulae, assisted by W.W. Campbell and Armin O. Leuschner. That night was the first occasion on which Keeler was able to directly compare the faint Hβ emission line in the planetary nebula NGC 6210 with the same line in a comparison source mounted on the telescope, and thus to measure accurately the radial velocity of the object. He also measured the wavelength of the green nebular line with respect to the magnesium and lead comparison lines, as he had on previous nights. Later they looked at the spectrum of the planet Jupiter, but it evidently did not make nearly as strong an impression on Hale as the planetary nebula had.[6]

The next morning Hale tested the daytime atmospheric conditions, observing the sun with a spectroscope on the 12-inch refractor. Edward S. Holden, the director of Lick Observatory, had offered him the chance to stay as a volunteer and use the 36-inch to continue his study of solar prominences. Hale was amazed at the unexpected opportunity. However, Keeler and Burnham advised him that the daytime seeing at Mount Hamilton was not nearly so good as at night, because of the heating of the nearly bare mountain slopes by sunlight and the resulting turbulent air flow. Hale's test with the 12-inch confirmed their fears, and he decided to reserve judgment on Holden's proposal. He had also been offered the use of the 23-inch refractor in Princeton by Charles A. Young, and had been considering another plan of going abroad with his new bride for a tour through Europe.[4] However, Hale's wealthy father solved the problem of where he should observe by ordering a special 12-inch telescope for his solar work, and starting to build a professional observatory for him in the back yard of the family mansion in Chicago.

This was a recurring motif in Hale's life until the day his father, Willliam E. Hale, died. Like Keeler, Hale was descended through both his parents from old New England pioneer families. He was born in Chicago on June 29, 1868, in a house at 236 North La Salle Street, a spot that is now in the center of the city's financial district. His parents' first two children had both died in infancy before young George was born and they treasured and guarded his life with special care. When he was three the great Chicago Fire destroyed much of the downtown area, but the Hales had moved to Kenwood, a suburb in the South Side, the previous year. As Chicago was rebuilt from the ashes, William E. Hale prospered as his elevator company provided the means of access to the new skyscrapers that were going up everywhere. George was a weak, sickly boy, always interested in experimenting in the laboratories and working with tools in the shops that his father provided for him in their various homes. His younger sister and brother, Martha and Will, were his willing, devoted assistants, but George was always the leader.[7]

Hale's parents wanted every educational advantage for him. He had a private tutor, Dr. E. F. Williams, who lived with the family for several years. After a disastrous start at Oakland Public School, Hale transferred to a private school, the Allen Academy, where he received special training and encouragement in science. It was at Wabash Avenue and Twenty-third Street, an easy street-car ride from his home. He also attended the Chicago Manual Training School, another private institution, located at Twelfth Street and Michigan Avenue. a few blocks south of the downtown area. Its Board of Trustees included Marshall Field, John Crerar, and George M. Pullman, all potent names in the Chicago power structure.[5]

When Hale was fourteen years old, he tried to make a telescope himself, using a single lens, but failed. It was then that he first met the middle-aged Burnham, who was working as a court reporter by day but observing double stars at night with his own telescope. He was friendly and helpful, and advised young George where an excellent Clark 4-inch refractor could be bought second hand. Hale was determined to have it, and pleaded with his father to buy it for him. He even had a deadline – he needed it to observe the transit of Venus on December 6, 1882. His doting parents could never refuse him, and Hale had his telescope before the crucial date. His career as a scientific observer and fund raiser was launched. [7]

From that day on, Hale was a dedicated astronomer. He observed often, visually at first, but soon had rigged up his telescope with a plateholder so that he could obtain photographs of the moon, sun, and planets. Burnham took him to visit Dearborn Observatory, then located at the old University of Chicago, in Douglas Park, on 35th Street near Lake Michigan. Its 18 ½-inch Clark refractor impressed him greatly. Burnham, a fascinating raconteur, told Hale stories of his astronomical work in Chicago and at the University of Wisconsin, and particularly of his visit to Mount Hamilton to test its skies for the Lick Trust.

But Hale was less interested in the old astronomy, of which Burnham was an outstanding practitioner, or the elevator business, which his father hoped he would take over, than in astrophysical research. He learned from a book he had received as a Christmas present how to make a spectroscope, and constructing one himself, he caught his first glimpse of spectral lines. Before long he had seen the solar spectrum, but he realized he needed a better instrument to make any serious study of it. Once again he appealed to his father, and his father bought a spectroscope for him. Soon the boy was reading everything he could find on astrophysics, and observing the solar spectrum at every opportunity. In the summer of 1886, Hale toured Europe with his parents. He met the noted astronomer and pioneer solar spectroscopist Jules Janssen at the Meudon Observatory near Paris, and also persuaded his father to let him have £40 to buy a professional-quality Browning spectroscope in London.[7] That same summer Keeler, eleven years older than Hale, started working for Captain Richard S. Floyd and the Lick Trust at Mount Hamilton, at a salary of $100 (approximately £20) per month.

On his return from Europe in the fall, Hale entered the Massachusetts Institute of Technology in Cambridge as a freshman. He was a good, but not outstanding, student. He was far more interested in astronomy, spectroscopy, and research than in his physics courses, which seemed dull and rigid to him. But in the summer vacation after his freshman year at M.I.T. he fitted up a laboratory for spectroscopy and solar research in the attic of the family mansion at 4545 Drexel Boulevard, and that fall, back in Cambridge, he started work as a volunteer assistant at the Harvard College Observatory. Its director, Edward C. Pickering, who had counseled Keeler when he first came north ten years before, allowed Hale to spend each Saturday at the observatory and help with the scientific work. Pickering was a pioneer in the field of stellar spectroscopy with an objective prism on the telescope, and Hale learned many valuable photographic and observational techniques from him. Even more important, he sharpened and developed his love of research.

Hale's intensity and driving interest in astrophysics, combined with his family's wealth and position, gave him the self-confidence that enabled him to approach famous scientists in a way no other young student would have dreamt of doing. Even before entering M.I.T., as a boy in Chicago he had corresponded with John A. Brashear, the Allegheny instrument maker who was Keeler's close friend, and had succeeded in obtaining a small grating from him for one of his early spectroscopes. Then Hale had traveled on his own by train to Allegheny and had introduced himself to Brashear, who was amazed that "Mr. Hale of Chicago" was so young, but was equally impressed with his knowledge of astrophysics. In preparation for his senior thesis at M.I.T., Hale went to visit Rowland at Johns Hopkins, as well as Young at Princeton. Rowland, who did not suffer fools gladly, at first tried to brush off the "infant," but the intelligent, persuasive Hale soon won him over. The genial Young was supportive from the start. These conversations confirmed Hale's desire to do a research problem on solar physics, and Pickering was willing to let him use the 15-inch refractor at the Harvard College Observatory to carry it out. Hale decided he would need a special photographic solar spectrograph for his project, and financed once again by his father, ordered one from Brashear. [7] It was a huge instrument, built to Hale's specifications, and loaded with extras. Brashear feared it would be too heavy for the flimsy old Harvard telescope, but went ahead with its construction as ordered. It was built on a rush time scale in less than four months from order to delivery, and cost Wiliam E. Hale $1,000. [8] The Lick Observatory spectroscope that Keeler had designed had cost only $100 more, but had been in the shop for over a year. [9]

Hale wanted to study solar prominences, the large irregular structures that early observers had seen on the limb of the sun at solar eclipses. Spectroscopists had discovered that the light of the prominences is concentrated in a few bright emission lines, particularly Hα, which gives them their red color. In 1868 Janssen and the English astronomer J. Norman Lockyer independently discovered that prominences could be observed visually even when there was no eclipse, by using a spectroscope to isolate the bright Hα line. This method cut down the diffuse light of the sky enough so that the observer could see a section of the prominence, defined by the slit of the spectroscope. If the slit were opened wide enough, the whole prominence could be seen, but if it were large, the slit had to be so wide that the contrast was greatly weakened. This made the prominence nearly invisible The problem was how to use a narrow slit to preserve the spectral purity, but yet let through the light of the whole prominence.

Hale conceived the brilliant idea of using photography, and scanning the slit of the spectrograph across the image of the prominence. This could be simply done by moving the telescope slowly across the sun. At the same time the photographic plate, in the focal plane of the spectrograph, was driven in the opposite direction, at an exactly equal rate, so that it remained fixed with respect to the solar image. The result was to build up on the plate a picture of the prominence, actually taken as the sum of a series of individual segments or slices. This is the basic idea of the instrument Hale invented for his senior thesis. He later named it the spectroheliograph. The principle had been suggested previously several times by various solar astronomers, including Janssen, but Hale had thought of it independently, and eventually became the first person to produce actual research results with it.[10]

Hale's senior thesis, however, although brilliant in concept, left a good deal to be desired in execution. The main problem, as in many of his later observational projects, was that he simply did not leave enough time to carry it out. His new Brashear spectrograph, a beautiful instrument, was delivered at Harvard in October and Hale was ready to put it on the telescope in November, but it proved too heavy for the old wooden tube and mounting, just as Brashear had forecast. Then Pickering let Hale use a fixed horizontal telescope, so that the spectroscope could be solidly mounted on a brick pier, rigidly set in the ground. A heliostat, or flat mirror driven to follow the solar motion across the sky, fed the sunlight into the telescope. Hale had the spectrograph adjusted by January, but problems with the mirror, which had to be resilvered, and uneven heating of the instrument plagued him until April, when he first attempted to photograph a prominence. He did not succeed. He had wanted to use the bright Hα line, but this required especially red-sensitized photographic plates, which he did not have at the time. Hence, he had to use instead the much weaker Hβ line, which can be recorded on ordinary blue-sensitive plates. Finally he did get a successful Hβ plate on April 14, 1890 which actually showed the "general form" of two prominences, although the picture was somewhat fogged by stray light. The rest of April and the first part of May was hazy in Cambridge, and on the few occasions on which the sky was clear, the sun refused to cooperate and produced no prominences. Nevertheless, Hale desperately took one exposure after another, but not surprisingly, none showed any prominences. He experimented with sensitizing plates for the yellow, red, or infrared, with the hope of recording Hα, but "reluctantly" gave up after a minor explosion and fire that occurred while his assistant was mixing silver nitrate with collodion. At that point, only a month before graduation, he had to stop experimenting and start writing his thesis. As he said in its conclusion, although he had not been able to produce any prominence photograph of research quality, he had demonstrated the feasibility of the method.[11]

After graduation came the marriage and the honeymoon trip to Lick Observatory. Less than two months after the wedding ceremony, Hale and his bride were back from California and had moved in with his parents. While Hale was still a student at M.I.T., his father had retained the famous Chicago architectural firm of Burnham and Root to design a small but handsome brick laboratory building for him, and had erected it in the back yard.[12] Now he ordered a 12-inch lens from Brashear, and added a large tower and a dome to the laboratory, all faced with stone matching the house, so the observatory would not look out of place in the affluent suburb of Kenwood. The telescope mounting, which he ordered from the Warner and Swasey firm in Cleveland, was designed to have a relatively long focal length, an advantage in solar research, and to be especially sturdy, so that it could carry Hale's heavy spectrograph without any flexure problems.[13]

While he was waiting for his telescope and observatory to be completed, Hale received a letter from Keeler, then still at Lick, who suggested some spectroscopic measurements to help clear up the problem of the nebular lines. Keeler asked Hale to measure accurately the wavelength of the MgO "fluting" which Lockyer had identified with the chief nebular line, and also the wavelengths of two nearby lines that occur in the spectrum of air.[14] Hale did not yet have the precision measuring instrument necessary for highly accurate wavelength determinations, and wisely left this problem for Rowland's laboratory at Johns Hopkins. However, the young Chicagoan's interest was whetted, and he did undertake a laboratory investigation of the MgO spectrum. Under ordinary conditions, the spectrum of the molecule contains a number of close clumps of lines called band heads. One of these was the feature that Lockyer claimed was observed as the chief line in the nebulae. According to the English spectroscopist, under certain conditions this single band head was brighter than any of the others, so that if it were barely visible, as in the nebulae, the other band heads would be invisible. Hale checked this statement carefully in his laboratory. He took over one hundred exposures of the MgO spectrum with various excitation conditions, experimental set-ups, exposure times, and slit widths, but never reproduced Lockyer's result. Hale found that whenever the band head near $\lambda 5006$ appeared, two other band heads, near $\lambda 4996$ and $\lambda 4985$, were also always present, only very slightly fainter. As Hale correctly concluded, if lines could be seen at these two wavelengths in the nebulae whenever the chief nebular line was observed, it would have been strong evidence in favor of Lockyer's identification. Since they were not seen in any nebulae, however, their absence cast strong doubt on the idea that the chief line was the strongest MgO band head.[15] This was not conclusive evidence, because it could be argued that the unknown physical conditions in the nebulae were so different from the conditions in Hale's laboratory source that the MgO relative intensities might be completely different from those he had measured. This interpretation did not seem particularly likely, however, so Hale's paper was valuable supportive evidence for Keeler that the chief nebular line was not the magnesium "fluting."[16]

Soon afterward, when Keeler received his definite job offer from Allegheny Observatory and accepted it, Hale in turn got a somewhat ambiguous proposal to come to Lick. Holden wrote him that he did not know where to turn to find Keeler's successor, and asked if Hale could come to Lick for a year and take over the stellar spectroscopic work as a volunteer. Holden had already hired Campbell in Keeler's position, and told Hale so, but he nevertheless implied that the whole field of radial-velocity work would be left wide open, which would have been surprising news indeed to the ambitious young man from Ann Arbor if he had learned of it. Very probably Holden expected that Hale, with his well equipped private observatory in Chicago nearly ready for use, would decline the offer again, but be doubly grateful in the future for having been asked. Hale, however, took the offer quite seriously and nearly accepted. He telegraphed Holden for more details, saying that he and his wife had engaged passage for Europe in the summer, but would postpone the trip if he were placed in charge of the spectroscopic work at Lick for three or four months. At this point Holden, who was at that very moment trying to hire yet another spectroscopist, Edward W. Maunder of the Royal Greenwich Observatory in England, backed off and telegraphed Hale that they might have difficulty finding suitable quarters for him and his wife at Mount Hamilton. The young Chicagoan, realizing then that he was rather far down in Holden's list of priorities, declined the offer and never came back to Lick again.[17]

By the spring of 1891, Hale's Kenwood Physical Observatory and its telescope were completed. He began working to try to get results with his spectroheliograph. He had given up using Hα, but found that the hydrogen lines in the blue spectral region were too faint to produce good prominence photographs. Then he noticed that the λ3933 emission line due to ionized calcium, Ca II, was particularly strong in prominences. This line, called K by astronomers ever since the solar spectrum was first systematically mapped by Josef Fraunhofer, is in the ultraviolet spectral region to which photographic plates are most sensitive. It was ideal for Hale's purpose, and he soon used it to take excellent photographs of prominences with his spectroheliograph, thus proving it in practice at long last.[7]

In June Hale staged a formal dedication of his observatory, bringing in Young from Princeton, Brashear from Allegheny, and Charles S. Hastings, formerly Keeler's physics professor at Johns Hopkins, from Yale for the ceremonies.[7] Keeler, who had just arrived at Allegheny, congratulated Hale "on the possession of such a splendid laboratory for research." He could hardly have avoided reflecting that Hale's private observatory was far better equipped than his own Allegheny Observatory or nearly any other American university observatory, and that Hale's father was far more supportive and accessible than the Allegheny trustees, who were all off to Europe or the Thousand Isles in Canada for their summer vacations.[18]

Just at the time that Hale was getting his Kenwood Physical Observatory into operation, the new University of Chicago, to be built a few miles away in Hyde Park, was being planned. Its founder and chief source of funds was John D. Rockefeller; its president was William Rainey Harper. This amazing bundle of energy had been born in New Concord, Ohio in 1856, one year before Keeler. From early boyhood Willie Harper had loved books, especially the Bible, and music, especially the trumpet. When he started school he was far ahead of the other boys, and he soon finished all his courses and was allowed to enter Muskingum College at the age of ten. There were seven other students in his college class, whose average age was 24, but Harper graduated at the age of 14. Among the subjects he studied at Muskingum his favorite was Hebrew. Yet he was a "real boy ... perfectly normal, jolly as a boy could be, free, friendly and well behaved." While he was in college he organized and led the New Concord Silver Cornet Band. After graduation he clerked in his father's store for three years while he studied languages on his own. Then he taught Hebrew at Muskingum for a year, before leaving to become a graduate student at Yale at the age of 17. There he studied Sanskrit, Greek and Chaucer, and earned his Ph.D. a month before his nineteenth birthday. Harper then married his childhood sweetheart, the daughter of the president of Muskingum, and after two years of college teaching, he became an instructor in Hebrew at the Baptist Union Theological Seminary in Morgan Park, a suburb of Chicago, in 1879. He was an enthusiastic, intensely earnest teacher, whose concentrated energy inspired his students. Harper never believed in wasting time on vacations, and he started a summer school in Hebrew at the seminary in 1881. It proved so popular that he started another at Chautauqua, New York in 1883, and then a correspondence school to reach still more students. In 1886 he returned to New Haven as professor at Yale University, the Yale Divinity School and Yale College. In addition he was president of the American Publication Society of Hebrew and principal of the American Institute of Sacred Literature (two correspondence schools) and principal of the Chautauqua College of Liberal Arts. In his spare time he edited two journals, one on Old and New Testament Studies, the other on Hebrew.[19]

Before he had left for Yale, Harper had met Rockefeller, who was a large financial supporter of the Morgan Park Seminary. The millionaire philanthropist originally was interested in founding a Baptist university in New York, but in the end it became the secular, but highly religiously oriented, new University of Chicago. Rockefeller and his advisers wanted it to begin its life as a four-year college with the hope that it would in time grow to become a university with graduate students, but they also wanted Harper as its president. He, on the other hand, insisted that Chicago should be a great university from the start, that would incidentally include undergraduates. A great deal of pushing and hauling ensued, but in the end Rockefeller agreed to Harper's basic position, and on February 16, 1891 the Bible scholar accepted the presidency. He was then 35 years old.[20]

Harper's appointment was not to become effective until July 1 of that same year, but he immediately began making short trips to Chicago to get things started. Hale was recommended to him for a faculty position at a very early stage, and Harper, who considered him a "young man of great promise," was anxious to get together with him.[21] They met on Harper's second flying visit to Chicago, and soon afterward he made the young Hale a tentative proposal. Harper's idea was that Hale's observatory would be moved to the campus, and thrown open to students and public visitors. It would be enlarged to include not only astrophysical research, but also the older astronomy of position. Another "young man" would be found to take charge of the astronomical work, as Hale developed the spectroscopic side of the program. The observatory would be endowed with enough money to pay the salaries of the two professors and the other necessary research expenses by a benefactor whom Harper did not name, but whom he clearly intended to be Hale's father. The entire operation would be under Hale's direction, and he would be promoted "as rapidly as your age and work might seem to you and to us to call for advancement."[22]

Hale, who received this offer just a few days before the dedication of his new Kenwood Observatory, rejected it out of hand. He wrote that he and his father could not agree to secure a position for him by giving or leasing the observatory to the university, and that if he were not competent enough to be appointed on his own merits, he would prefer to gain experience by further study and hope for the best in the future. Harper replied that Hale had misunderstood him; he claimed he had had no intention of making the gift of the observatory a condition of appointment.[23] In fact it was exactly his intention, and there was little basis for further discussion. Soon after the dedication Hale went abroad, traveling around Europe with his wife and meeting many English and continental astronomers and astrophysicists. After his return to Chicago in the fall, he buried himself in solar research.[7]

As soon as Harper moved to Chicago, he began recruiting faculty members in earnest for the university that was to open its doors for students the following year. He wanted only the best, and financed by Rockefeller, he set the standard salary for the head professor of a department at $6,000 per year. This proved not high enough, and in December 1891 it was raised to the previously unheard of figure of $7,000. Harper's own salary was $10,000. With this kind of money as bait, he persuaded Thomas C. Chamberlin to leave the presidency of the University of Wisconsin to become head of the geology department at Chicago, and Albion W. Small to resign as president at Colby College to take over the sociology department. In a daring raid on Clark University, Harper carried off nearly half its faculty including two full professors, one of them, Albert A. Michelson, later to be America's first Nobel Prize winner in physics.[20]

Still Harper had time to try again to land the young Chicagoan with the wealthy father and the well-equipped observatory.[24] Asaph Hall, the noted American astronomer who had discovered the two satellites of Mars, was among those who recommended Hale to Harper for a post at the university. He wrote Harper that the low flat campus not far from Lake Michigan with its frequent fog, mist, and smoky skies would not be a good site for classical positional astronomy, but that it might be adequate for the newfangled astrophysical work. He added significantly:

> Again it was said that you have in Chicago a young man, Mr. George E. Hale, who is devoted to his branch of astronomy, and who has already shown good ability. It was said that Mr. Hale's father is a man of great wealth, and is very generous in the support of his son's investigations. It may be well for you to know Mr. Hale.[25]

Soon after receiving this letter, Harper met again with Hale and, after some negotiations, signed him up for the faculty. The terms were not very different from those Harper had first proposed. Hale agreed that the university would have the use of his Kenwood Observatory, but only for graduate students under his own personal supervision. He was to be appointed associate professor of astrophysics and director of the observatory, but would receive no salary for the first three years, or "until such time as the University has the necessary funds for this purpose." Since Harper consistently maintained throughout his presidency that he never had enough money to run the university, Hale could have had few illusions on this point. Part of the agreement was that Hale would be in charge of all the work done in the observatory but would not be required to teach. The university would pay the operating expenses, but Hale's father agreed to give the entire observatory, telescope, instruments, and all, to the university if his son was retained and if Harper could raise $250,000 for a larger observatory.[26] Each side got what it wanted from this arrangement: the University of Chicago an observatory and a young research professor, Hale an associate professorship at the age of twenty-three, all financed by William E. Hale's fortune. Michelson strongly favored the move into astrophysics, and congratulated Harper on "the acquisition" when he learned that Hale had accepted the position and that his observatory would be part of the university.[27]

An important part of the agreement, added by Harper himself, was that although Hale would be in charge of the observatory and astrophysical research, he would not direct the work of any classical astronomers who might be added to the faculty. Harper undoubtedly inserted this clause into their agreement because he was at that very moment negotiating with Thomas Jefferson Jackson See, a young Missourian who was just completing his Ph.D. thesis in astronomy in Berlin. See, three years older than Hale, had met Harper in Germany the previous summer and was well acquainted with Eri B. Hulbert, a close friend of the president from Morgan Park days. See's educational credentials were far more in tune with Harper's ideas than Hale's were; he knew Latin, Greek, French, and German, had traveled widely in Europe and the Near East "for purposes of historical, antiquarian and artistic study, and for general culture," and was a classical astronomer to boot.[28] See was expansive in his ideas and had no doubt that "an enthusiastic capable man [himself] would soon be able to raise money for an observatory equal to any in the world."[29] Harper offered him an assistant professorship at $800 a year, and See snapped it up.[30] An unpleasant contretemps ensued when a newspaper back in Columbia, Missouri published a story that See had been appointed to the University of Chicago faculty at a salary of $6,000, and one of his enemies forwarded it to Harper. The president now learned that See was a highly controversial character who as a student had been one of the leaders in a rebellion that had resulted in the ousting of the president of the University of Missouri. See defended himself against his enemies' charges that he was "a thoroughly unscrupulous ... intriguer" and a plagiarist "largely devoid of moral principle" in a series of long, near hysterical letters to Harper, who went through with the appointment, no doubt largely at the urging of Hulbert.[31] This episode was only a mild foretaste of the battle to come between See and Hale.

Immediately after his appointment to the University of Chicago faculty, Hale drew up a plan of instruction for graduate work in astrophysics. It was almost completely concerned with solar research – photography of prominences, spectroscopy and laboratory work of the kind Hale was doing. There were no theoretical courses and no work in orbit determination, the mainstays of the old astronomy.[32] Hale was prepared to lead a complete break with the past.

However, his main contribution to the University of Chicago was not in teaching, but in helping to secure the funds with which Yerkes Observatory was built. The wheels were set in motion less than a month after Hale's appointment. He went with his wife to a resort in the Adirondack Mountains for a summer vacation, but interrupted the holiday to go to Rochester for the meeting of the American Association for the Advancement of Science. There he heard that two unfinished glass disks, intended for an achromatic telescope lens 40 inches in diameter, were unexpectedly available for purchase. They had been ordered for a projected Southern California observatory, planned for erection on Mount Wilson. A long-time civic booster and trustee of the University of Southern California, Edward F. Spence, had publicly announced in 1887, as Lick Observatory neared completion, that regional pride demanded that there be an even larger telescope in the Los Angeles area. A land boom had brought sudden wealth to the Southland, and nothing seemed impossible. Spence launched a campaign to raise money for the observatory with a $50,000 contribution of his own, and a rash of enthusiastic newspaper editorials supported his plan. Railroad companies vied for the franchise to build a line to the summit, and developers planned hotels to serve the scientists and hordes of tourists who they believed would flock to the observatory on excursion trains from the East. The Spence Observatory Trustees commissioned Alvan Clark and Sons to make a 40-inch lens, and, after numerous attempts, a French glass maker cast the crown and flint disks and Clark began grinding and figuring them. But in 1891, with the lens only partly completed and only partially paid for, the Los Angeles land bubble burst. Spence and the other promoters were wiped out, and in 1892 they defaulted on the payments on their contract with Clark. He was being pressed by the Parisian glass maker for $16,000 due on the disks, and was anxious to find another buyer for the lens.[33]

Hale heard the story with growing excitement, and hurried back to Chicago to pass on the news of the great opportunity to Harper. They could build the largest telescope in the world if they could only raise the necessary money. At the suggestion of Charles L. Hutchinson, a wealthy Chicagoan who was one of the trustees of the university, they approached the street-railway magnate Charles T. Yerkes. Harper had been trying to get a large contribution out of him, but had failed to rouse his interest in a hoped-for biology building. Yerkes was an unscrupulous, flamboyant character who would clearly be interested in a telescope or in anything else that could be said to be the biggest in the world. He had recently married a chorus girl whom Harper described to Frederick T. Gates, another Baptist minister who had become Rockefeller's personal educational adviser, as "the most gorgeously beautiful woman I have seen for years."[34] At Harper's suggestion, Hale drew up an eight-page letter describing the exceptional opportunity to get the 40-inch lens for Chicago. He spelled out in simple terms how much more light a telescope built around it would collect than the Lick refractor, and how the new instrument would attract visitors to Chicago and fame to the University. He estimated the total cost of the projected observatory as $300,000, and stated that he already had a subscription from a resident of Chicago who wished to remain anonymous (his father) for $30,000 (Hale's estimate of the value of the Kenwood Observatory) on condition that the university raise the total amount.[35]

The letter awakened Yerkes' interest, and on October 2, 1892 Harper and Hale went to call on him at his North Clark Street office. After only a little persuasion on their part, he agreed to build "the largest and best [observatory] in the world." He savored the name "The Yerkes Telescope," and recalled that in his school days he had predicted that he would be the "owner" of the largest telescope in the world.[7] Hale and Harper were overjoyed. "The whole enterprise will cost Mr. Yerkes certainly half a million dollars," Harper reported. "He is red hot and does not hesitate on any particular. It is a great pleasure to do business with such a man."[36] The news, when released, created a sensation, particularly in Yerkes' own newspaper, the *Chicago Daily Inter Ocean*. According to it, the estimated price of the complete observatory would be $500,000. "The cost, however, cuts no figure as far as the university is concerned," it added. "Mr. Yerkes, when he took the matter in hand, simply stipulated that the observatory and its telescope should beat everything of its kind in the world." Hale and Burnham were quoted as attesting that the climate and atmosphere of Chicago, where everyone imagined the telescope would be located, were well suited for astronomical work. The next day, in a follow-up editorial, the *Inter Ocean* compared Yerkes and Sidney A. Kent, who had just given the University the money for a chemistry building, to Lorenzo the Magnificent, saying that they were doing for science in Chicago what he had done for art in Florence. The *Tribune,* almost as respectful of Yerkes, called the telescope a "Princely Donation to Chicago University. No Limit as to Cost," and added "The Lick Telescope will shortly be licked."[36] Keeler was more skeptical of the site, and wrote Henry Mathews, the secretary of the Lick Trust, "Our Chicago friends will have a bigger telescope, but 36 inches on Mt. Hamilton will beat 40 inches in Chicago. Still I am sorry to see the L[ick] O[bservatory] lose the prestige of having the biggest telescope in the world, as it counts for a good deal with the unreasoning public."[37]

Keeler was becoming better and better acquainted with Hale. In the fall of 1891, only a little more than a year after he had graduated from M.I.T., the young Chicagoan had decided to launch a completely professional journal of spectroscopy and the new astronomy. He planned originally to call it *The Astro-Physical Journal,* and asked Keeler for a paper on the Lick Observatory spectroscope for the first issue, together with a statement of support to be published with similar messages from other eminent astrophysicists.[38] The new Allegheny director was glad to oblige the would-be editor, but Hale soon found that he could not raise enough money to buck the well established *Sidereal Messenger,* edited by W. W. Payne at Carleton College in Minnesota. Even though Hale was being supported by his father, while Payne depended on the income from the *Sidereal Messenger* for his livelihood, the popular magazine seemed too well entrenched for Hale to compete with it. The only alternative was to join it, so when Payne proposed they combine forces, Hale accepted.[39] Thus, the *Sidereal Messenger* was transformed into *Astronomy and Astro-Physics.* Each issue of the new journal contained two well-defined sections, one on astronomy, mostly from the amateur point of view, edited and in large part written by Payne, the other, on astrophysics from a more professional point of view, edited by Hale. Keeler's paper on the Lick spectroscope, containing a photograph and a carefully made scale drawing, appeared in the second issue.[40]

A few months later, just after Hale had been appointed to the University of Chicago faculty, Keeler visited his Kenwood Observatory for the first time. The Allegheny director was on a vacation trip to the upper Great Lakes with two friends, one of them Brashear's son-in-law and shop foreman, James McDowell. They stopped off in Chicago, and Keeler inspected Hale's telescope, spectrograph and other instruments with close attention. No doubt he was impressed not only with the observatory and apparatus, but also with the wealth of Hale's father, who invited him to dinner with the family. Young George, on his part, could see that Keeler was not only a first-class research scientist, but also a very handy instrumental expert and extremely personable to boot.[41]

Hale visited Keeler at Allegheny for the first time in November of that year, and saw his friend's new spectrograph.[42] Later that month Keeler went back to Chicago again, to give a talk to the Section of Astronomy of the Chicago Academy of Science, one of the first of the many organizations that Hale, throughout his career, found moribund but useful to his purposes, and pumped into life, at least for a time, with his own sheer energy. Keeler had been preparing a lecture on "Great Telescopes," drawing on his results and experiences at Lick Observatory, for delivery at the Western University of Pittsburgh the next winter; with some added touches on the kind of results that might be expected from the Yerkes telescope it was perfect for Chicago.[43]

Each time they met the two friends undoubtedly discussed their own research, the latest news they had of the work of other astronomers, and the ideas Hale had for organizing and rejuvenating American astronomy. One fascinating topic was Hale's plans for what became the first international astronomical meeting in the United States. It was the World Congress on Astronomy and Astro-Physics, ultimately held in Chicago in August 1893. That was the year of the World's Columbian Exposition, or World's Fair, designed to exhibit the material progress and richness of America four centuries after its discovery by Christopher Columbus. The organizers did not get started soon enough to bring it off by 1892, but when the Fair finally did take place the next summer in Chicago, close to Hale's home and the university, it captured the imagination of a generation of Americans.[44]

Soon after the Fair was announced, Charles C. Bonney, a leading Chicago lawyer and reformer, who had been a school teacher and administrator in downstate Illinois in his youth, conceived the idea of holding a series of World Congresses, or meetings, in conjunction with it. He felt that however magnificent a display of industrial achievements and material progress the Fair itself might be, something still higher and nobler was needed, even demanded, by the enlightened and progressive spirit of his time. He therefore proposed a series of meetings in which the leaders of government, science, religion, finance, and literature could discuss their fields and describe the problems of the age and their solutions to them. He set up an organization, the World's Congress Auxiliary, to plan these meetings; it adopted as its motto, used on all its publications, "Not things, but men." Over a hundred congresses were held under its auspices, many simultaneously, during the months from May to October 1893.[45]

The World's Congress Auxiliary named local committees to decide on the topics, programs and speakers in each subject. Hale, his father an important figure in Chicago business circles, had no trouble getting himself appointed to the General Committee of the World's Congress Auxiliary on Scientific and Philosophical Congresses, as a member of the Special Committee on Astronomy.[46] At the time these first local committees were set up, in the fall of 1890, Hale was only twenty-two years old; yet he was the main force responsible for organizing the World Congress on Astronomy and Astro-Physics in the form it eventually took. He saw the congress as an opportunity to advance the new and somewhat suspect subject of astrophysics, and incidentally to gain recognition for himself and his observatory. By the spring of 1891 he had arranged for the dignified and elderly George W. Hough, professor of astronomy at Northwestern University, to be appointed senior member of the Committee on Astronomy.[47] Hough was a classical astronomer of the old school, whose life work was making visual measurements of Jupiter and of double stars with the Dearborn Observatory refractor.[48] He was well known in astronomical circles, and provided the scientific respectability that Hale lacked at that time.

In March 1892 the subject of the Congress was broadened to mathematics and astronomy, a combination that made more sense to the classical scholars who dominated the General Committee on Science and Philosophy than to Hale. Mathematician E.H. Moore, was added to the organizing committee, along with astronomers George C. Comstock of the University of Wisconsin, Payne, and Malcolm McNeill of Lake Forest College, near Chicago, as well as several Chicago amateur astronomers. The subjects had been divided into three "chapters", pure mathematics, astronomy, and "astro-physics", the latter including spectrum analysis, astronomical photography and stellar photometry, while "physical astronomy", as the study of planets was then called, was included under astronomy. A preliminary "address" or announcement was sent to a long list of advisory-council members, including most of the foremost astronomers of Europe and America, as well as to all known astronomical societies and many other astronomers, soliciting suggestions for subjects to be discussed and people to be invited.[49] Hale, as secretary of the organizing committee, was particularly active in encouraging leading astronomers and astrophysicists to attend, and in using *Astronomy and Astro-Physics* to publicize the Congress.[50] In the summer of 1892 Burnham resigned his position at Lick and came back to Chicago. He had become increasingly disenchanted with life on the mountain under Holden, and with a large family to support he jumped at the chance to return as clerk of the United States Court for the Northern District of Illinois, at a salary more than double the $3,000 he had been getting in California.[51] Burnham's membership on the organizing committee provided great prestige to the congress. He and Hough organized the classical astronomy part of the program, while Hale organized the astrophysical sessions.

By the spring of 1893, a striking "White City" of large and imposing buildings, grouped attractively around the lagoons, canals and inlets of Jackson Park was nearing completion to house the Fair. With the World Congress on Astronomy and Astro-Physics only five months away, Hale suddenly decided to mount an astronomical expedition to Pike's Peak in an attempt to photograph the solar corona without an eclipse. He had been having good success in photographing solar prominences with his spectroheliograph, and had seen how effectively it could cut down the diffuse sunlight scattered by dust particles in the earth's atmosphere and on the surface of the lens. For the expedition he ordered a special long-focus reflecting telescope from Brashear, and by putting a blackened diaphragm just the size of the solar image at its focus he could eliminate most of the direct sunlight; then he hoped that a spectroheliograph behind it, working in the ultraviolet spectral region where the corona might be especially bright, would do the rest. Hale soon realized that if he worked in the strong K absorption line of the solar spectrum, the sunlight would be particularly faint and he would have the best possible contrast for photographing the weak radiation of the corona.[52] It was a good idea, building on the spectroheliograph technique that Hale had developed himself and knew well. To make it work, he would certainly have to get away from the hazy, dirt-laden skies of Chicago, and the clear thin air of the western mountains offered the best hope of success. He quickly selected Pike's Peak, which was easily accessible by railroad and has very little atmosphere above it. In deciding to go on the expedition before the Congress met, Hale clearly hoped that he would be able to come back with photographs of the corona that would prove that he had succeeded where William W. Abney, William Huggins, Samuel P. Langley, and the other giant figures of his youth had failed, and lay his pictures before the assembled savants of the world as a proof of his own skill and ability.

Although Hale was very good at visualizing important new research programs, he generally depended on others to do much of the observational work for him, beginning with his sister and younger brother in his Kenwood days.[53] For the corona expedition he wanted to take Keeler, Brashear, Burnham, Henry Crew, and McNeill, "a jolly company", to help out.[54] Keeler thought Hale's scheme might not succeed ("I am a little inclined to be skeptical", he confessed) but deserved a trial. He expressed some interest in accompanying Hale, but said that as he planned to attend the Congress and "take in the World's Fair", the expenses of the additional trip to Colorado would be too much for him.[55] Hale, working through Harper, at once requested passes on the Chicago, Rock Island, and Pacific Railroad for the entire group. Free transportation on the railroads was one of the frequently available fringe benefits of a professor's life in those days; Keeler apparently always had a pass on the Pennsylvania Railroad for his travels in the East. For the Colorado trip the president of the Rock Island personally paid for the tickets, and Harper then had to forbid Hale to apply to George Pullman for free sleeping-car berths, arguing that such a request might interfere with getting a bigger contribution for the university from him.[56]

June was the best summer month for clear weather at Pike's Peak, according to the records Hale had consulted, so in the middle of that month they set off for Colorado. Brashear was only able to finish the prism for the spectroheliograph they were to use at the very last moment, and could not find time to go with them himself.[57] All the rest of the projected party, and two more astronomers whom Hale had later invited, all declined, so in the end only Keeler and Hale went, accompanied by Evelina Hale, who was supposed to act as an assistant to justify her pass. She was almost constantly sick or suffering from severe headaches all the time they were in the Rockies, and actually was more of a hindrance than a help, as Hale often had to take her down from the Peak to Manitou Springs, or stay with her there while Keeler worked alone. The two men were not much troubled by the altitude, except that they found that prolonged work on the Peak was very fatiguing. The weather was a disappointment. For the first few days it was clear early in the morning, but by 9 a.m. cumulus clouds began to form, and before noon the sky was usually completely overcast. Only a few days after their arrival they awoke to find the silvered glass telescope mirror, which had been in perfect condition a few days before, badly tarnished. They dashed off a telegram to Brashear to send the chemicals they needed to clean and resilver the mirror; he rushed them out by express and they received them five days later.[58] Keeler, who had watched Brashear resilver mirrors many times, did the job himself this time on Pike's Peak. Then there was a sudden snowstorm, but on July 1 they were able to take a few exposures through a milky haze with some clouds passing over the sun. The next day they had more clouds, and on July 3 smoke from forest fires. None of their plates showed any trace of the corona; all they saw was the uniform sky close to the sun. Their railroad passes expired on July 4, so they had to leave with no results. The method looked promising, but it had not had a fair trial.[59] They had done their best, and Keeler's skill and experience with astronomical experiments had been invaluable, but Hale had allowed far too little time for a serious test of the method.

Nevertheless, the trip to Pike's Peak greatly strengthened the bonds of friendship between Keeler and Hale. It had been much like a hunting or camping trip, and for years afterward in their letters to one another they referred humorously to "varmints", the name they had given to the bedbugs that infested their blanket rolls on Pike's Peak. Soon after their return, in urging Keeler to plan to take part in the discussion of stellar spectra at the upcoming World Congress in addition to presenting a paper, Hale wrote "I suppose by that time your nose will have ceased to *radiate* light of long wave-length, your ears will have taken on once more their normal ultra-violet (plus red) hue and you will gladly appear before the public eye."[60]

Keeler had long before decided that he would attend the congress, present a paper, participate actively in all the scientific discussions, and learn as much as he could about the astronomical and astrophysical research going on in the country.[61] He had hoped to have enough new results on stellar spectra from Allegheny for his paper, but as the bad weather had prevented him from accumulating enough data, he decided to present the final results of his nebular work at Lick Observatory instead.[62]

On August 19, 1893, when Keeler, accompanied by his wife and infant son "Kidlums", took the train from Pittsburgh to Chicago, the United States was in a serious financial depression. Many workers were unemployed, the steel mills in South Chicago were laying off men, and mass meetings of the jobless were being held. Some of the finest hotels in Hyde Park had gone bankrupt. Yet attendance at the Fair, after a shaky beginning, was picking up by August, and the hotels were open and expected to pay off their creditors. On the national scene President Grover Cleveland was reported ill with a kidney infection, and newspaper attention shifted to Vice President Adlai E. Stevenson, who was spending August in Washington, but a few days later Cleveland's friends told reporters that he was only tired, not seriously ill, and besides he was getting better. He issued a proclamation opening the Cherokee Strip in the Indian Territory to homesteaders effective September 16, 1893 at 12 noon, setting off the last large-scale land rush in American history.[63]

In Chicago the Keelers stayed with his mother and sister-in-law in a small cottage at 63rd Street and Woodlawn, which the U.S. Geological Survey had rented for the summer near the Fair.[64] Keeler's brother-in-law, David T. Day, was a member of the Survey, and apparently the fringe benefits for a government scientist included housing at the taxpayers' expense for members of his extended family. The Brashears were nearby in the Tasmania, a small hotel, and Leuschner, the young California graduate student who had spent a month at Allegheny with Keeler earlier in the summer, was not far away in another hotel.[65] The astronomers were congregating in Chicago.

The week of August 21 was set aside for the ten Congresses on Science and Philosophy. They met, as all the congresses did that summer, not at the Fair, but in the Art Institute on Michigan Avenue in downtown Chicago.[66] The original plan had been to locate the Fair there, in the narrow strip of land along Lake Michigan, but it was too small, for the lake had not been filled nearly so far out as it is today. Hence the decision had been made to hold the Fair in Jackson Park, but the construction of the Art Institute was hurried near the north end of Lake Front Park, so that it could be used for the World Congresses that summer, and then afterward as a permanent art museum.[67] Workmen were still hammering inside the building, when the Congress of Representative Women began in mid-May, with speeches by Julia Ward Howe, Susan B. Anthony, and Clara Barton, but by August the structure had been completed.

On Monday morning, August 21, all the Scientific Congresses met together for the opening ceremonies in the huge Columbian Hall on the ground floor of the Art Institute. There was a tremendous ovation for Hermann von Helmholtz, the great German physicist, under whom Keeler had studied in Berlin. Helmholtz was seventy-three years old, white haired, stooped and somewhat frail, and the audience stood and applauded and waved their handkerchiefs and hats after he had been introduced and given a little speech, in lightly accented English. Bonney, the president of the World's Congress Auxiliary, delivered a long and uplifting speech of welcome.[68] No record of his words on this occasion has been preserved, but he probably followed fairly closely the lines of his speech opening the World Congresses in May, when he said:

> The New Age has dawned. A new leader has taken command. The name of this leader is Peace ... In the service of this new commander we proclaim a Universal Fraternity of Learning and Virtue, as the best means by which ignorance, misunderstanding, prejudice and animosity can be removed; and intelligence, charity, productive industry and happiness be promoted. For these high purposes the World's Congress Auxiliary of the World's Columbian Exposition was organized, the leaders of progress invited, and the arrangements made for the World's Congresses of 1893.[69]

The chairmen of all the congresses responded, including Hough on behalf of the astronomers, and Felix Klein, the well known Göttingen professor, on behalf of the mathematicians. Klein, an official foreign delegate, spoke in German, and his words were translated into English by the Russian Prince Wolkonsky, who tacked on at the end a happy little speech of his own, for which he received nearly as much applause as Helmholtz.

After the opening ceremonies, the Congress on Mathematics, Astronomy and Astro-Physics, as it was called on the program, split up into two sections, one on Mathematics, the other on Astronomy and Astro-Physics (referred to by the *Chicago Tribune* as the Astronomological Section), which from then on met separately.[70] The astronomers had a classroom-sized room on the third floor of the Art Institute, with the mathematicians in an even smaller room next door. Just beyond the Mathematics Congress, in a considerably larger room, the Congress on Psychical Science was going on. In 1893 this subject had somewhat the same fascination that the study of U.F.O.s has today, and papers with titles such as "Elementary Hints on Experimental Hypnotism," "Experiments with the So-called Divining Rod," "Possibilities of a Future Life," and "Exhibition of Spirit-Photographs Known to be Spurious, and Others Which have been Supposed to be Genuine, with Remarks," attracted large crowds of interested people.[71]

Down the hall, attendance at the Congress on Astronomy and Astro-Physics was disappointingly small, but the astronomers and astrophysicists who were there found the papers and the personal contacts valuable. The foreigners present included Max Wolf of Heidelberg, Pietro Tacchini of Rome, Egon von Oppolzer of Vienna, and Eugen von Gothard of Hereny, Hungary. In addition two American astronomers were present reporting on work they had done in South America: William H. Pickering, who had recently returned from two years at the Harvard Observatory southern station in Arequipa, Peru, and John M. Thome, the long-time director of the Argentine National Observatory in Cordoba. Several other astronomers and astrophysicists had sent papers from abroad, which were read for them by proxy. Rowland divided his time between the theoretical sessions of the Congress on Electricity, which was meeting in three large rooms on the ground floor, and the more astrophysically oriented sessions of the Congress on Astronomy and Astro-Physics. Telescope makers were well represented for Alvan G. Clark and W.R. Warner, whose firms had made the lens and mounting for the Lick 36-inch refractor a few years before and were now working on the Yerkes 40-inch, were present, along with Brashear.

Hough gave the first paper on Tuesday, describing his observations of Jupiter and his conclusion that it is gaseous, with a structure and features strongly affected by its rapid rotation.[72] Frank Bigelow, a Weather Bureau scientist, spoke on his theory that the structure of the outer layers of the sun, particularly the corona, were strongly influenced by magnetic

194

Fig. 15. Group picture at the World Congress on Astronomy, Astro-Physics and Mathematics, held in the Art Institute in Chicago. John A. Brashear (with beard) is seated at the far left in the front row on the platform. Keeler is seated next to him, and Armin O. Leuschner next to Keeler. T.J.J. See, wearing a light suit and holding a soft hat, is seated to the right of the center of this row, and Henry A. Rowland, holding a straw hat, is at the far right of the same row. Hale is seated at the left end of the row just behind Brashear, and Edwin B. Frost is seated next to Hale. The two men with white beards in the second row from the back are (left to right) George W. Hough and Alvan G. Clark (1893). (Reproduced by kind of permission of Yerkes Observatory.)

forces, and tried to trace its magnetic field.[73] Rowland, who was present for this paper, said in the discussion that he believed the sun had a magnetic field, and thought its effects on the earth's magnetic field were well worth studying. Most of the papers were summaries and reviews of the speaker's work, rather than new research results.

On Wednesday the astronomers, astrophysicists and mathematicians went out to the Fair, and saw the opening of the exhibit of the mounting for the Yerkes 40-inch telescope, which had been erected in the huge Manufacturers and Liberal Arts Building. Hale, Clark, and Warner gave short talks at the base of the telescope, and the Warner and Swasey resident representative demonstrated its motions and controls. There were many other astronomical exhibits in the building.[74] The German educational exhibit included Gustav Kirchhoff's original spectroscope and samples of Jena optical glass. Warner and Swasey had other telescopes besides the 40-inch on display, and Brashear had several instruments, including the spectroscope for the 40-inch, and a selection of mirrors, prisms, and gratings. He was especially proud of a 20-inch telescope lens he was exhibiting, "made from American glass by an American Optician – calculated by an American Physicist."[75]

There were two very complete exhibits of astronomical photographs, from Harvard and from the Royal Astronomical Society, and smaller displays from many other observatories and astronomers. Keeler's excellent drawings of Jupiter and Saturn, which he had made at Lick, were hung in a "splendid place" in the Warner and Swasey exhibit.[76]

The astronomers strolled through the other buildings and exhibits at the Fair, and Keeler must have visited the California Building, described as "second in size only to the great Illinois State Building." It was a large, mission-style building, with many domes and arches, surrounded by palms and orange trees, designed to display all the best of California, a largely agricultural state at that time. Three sides of the building were reproductions of the Santa Barbara, San Antonio de Padua, and San Juan Capistrano missions, while the belfries and towers were copied from Carmel and San Luis Rey. The centerpiece inside the building, in the great central dome, was a date palm tree 50 feet high, which had been dug up in Mission Valley, near San Diego, where it had been planted as a seed by Father Junipero Serra 123 years before, and transplanted to Chicago.[77] Part of the California educational exhibit in this building was a large collection of photographs from Lick Observatory, of the telescopes and buildings, and of the moon, planets, bright nebulae and other astronomical objects. Unfortunately, the pictures were poorly arranged and lighted, so they did not make a very good impression.[78] Santa Clara

County, where Lick Observatory is located, had an exhibit built around its boast that in 1891 it had produced twenty million pounds of prunes, while all the rest of the United States together had produced only nine million pounds. The center of this exhibit was a life-sized statue of a horse and rider, clad in armor, completely covered with dried prunes, selected so that their colors enhanced the design. If Keeler examined the guest book in the California Building, he must have agreed with at least some of the comments recorded there: "California, we are proud of you", "Surpasses anything the world has ever seen", and "Why can't I go and live there permanently? Magnificent show".[79]

Probably the most interesting session of the Congress on Astronomy and Astro-Physics was on Thursday, when the meeting was moved to a room in the basement so that slides could be shown.[80] Keeler described his measurements of the wavelengths of the two principal lines in the spectra of nebulae. He explained how he had been able to use a high-resolution Rowland grating and had thus achieved unprecedented accuracy, enabling him to prove conclusively that the green lines are not due to magnesium, as Lockyer had suggested, and do not coincide with any solar absorption lines, or with any lines measured in the laboratory.[81] Hale described his photographic work on the sun, and Edwin B. Frost read a paper from Campbell at Lick, describing the spectrum of Nova Aurigae and emphasizing its similarity to the spectra of nebulae. Spectra of this same nova taken by von Gothard with an objective prism on a 10-inch telescope were also shown, as well as photographs of the corona, taken at the eclipse of April 16, 1893 by Schaeberle.

A paper with many modern resonances, by Williamina P. Fleming of Harvard College Observatory, was read on Friday. She was in charge of spectral classification on the Harvard objective-prism photographic plates, and her paper, "A Field for Women's Work in Astronomy," described this work. She began by mentioning the early woman astronomers, Caroline Herschel and Maria Mitchell who, because of their family connections, had the chance to work in astronomy. She said there must be other women with similar aptitudes and tastes, who could do just as well if only they were given the chance. The United States, she continued, is a liberal country, inhabited by large-hearted people. Immigrants come from all over the world and get a fair, open chance and equal opportunities in pursuing their careers. Likewise there is no other country in which women have advanced so far as America. There is very little jealousy or narrow-mindedness in research against women, so they "need not be discouraged ... even if one or two [men] ... in their superior judgment refuse to give credit to their work." The main body of her paper, about photography as applied to objective-prism spectroscopy, emphasized the support of wealthy women such as Anna Draper and Catherine W. Bruce, and the work done by women at Harvard under the benevolent direction of Edward C. Pickering. It contained a good shot at the "old-time astronomer" who clings tenaciously to his visual observing, but is left far behind by the new and better methods of photography. Her paper concluded, "While we cannot maintain that in everything woman is man's equal, yet in many things her patience, perseverance and method make her his superior. Therefore, let us hope that in astronomy, which now affords a large field for woman's work and skill, she may, as has been the case in several other sciences, at least prove herself his equal."[82] Fleming was not present at the Congress when her paper was read, but it was highly praised by Brashear, and William H. Pickering testified to the efficient work of the women on the Harvard Observatory staff.

On Saturday, the last day of the Congress, Pickering gave his paper, "Is the Moon a Dead Planet?", in which he claimed that it was not, and that his observations provided evidence for the existence of a lunar atmosphere and water. A paper by his brother, Edward C. Pickering, on the Harvard system of spectral classification and its implications for the constitution of the stars, was read. Keeler said nothing after the first paper, although he must have been skeptical, for he considered William H. Pickering's published drawings and descriptions of Mars crude and unintelligible.[83] Keeler started the discussion of the second paper with a review of the other systems of spectral classification then in vogue, and emphasized that no single system completely described all the stars, so that all of them had still to be regarded as provisional. The theory was too speculative, he said, and what was needed were more observational facts on the spectra of the stars.[84]

At the close of the session Wolf announced that, in accordance with a custom in Europe at the end of an astronomical meeting, he would name one of the asteroids he had recently discovered in honor of the Congress. He suggested either Chicago or Illinoa as appropriate titles, and since the vote was for the former, this minor planet is now and forevermore known as 334 Chicago. Brashear proposed a vote of thanks to the foreign delegates, Keeler proposed a vote of thanks to the officers of the Congress; both motions were adopted unanimously, and the meeting was adjourned.

Along with many other visitors to the Fair, Keeler had spent more than he could afford, and had to borrow $50 from Hale to get back to Pittsburgh, but he was glad he had gone and thought the Congress on Astronomy and Astro-Physics had been a real success.[85] It was estimated that the total attendance at all the Congresses in Chicago that summer was nearly one million people. All the Columbian Exposition buildings in Jackson Park had been built as temporary structures and were torn down after the Fair was over, except for the Fine Arts Building, which remains as the present-day Museum of Science and Industry. The Art Institute where the Congresses met still stands on Michigan Avenue, one of the great art museums of the world.

For Keeler as an astrophysicist, the World Congress on Astronomy and Astro-Physics was very important. It was a spectacular breakout from his scientific isolation in Allegheny.[86] His own paper was one of the most important presented at Chicago and undoubtedly captured the interest of Rowland, the great physicist who had been too busy to pay any attention to Keeler in his student days at Johns Hopkins. The Allegheny spectroscopist had taken an active part in the discussions and demonstrated his knowledge and expertise. Keeler had arrived as a leader in his field at Chicago, and from that time on he was widely considered an expert whose opinions were valuable and important.[87]

Important as was Keeler's participation in the Congress, even more important was his collaboration with Hale in establishing firmly a professional journal of astrophysics. The young Chicagoan, once he had started publishing *Astronomy and Astro-Physics* with Payne, found that keeping his end up was a heavy burden. Editing the manuscripts that were submitted was only a small part of the work; beating the bushes to get more papers, translating articles from foreign journals, and writing reviews of published papers or short notes on recent discoveries all added up to a never-ending job. He needed help, and the ideal person in his mind was Keeler. Thus in September 1892, nearly a year before the World Congress met, Hale wrote Keeler and asked him to become an associate editor of *Astronomy and Astro-Physics*. Hale explained that he considered his alliance with Payne only temporary, and that as soon as the University of Chicago could come up with the necessary money he planned to get the whole journal away from the Carleton College professor and make it a real professional journal, with a "first-class mathematical astronomer" editing the "general astronomy" part. This might not occur for some time, "even a year or more", and in the interim he personally did not have enough time to make the journal all it could be. Hence he wanted to have one or more associate editors for astrophysics, and he asked if Keeler could not help him out.[88]

Keeler, after considering it briefly, politely declined. He said that Hale had already greatly improved the journal and that if Payne's department could be similarly upgraded, *Astronomy and Astro-Physics* would become a journal of "the highest class". But Keeler claimed that he himself was not "a ready writer like Holden" and feared he did not have the qualifications needed for the position.[89] Hale would not accept this answer, and begged Keeler to reconsider. He said he needed the Allegheny director's name in the list of editors to attract readers. He could see that Keeler imagined that he would have a bit of work saddled off on him, but this would not be the case at all, Hale assured him. It would take practically no effort at all. He wrote:

> I sincerely hope you will not let your modesty in regard to writing keep you from [accepting]. You know as well as I do that you are as good an "ink-slinger" as the best of them, and you also know that your name would assist very greatly in making the Journal a real success. I shall be *very greatly disappointed* if you do not help me out in this matter, for I have really counted more on you than on all the others together. Burnham and Crew join with me in urging you to consider how greatly you could help the journal and the subject of astro-physics by a very small amount of work. If you do not agree to help me I shall be greatly inclined to give up my whole scheme of associate editors, and this would keep the journal where it is, and thus injure it. I know that you will be willing to aid the progress of science in this easy way, for there is no question that a journal of the highest class assists very materially in advancing the science which it represents.[90]

How could Keeler stand in the way of the progress of science? He accepted, and became associate editor for stellar and nebular spectroscopy, while Crew, now at Northwestern University, and Joseph S. Ames of Johns Hopkins became associate editors for laboratory spectroscopy. Keeler had suggested Campbell as an additional associate editor for the West Coast, but Hale, although he enthusiastically urged his friend to criticize anything he did not like about the journal, did not accept this suggestion.[91]

Keeler plunged into his editorial duties with an amusing review of an article in the *Boston Commonwealth* by John Ritchie, Jr. entitled "Do Large Telescopes Pay?" Ritchie, an acerbic amateur astronomer who devoted much of his time to trying to make life difficult for Edward C. Pickering at Harvard, argued they did not. A defense of big science against this critic in the center of New England culture was clearly required, and Keeler provided it. He pointed out that telescopes were not used only for visual studies of planets, and their satellites, as Ritchie had seemed to imply. "[A]s for celestial spectroscopy, and the whole modern development of astronomy in the direction of astrophysics, Mr. Ritchie either does not consider them worthy of mention, or he is unaware of their existence ... [T]he spectroscope is not only the key which unlocks the constitution of the stars; it has taken its place among the time-honored tools of the astronomer as an instrument of precision. For the purposes to which it is applied, the demand is always for more light, and light can be obtained only by constructing large telescopes." On the other hand, Keeler agreed completely with "[t]he part of Mr. Ritchie's article which relates to the insufficient endowment of large telescopes ... An observatory without the means of carrying on observations is as useless as a manufacturing plant without workmen."[92] Truer words were never written. Keeler could always find a way to emphasize the positive at the same time that he diplomatically corrected the errors.

The astrophysical part of *Astronomy and Astro-Physics* was often short of material, and Hale would press Keeler to write a short paper to fill it out, or to provide quick translation from a French or German journal that they could republish.[93] Payne was a constant source of irritation to the two friends, with his tedious sermons on the subject of meeting his copy and proof deadlines, together with his complete lack of understanding of any of the finer points of astronomy. Keeler, normally exceedingly polite in his correspondence, was often brief and peremptory in his replies to Payne's editorial demands.[94] Yet all in all, they felt they were accomplishing something for astrophysics, even though there were only 520 paid subscriptions. Of these, nearly 100 were outside the United States. The total receipts for the first year of existence of *Astronomy and Astro-Physics* were approximately $2,800; the deficit was only $200.[95]

But these encouraging results soon proved to be largely due to the newness of the journal and to the fact that many former readers of the *Sidereal Messenger* had signed up for a year of *Astronomy and Astro-Physics* to see what it was like. As their subscriptions ran out, many of them did not renew, as Payne wrote Hale:

> We are losing so many of our old subscribers who were amateurs that we have thought it necessary to do something. Our new plan and matter is excellent, of course, but it goes above the heads of eight-tenths of the persons I wish to reach and the amateurs are leaving us ... All in kindly mode but never-the-less going.[96]

The trend continued, and a few months later, when Hale and Keeler wanted to add more space for astrophysical research papers, Payne replied in the same vein:

> I am sorry to say that we are loosing so many of our amateur subscribers that we will need to hold closely to our present size of *Astronomy and Astro-Physics*. I have seen no other way to save those members from deserting us entirely than to undertake another form of

publication which is less rigorous ... If I can keep the interest of the amateur line until they are able to do harder work and appreciate more difficult subjects we may be able in time to increase our list so as to make it self-sustaining and also to improve its matter and illustrations.[97]

That was Payne's credo, but not Hale's and Keeler's. So, in the fall of 1893, the Carleton professor, on his own, founded a new magazine, *Popular Astronomy*, aimed specifically at amateur astronomers and school teachers.[98] Although Keeler did not want to have anything to do with editing it, he was glad to write for *Popular Astronomy*, and his series of articles on astronomical spectroscopy began in the first issue. Payne was delighted with them as "brief, clear, plain statements of facts for the popular readers, the teachers, and those beginning the study of astronomy," and assured Keeler that "they are being appreciated as nothing else of the kind has been since we began our work."[99]

Payne's new *Popular Astronomy* did not finish off *Astronomy and Astro-Physics;* it continued much as before, but with more serious research papers in the traditional astronomy section. Then suddenly, two months after the World Congress, Hale decided to go to Germany for the postgraduate study he had never had time for after his graduation from M.I.T. Work on the 40-inch lens was progressing at the Cambridgeport shop of Alvan Clark and Sons, but otherwise little could be done to hasten the building of Yerkes Observatory, and Hale probably felt that as director-to-be of the world's largest refracting telescope, he could use a year profitably to learn a little more astronomy. There remained the problem of editing the astrophysical part of *Astronomy and Astro-Physics,* and Keeler soon learned that all Hale's protestations just two years previously that the work would not be "shoved off" on him if he became associate editor had become inoperative. The entire responsibility now in fact rested on Keeler. Hale did not even have time to stop at Allegheny to discuss the journal, as Keeler had "insisted," but went straight to New York and boarded the ship for Europe. All the arrangements had to be made by mail. At least Hale managed to get Crew, at Northwestern, to read the proofs, instead of See, whom neither of them trusted. Hale's observing assistant at Kenwood, Ferdinand Ellerman, became the contact man in Chicago. He forwarded all the manuscripts to Keeler, who made the necessary editorial decisions and sent the accepted copy on to Payne at Northfield.[100]

Hale gave a lecture in Boston before he sailed, and when he first arrived in Europe he had a "glorious time" visiting astronomers and observatories in England and France.[101] Once he settled down in Berlin, however, life became less interesting for him. He could not understand German very well, and never seemed to find time to put much effort into learning it. Hale was not really interested in organized study, and soon was skipping the physics lectures to stay in his apartment and work on plans for the Yerkes Observatory building and on schemes to help Harper persuade Yerkes himself to provide more money for spectroscopes and for an additional small telescope.[102]

Keeler meanwhile was coping with *Astronomy and Astro-Physics.* Sometimes there was an agonizing dearth of material, as in the December 1893 issue, when absolutely no new manuscripts came in, and Keeler had to eke it out with translations from European journals and selections from other English-language magazines. He used this experience as an opportunity to encourage Hale to practice his German by doing some translations himself. A month later there was a flood of new manuscripts and Keeler had to put some of them over to the next issue. Always he was irritated by Payne's complete inability to understand any of the astrophysical material he was publishing. On one occasion Payne reprinted an item from the *Publications of the Astronomical Society of the Pacific* in his section, without realizing that it was a translation by Keeler which that journal had itself lifted from an earlier issue of *Astronomy and Astro-Physics.*[103] Yet Keeler kept his sense of proportion and was still able to laugh at his own editorial efforts, as when he wrote his friend Ames,

> I had no idea so many "crank" articles were sent in for publication. Probably you think that they have not all been kept out of the journal, and with some reason; but if you saw some of the stuff that is sent in, you would realize how virtuous we are by comparison with the sins that we might commit.[104]

Meanwhile, in Europe, Hale had been consulting with as many astronomers and spectroscopists as he could on the prospects for support if he and Keeler broke away from Payne and founded a new completely astrophysical journal. Hale's idea was to adopt a very wide definition of astrophysics, including not only astronomical and laboratory spectroscopy, but also photometry and physical studies of planets. He planned to set up an editorial board that would include one leading astrophysicist from each of the larger European countries, and his contacts assured him that if he jettisoned the traditional astronomy and popular articles, more than enough papers would come in from Europe to fill the void. Keeler and Hale comforted themselves with the thought that Payne was most interested in *Popular Astronomy* anyhow, and would not really mind if they killed *Astronomy and Astro-Physics.*[105] Certainly many astronomers they knew considered Payne dangerously unprofessional, a "rooster" as one of them wrote, and thought that it was only a matter of time before he would be eased out of the picture.[106]

Thus, when Hale returned to America after another unsuccessful attempt to photograph the corona without an eclipse, this time from Mount Etna, in Sicily, he arranged to see Keeler at the meeting of the American Association for the Advancement of Science in Brooklyn in mid-August 1894.[107] They agreed that the time was ripe and that they would go ahead. The natural place to publish the new journal was the University of Chicago Press, so as soon as Hale got back to his Kenwood Observatory he arranged to see President Harper. The conference was a success, and Harper told Hale to open negotiations with Payne and find out how much money it would take to buy him out.[108] Thus just at the end of August, Hale met with Payne over dinner at a Chicago restaurant, together with Burnham and Crew. Payne was not pleased to learn that Hale and Keeler wanted to sever their connection with him and start publication of a professional-level, purely astrophysical journal on their own, but he agreed to turn *Astronomy and Astro-Physics* over to them for a fair price. He said he would not compete with them by continuing publication of his part of it on his own, although he of course intended to go on with *Popular Astronomy*. Burnham, an apostle of the old-time astronomy who had no use for spectroscopes or other astrophysical gadgets, encouraged Payne to fight the breakup, but the wily Minnesotan knew that if Hale and Keeler withdrew their astrophysical section, *Astronomy and Astro-Physics* would lose half its subscribers. It would be wiser for him to sell out than to go broke. Crew, basically a laboratory physicist himself, naturally sided with Hale, and after Payne had grudgingly agreed, on the way home after dinner the enthusiastic young Chicagoan managed to persuade Burnham too, that it was the right decision.[109] Crew, in describing this conference to Helen Wright many years later, said that "Hale had the ability of putting a knife in your side, and of making you think he was doing a favor."[110]

Payne, although he had agreed to the sale, began having second thoughts as soon as he got back to Northfield. He wrote Keeler, saying that "Hale and Crew" were planning to break away and publish a new purely astrophysical journal. He asked if Keeler would stick with him as editor of the astrophysical part of *Astronomy and Astro-Physics,* perhaps in collaboration with Ames. Keeler at once warned Hale that Payne might try to go on with *Astronomy and Astro-Physics* on his own, and that they had better sign up all the important contributors (such as Ames) for their new journal before the Northfield editor got to them. Keeler also wrote Payne himself, gently but firmly informing him that he was committed to Hale. He urged Payne to go on with *Popular Astronomy,* perhaps including more professional papers. *Astronomy and Astro-Physics* had been a happy combination, Keeler said, and he was sorry to see it come to an end.[111]

Nevertheless, Payne proposed a new scheme that would give Hale complete control of the astrophysical side, and allow him to expand it as much as he wanted, if he would stay with *Astronomy and Astro-Physics.*[112] Hale's mind was made up, however, as Payne finally realized after another conference, this time at Burnham's office, and he then got down to the business of negotiating the sale. Payne agreed to sell his share of the magazine, including the 420 active subscriptions they still had, all the advertising, seventy-five sets of unsold back issues, and half the engravings (from which illustrations were printed) used in the past three volumes, for $600.[113] Apparently Hale's father was not willing to finance the purchase entirely himself, although he was probably the largest contributor with a check for $150. Keeler got a contribution of $100 from Mrs. William Thaw, and added $50 of his own, but even so Hale could not raise the entire amount. He was forced to renegotiate the agreement. Payne now agreed to keep the back issues and to reduce the price to $500, of which he would actually get only $400, all the cash Hale had been able to lay his hands on, when the transfer was made in late December, with the balance due in 60 days.[114] Relations became very strained at the end, but finally the deal was consummated and the *Astrophysical Journal* was born.[115]

This was the name Hale and Keeler had always wanted to use for the new journal, which took the place of *Astronomy and Astro-Physics,* and although they usually wrote astro-physical with a hyphen, the University of Chicago Press decided that it was one word, and so it became. Hale and Keeler were the editors (every issue since Hale retired in 1924 has borne the words "Founded in 1895 by George E. Hale and James E. Keeler" on the cover or the title page), and Hale had assembled an outstanding board of associate editors. The European members were Huggins, Tacchini, Hermann Vogel, Nils P. Duner of the Uppsala Observatory in Sweden, and M. Alfred Cornu of the Ecole Polytechnique in Paris, while the Americans were Rowland, Young, Edward C. Pickering, Michelson, and Hastings. Cornu, Rowland, Michelson and Hastings were primarily physicists, while the others were basically astronomers. These men were to function as members of an advisory board as well as unofficial ambassadors of their countries or institutions to the journal. There were also five assistant editors, Crew, Ames, Campbell, Frost and Frank Wadsworth, who were expected to help in the actual work of editing the journal.[116]

Keeler and Hale had debated longest over Campbell. Keeler wanted him as one of the assistant editors, arguing that Lick Observatory should be represented and that Campbell was one of the most active research workers in America. Hale was more doubtful. He recognized Campbell's scientific credentials, but considered him too quick to criticize his elders, men like Huggins and Kayser, "who should be treated with great respect."[117] In the end Hale agreed that Campbell belonged on the editorial staff, but the young Lick astronomer then declined to serve. He claimed that he needed all the time he could find for his own research, and that whatever moments he could spare for editorial work belonged first to the *Publications of the Astronomical Society of the Pacific*. Even more important, however, he did not want to be muzzled by accepting the assistant editorship. "Now the truth is Kayser and Huggins rode rough-shod over some of my work," he wrote.

> They did not treat it *on its merits* ... Kayser paid no attention to my comparisons, except to say I had committed some mistakes, which after all were not mistakes ... Huggins was extremely unjust to me in his article on the present spectrum of *Nova Aurigae,* in that he misrepresented my description of the principal line. His recent note in A. & A. P. is not at all candid. His ignoring Lockyer's February 1890 photographs of the Orion Nebula spectrum was outrageous. In my opinion *age* is no excuse for injustice: it only adds to the culpability.[118]

But Keeler and Hale would not accept this answer; by now they both wanted Campbell on the staff of their journal. They wrote him persuasive letters urging him to reconsider. Keeler agreed with Campbell that Kayser and Huggins had been unjust to him, but ascribed it "more to the way of the world in general than to any peculiarity of theirs." Like Campbell, Hale did not get the credit that was due him, because of his youth, Keeler wrote. "We younger fellows can console ourselves with the reflection that every dog has his day, and some time we may get big credit for little work ourselves ... In the meantime I suppose it is better policy, and really more effective to oppose the older men gently, even when we are right."[119]

Campbell could not resist this appeal, particularly because of the warm personal friendship and respect he had felt for Keeler ever since he had worked with him in the summer of 1890. Campbell accepted the assistant editorship, though it really meant little more than adding his name to the list on the title page.[120] In California he was almost as far from the scene of the editorial action as if he had been in Europe. A few days after becoming an assistant editor, Campbell received the November issue of *Astronomy and Astro-Physics,* and found that it contained a criticism of his spectroscopic work on Mars by Huggins. This was too much for

Campbell; he wrote "in just as mild a way as I can" a reply explaining that he was right and Huggins was wrong, and fired it off to Hale. A few days later the young Lick spectroscopist cooled off a little and realized that "my reply did not do either the subject or myself justice," and he wrote an even milder (but still stinging) reply, and telegraphed Hale to substitute it for the earlier version.[121] Slowly but surely the combative young Campbell was being co-opted by the establishment.

Wadsworth, a young physicist who had been Michelson's assistant at Clark University, and then Langley's assistant at the Smithsonian, was now an assistant professor of physics at Chicago. His main expertise was in instrumentation, on which he tended to be somewhat dogmatic, and Keeler was not enthusiastic about taking him on as an assistant editor. Hale, however, said he wanted Wadsworth on the staff partly to keep Michelson happy "and also because he can be very useful in helping us on drawings of apparatus, etc." He was appointed and soon Hale's letters were full of admiring references to Wadsworth's ideas, Wadsworth's opinions, and a projected book by Wadsworth.[122]

Even before the first number of the *Astrophysical Journal* was published, Hale and Keeler called a meeting of the editorial board. It was held at the Fifth Avenue Hotel in New York on November 2, 1894. None of the Europeans could attend, but Young, Hastings, Rowland, Michelson and Edward C. Pickering were there, in addition to Hale and Keeler. J.K. Rees, professor of astronomy at Columbia University, was also present as an invited visitor. Each of the board members paid his own traveling expenses, except that Hale and Keeler between them picked up the check for lunch for the entire group plus dinner for those who remained in the evening; their total bill was $35. Pickering was elected to preside at the meeting, on Hale's motion (he was considered too young to chair this august group himself); Keeler was chosen secretary and kept the minutes. They discussed general matters such as the importance of astrophysical research and the need for a journal specializing in it. When they got down to specifics there was some argument about the name of their publication. Rowland and Young wanted the word "spectroscopy" in its title. Finally they agreed to stick with *The Astrophysical Journal*, but use as an interpretative subtitle, *An International Review of Spectroscopy and Astronomical Physics*. After lunch they settled several matters of style, among them to print spectra in figures with blue on the left and red on the right, to list spectral lines in tables in order from short wavelengths to long, to use ångström units for wavelength, to use km/s for velocity, to use Vogel's new notation for the hydrogen lines, in which they were listed as Hα, Hβ, Hγ, etc., and to use DM as the designation for stars listed in either the Bonner Durchmusterung or the Cordoba Durchmusterung.[123] All these questions were controversial at the time; all the decisions adopted on that day except the last have become the accepted standard used by astronomers everywhere.

Before returning to Allegheny after the editorial board conference, Keeler stopped off for a short visit at New Haven where he had many friends and relatives. He had recently developed an interest in his ancestry, sparked by correspondence with an author who was preparing a genealogical book on the Keelers in America. Connecticut was the center of the Keeler territory, and the Allegheny astronomer went out to Ridgebury, the little village where his great grandparents, and their parents and grandparents, had lived back to the days of the American Revolution and the French and Indian Wars. Their houses were all gone, but he saw their graves in the churchyard and read their names in the marriage registry book.[124]

The first issue of the new journal came out in January 1895. It contained articles by Michelson, E.E. Barnard, Rowland and one of his young collaborators, and several other well-known scientists. In addition to the original papers there were, as in subsequent issues for many years, reviews of important papers that had appeared in other journals, as well as notes on the progress of astronomy and spectroscopy.

Although Hale saw faults in this first issue, Keeler was very pleased with it. To its readers the *Astrophysical Journal* was an instant scientific success.[125] Payne, his bitterness at the rupture quickly evaporating, congratulated Hale "most heartily on the neat appearance of your first number." Hale, no matter what he felt inside, was quick to thank him and to praise *Popular Astronomy* in turn, and as Keeler had always managed to stay on good terms with the elderly editor, relations between the two journals were soon on an even, friendly keel.[126]

Hale was discouraged, though, that new subscriptions did not come in faster. The first number had cost $600 to produce and mail, the second, the February issue, would cost $500; they would have to cut expenses by reducing the number of pages in subsequent issues. They had only about 200 subscribers, at $4 per year, and there was no other source of income; the projected deficit, even with the reduction in size of the journal, was several thousand dollars per year. Hale could not even raise the $100 final installment on the purchase of the journal that was due Payne by the end of February. "It worries me constantly but I don't know how to remedy matters," he complained. He knew that he would have to go out and approach his father's wealthy friends for money, and he hated to do it.[127]

Keeler, at a distance and not so directly involved, could afford to adopt a more philosophical attitude. He advised Hale to save all the favorable notices about the journal he was receiving, and at the end of the year use them in a fund-raising campaign among a few of the prominent citizens of Chicago. Although he did not say it, he undoubtedly realized that if all else failed, Hale's father would certainly pay off the deficit.

> In the mean-time if we keep the Journal up to its present mark it must be taken by institutions that wish to keep up with the times. The university ought to recognize the fact that it can afford to publish the journal at a loss, as the best possible form of advertisement. The Johns Hopkins does the same thing right along.
>
> Finally, if we fail at the end of the year it simply means, as I said, that the world is not big enough yet for such an undertaking. If we have done our best we can bust up with a clear conscience, and regard even the jeers of Payne with equanimity. My principle is, do the best you can, and then laugh if you fail. Above all, don't worry about the matter. You can't afford to lose your health for fifty *Astrophysical Journals.*[128]

The journal did grow and eventually prospered, although there were many crises along the way, not all of them financial. One of the continuing problems was Percival Lowell, the wealthy Bostonian who had built his own observatory in Flagstaff, Arizona, to study Mars. To Keeler and Hale, as to professional astronomers in general, he was the epitome of an amateur enthusiast, with no real astronomical training, who had no hesitation in rushing in where brave men feared to tread. Actually he had a Harvard degree with honors in mathematics. Lowell had become convinced in his own mind that there was intelligent life on the red planet; his observations were intended chiefly to prove this. Although his writings had a tremendous public impact, he never made much impression on the skeptical professionals of his day.[129]

As soon as Lowell learned that Hale and Keeler were going to publish the *Astrophysical Journal,* he submitted a manuscript recounting his latest visual observations of Mars and the conclusions he drew from them. Initially Hale favored rejecting it, because Lowell had included too many statements about "oases" and the "advancing wave of vegetation," which were speculative interpretations rather than observational facts. Keeler was more sympathetic, although he disliked as much as Hale did Lowell's "dogmatic and amateurish" style and his tendency not to distinguish "between what he sees and what he infers." But, Keeler thought, Lowell had spent a lot of time at the telescope working on Mars, and his observations must have had some value, so they should publish them. Besides, they had already decided to accept another paper on Mars from A.E. Douglass, Lowell's assistant at Flagstaff, and it would be awkward to publish it but reject the director's paper. So Keeler composed a "disclaimer" to be printed along with Lowell's paper, stating in carefully chosen language that in publishing it the editors were not endorsing the correctness of his interpretations, but only providing a vehicle for placing the data before their readers. Hale thought the disclaimer was excellent, but Lowell then asked for his manuscript back, to make a few changes, and to add another page of figures. They were quite expensive, but Lowell was willing to pay for them himself. When he returned the rewritten manuscript, the changes he had made were too much even for Keeler. In the new version the elegant Bostonian wrote of "vegetal life" on Mars as if it were a certainty, and claimed that the observational "facts" pointed to the conclusion "that intelligent beings of some sort or other exist at this moment upon the planet which is our own next to nearest neighbor in space." This would not do.

Keeler wrote a diplomatic letter to Lowell, saying they could not publish his paper because "the line between observations and inference is hardly drawn with sufficient clearness, and ... the conclusions are scarcely justified by the comparatively small amount of observational material." He added that although he and Hale were fairly conservative, many other editors felt less responsibility for the opinions expressed in their journals – in other words, that Lowell was free to take his paper elsewhere. Hale was ecstatic. "You were certainly cut out for a diplomat!", he wrote Keeler. "Your letter to Lowell put the thing exactly right. I could not have done it to save my life ... Percival may rise up in his wrath to defend his beloved 'denizens' [of Mars], but he will waste his ink, and might just as well find some other editor at once."[130] Lowell then submitted a completely factual observational paper, "On Martian Longitudes," without a word on interpretation in it. They accepted it; Keeler assured Hale that it was "all right," if a bit "amateurish" and with a few places in the text "that seemed a little bizarre." However, when Lowell made extensive revisions in proof to this paper, changing some of the numerical values of the measurements that he claimed to have made, or at least substituting results he had evidently not included before, it was too much for both of them. They published the paper, but Hale wrote Lowell that "on account of the expense, trouble and uncertainty which always seem to attend the publication of his articles, we have decided not to accept any that may be sent in the future."[131] Lowell soon afterward founded the *Annals of the Lowell Observatory,* a series which he personally controlled, and which, with his books, became the main medium in which he dispensed his results.

As the *Astrophysical Journal* continued, Hale, in Chicago, actually

functioned as the managing editor. He made most of the editorial decisions, and handled all the relations with the printer. These were often
stormy. On one occasion Young told Hale that instead of his subscription
copy of the *Astrophysical Journal* he had received the *Journal of Geology,*
another magazine published by the University of Chicago Press, bound
up in the *Astrophysical Journal* cover. He could understand the mistake,
but the Press ignored his request for the right journal for weeks. Even
worse, a subscriber in Ste. Catherine's, Ontario, had received a copy of
the *Biblical World,* another University of Chicago Press journal edited by
Harper himself, instead of his *Astrophysical Journal.* Hale complained
bitterly to the president that the Press was slow, error-prone, and generally incompetent. He said that he was sorry now that he had ever raised the
money to buy out *Astronomy and Astro-Physics,* and that he would not
have done so if he had known how the Press operated. This outburst
brought a severe chastisement from Harper. He had always regarded
Hale as a fairly moderate man, he wrote, but he would have to change his
opinion after reading these letters. "Patience is needed in all these things
and I am sure that we shall all have more of that desirable article as we
grow older." Hale did not take this sermon lying down, but fired back
accounts of several more instances of what he considered unnecessary
delays by the Press, and challenged Harper to submit the whole matter to
a committee composed of "those members of the faculty who through
their own editorial work are fully qualified to form a correct opinion of
the efficiency of the University Press."[132] Harper wisely let the matter
die.

On all difficult editorial questions, Hale consulted Keeler, and they
always discussed general policies as well as their latest ideas and information on recent astronomical research. In addition, Keeler's role was to
supply material, not only his own research papers, but also reviews of
important articles that had appeared abroad and especially translations of
significant papers from the German journals. He took pride in his efforts,
particularly as he sometimes saw botched translations of the same papers
in the British journals. Once he even saw the meaning of the original
German author exactly reversed in an English translation in *Nature.*[133]

The translated articles were especially needed when there were not
enough original papers in hand to fill out the pages of the journal. On one
occasion in early 1896, Hale wrote Keeler that "the dry season" was upon
them, and begged him for some translations of "notes, reviews – *anything*
(barring Lowell, Flammarion et al) will be welcome." Keeler replied that
"the dry time to which you refer has struck me in an unfortunate season
with respect to my capacity for irrigation." He was teaching a class,

preparing several public lectures, and entertaining a visiting aunt, and had no time for translating anything. Hale was insistent. A long paper he had counted on from Wadsworth had turned out to have an error in it and had to be revised. Another manuscript contained incorrect wavelengths, and had to be returned to its author, Hale was "forced" to telegraph Keeler to translate a German paper by Bernhard Hasselberg on the arc spectrum of titanium. Keeler began at once, and in spite of his class, his lectures and his aunt he was able to send off the first half of the translation to Hale just four days later. By that time, however, the situation had changed completely. So much unexpected material had come in that they could not use it all. Hale was philosophical, writing, "I am not sorry you are translating Hasselberg, as it will be a good thing to use in a future issue in any case." When this letter arrived, Keeler had nearly completed the task, but he put the paper aside for another day and went back to his class and public lectures. He did not finish translating Hasselberg's paper until months later, but it finally did appear in the journal.[134]

When Keeler took the spectrograms that demonstrated observationally that the rings of Saturn are composed of a myriad of small particles in orbit about the planet, as described in the previous chapter, the fact that he was one of the editors of the *Astrophysical Journal* undoubtedly contributed to the speed with which his paper was published. On the other hand it was exactly the type of paper, at the forefront of current astrophysical research, that the journal had been founded to publish, and that its subscribers wanted to read.

Although he worked very hard first on *Astronomy and Astro-Physics*, and then on the *Astrophysical Journal*, Hale's main interest since 1892 had been getting Yerkes Observatory completed and into productive research. It was agonizingly slow business. The telescope tube, mounting and controls had been finished in time to make a wonderful display at the Chicago World's Fair, but little else had been accomplished at that time. Yerkes himself, although he wanted nothing less than the largest and best observatory in the world, found it difficult to bring himself to pay for it. He seemed to think that the telescope was all that was needed, and strongly resisted providing any funds for auxiliary instruments, such as spectrographs. Harper was building a whole university, and could not see his way clear to putting any appreciable amount of general funds into the observatory, particularly as Rockefeller, his principal source of support, specifically disapproved of spending money for astronomical research rather than classroom instruction.[135]

At least the site question was settled. It was obvious to every astronomer that the low, flat, smoky Chicago lakeshore would never do as the site of the largest telescope in the world. Harper wrote to a long list of famous astronomers, including Keeler, soliciting their opinions, and they all advised building the observatory away from the city and its smoke. Many sites were considered, nearly all of them small towns and suburbs in the immediate vicinity of Chicago. Hale favored Lake Forest, Illinois, a wealthy North Shore suburb, where his friend McNeill was professor of astronomy in the local college, but in April 1893, the Board of Trustees decided to locate the observatory at Williams Bay, near Lake Geneva, Wisconsin. Chamberlin, who knew the region well, recommended it strongly to Hale for its "surrounding high hill, clear crisp atmosphere without fogs or vapors, no night or Sunday trains, easy access to Chicago, ... quietness as a place of residence, [and as a] resort for the choicest people of Chicago." One of the trustees, George C. Walker had a summer home on Lake Geneva; two others, Martin A. Ryerson and Hutchinson, were soon to follow him there. Several other friends of the University lived nearby, including John Johnston, Jr., who donated the land where the observatory was to be built. The only possible flaw was the site's proximity to the lake, but Burnham gave his opinion, based on his experience at the Washburn Observatory on Lake Mendota in Madison, that the small amount of open water would not present a problem, and the choice was made. Hale agreed that Williams Bay was a beautiful place that had every advantage as a site except that its distance from Chicago, seventy miles, was a little extreme. Keeler, like most astronomers, was pleasantly surprised that the Chicagoans had let the telescope get out of the city, and agreed that the site was "a very suitable one."[136]

When Hale went off to Germany after the World Congress, he spent much of the winter of 1893-94 working on his plans for Yerkes Observatory, rather than studying physics as he had said he planned to do. The overall design was clearly influenced by his memories of Lick Observatory, which had impressed him so greatly just a few years before. Like Lick, Yerkes Observatory was built with a large, elevated dome at one end of a long hall of offices; the main difference is that at the other end of the hall at Yerkes there are two smaller domes, connected by a shorter, perpendicular hall, while at Lick there is only a single small dome. Inside, however, Hale's own ideas came to the fore. Yerkes Observatory was specifically designed for astrophysics, not the old astronomy, and Hale regarded it as the first large observatory that was adequately laid out for both the astronomical and physical sides of the science. The large basement was full of photographic darkrooms, spectroscopic laboratories, and instrument shops, while the second floor of the short hall between the two small domes was outfitted with sectional roll-away roofs that could be opened for solar work. Hale sent the general design he had drawn up to Henry Ives Cobb, the university architect in Chicago, who converted it into detailed plans for the building. Harper, Cobb and Burnham went out to Lake Geneva on a wintry December day, met Walker, and picked out the exact spot where the observatory would be built.[137]

Finally actual construction work began in the spring of 1895. Hale had held down his budget requests, as Harper had demanded, while the observatory was only a distant, theoretical possibility, but now the work was about to begin, he had to let the trustees know what it would really cost them to operate Yerkes as a research institution. Although he maintained that he needed only one third as much as the Harvard Observatory budget, or half as much as the Paris Observatory or Pulkova appropriations, the sum he requested for the year, $22,000 was nearly four times as large as his earlier estimate. Harper was appalled and insisted on strict economy, but approved a $15,000 budget, probably far more than Hale had expected.[138]

Once work began at the site, Hale was anxious to move to Williams Bay, at least for the summer. He had conceived a new scheme for detecting the solar corona without an eclipse, this time by measuring its expected heat radiation with a bolometer, the tricky, energy-sensing device that Langley had developed years before. Wadsworth felt certain this method would succeed, but Hale wanted Keeler's candid opinion of the idea. The practical Allegheny astronomer's reply was scarcely encouraging; the basic idea was sound, he thought, "but knowing the cussedness of the bolometer (at least of the aboriginal breed) my expectations of your success are not very high". But he hoped Hale would win out, and told him that if perseverance counted, he deserved victory.[139]

At Yerkes progress on the building and on the corona was maddeningly slow. Hale tested the bolometer in the laboratory in Chicago; it proved insensitive and had to be torn down and rebuilt. Much of the work was done by Wadsworth. Hale did not get out to Williams Bay with it until August, when he set up the instrument with a small telescope in a shed, but he was constantly plagued by clouds, haze, smoke from a forest fire in Michigan, and failures in the apparatus. In the end he got no results, and had to put off a real trial of the method until another year. It was Pike's Peak all over again, as Hale himself admitted.[140]

At about the same time Hale did make an important research contribution that helped solve an outstanding astrophysical problem. This was the origin of the spectral line at $\lambda 5876$, which had first been discovered in emission in the spectrum of the chromosphere at the 1868 solar eclipse. Spectroscopists at that time already realized that each chemical element has its own characteristic lines, but the element that emitted this particular line, which can always be observed in the chromosphere and in solar prominences stubbornly resisted identification. The line was given the name D_3; the hypothetical element that emitted it was called helium, to indicate that it is present in the sun. The same line was found in emission in the Orion nebula and in some planetary nebulae, proving that helium is present in them as well. The German spectroscopist, Julius Scheiner, soon noticed that another emission line observed in gaseous nebulae has a wavelength, near λ4471, approximately the same as the wavelength of an absorption line in some of the stars in Orion. Keeler's and Campbell's accurate wavelength measurements of the spectra of nebulae, described in the previous chapters, then conclusively proved that D_3, λ4471 and several other lines occur both as emission lines in these objects and as absorption lines in the Orion stars. Furthermore the wavelengths the two spectroscopists measured showed that not only D_3 but also λ4471 occur as emission lines in the chromosphere and prominences.[141] Thus all these lines presumably resulted from helium, and they demonstrated the presence of this element in stars, as well as in nebulae and the sun.

Then in 1895 William Ramsay and William Crookes in England found the λ5876 line in the spectrum of the gas given off by cleveite, a uranium-bearing mineral, when it was dissolved in weak acid. They had thus discovered helium as a gas on the earth. Today we know that terrestrial helium is the neutral form of α particles given off in the decay of uranium, thorium, radium and other naturally occurring radioactive elements. At the time, however, there was a slight discrepancy between the wavelength measured by Ramsay and Crookes in the terrestrial sample, and the solar value. Lockyer, after hearing of their discovery, heated some uranite, another uranium-bearing mineral, himself, and saw not only λ5876 but λ4471 as well in the spectrum of the gas that it gave off. This really clinched the identification. But Friedrich Paschen and Carl Runge in Germany studied the spectrum of a terrestrial helium sample with very high dispersion and found that the laboratory D_3 line was double, not a single line as solar D_3 was believed to be. Runge therefore concluded that terrestrial helium had not actually been found, unless the solar line were in fact double. Here Hale entered the picture. He at first doubted the identification himself and tried to get a sample of the purported "helium" for his own experiments. Then, after reading Runge's short paper, he immediately observed a solar prominence with his high-dispersion spectroscope at Kenwood, and saw that the yellow line was in fact double in the sun, just as in the laboratory. He and Ellerman measured the separation of the two components, and found it essentially the same as Paschen and Runge had measured. This absolutely confirmed the identification.[142]

Meanwhile, Clark had been making real progress on the 40-inch lens in his optical shop in Cambridgeport. He pronounced it finished in September 1895, just as Hale was giving up on the corona for the year.[143] According to Clark's contract, the lens was to be tested and declared satisfactory by an independent expert before Yerkes would pay for it. Hale recommended that Keeler be asked to act as the expert, and Yerkes and Clark both agreed to this choice. Hale hoped that Burnham could participate in the test too, but in the end it proved impossible, for his travel plans did not mesh with Hale's and Keeler's. The test was scheduled for mid-October, when the two of them were to be in Cambridge for the second meeting of the Board of Editors of the *Astrophysical Journal*. It was held at the Harvard College Observatory; Keeler and Hale were by now both sufficiently important figures to be invited to stay with the Pickerings at the director's house.[144]

Thus, on the evening of October 17, 1895, after a long day around the conference table and no doubt a good dinner, Keeler, Hale and some of the other board members trooped out to Clark's establishment to look at the 40-inch lens. It was installed in a temporary telescope on the same outdoor brick pier that previously had been used, built up a little higher each time, to test the Princeton, Naval Observatory, and Lick objectives. In this arrangement it was possible to point the lens at a star and hold the eyepiece rigidly at the focal point, but not to set it accurately nor to follow the diurnal motion of a star. The seeing was very bad, for the atmosphere was quite unsteady, but even so Keeler and Hale could see a large, out-of-focus "wing" on one side of each star image. No doubt Clark assured them that the lens was not really that bad, but that it must have been temporarily distorted by uneven heating and cooling. Perhaps the lens was pinched in its mounting and out of shape. The next night they tried again. The seeing was even worse, but they could still see the wings. Then, mercifully, the sky clouded over. The following night the wings were still there. With several of Clark's workmen helping, Keeler and Hale found that they could lift the front element of the lens and rotate it in the cell. The wings rotated with the glass disk, showing that they were not an effect due to the ground or to the pier, but were definitely associated with something that turned with the front element. As they rotated it, the image became worse, more and more elongated. They went to bed.

The next day was Sunday, and evidently Alvan Clark and Sons did not do business on the Sabbath, for no tests were made that night. Finally, on Monday Keeler and Hale tried again. This time they decided to rotate both elements of the lens. To accomplish this, they had to take the lens completely out of the cell. They rotated the two elements ninety degrees, put the cell back together and turned the telescope to a star. The seeing was good. To their great joy they saw that the star disks were small and round, and that the wings had vanished. Very probably the second element of the lens had been pinched in its cell, and in taking it apart and putting it back together they had relieved the stress. It was a good lens. Clark had done his job well.

Keeler tested the lens on stars at several different altitudes, from the zenith nearly to the horizon. In his opinion the definition was as good as the definition of the 36-inch Lick telescope. He looked at Sirius, a very bright star, and could see that the lens was free of "ghosts" or extraneous faint images. He and Hale observed the Trapezium, the multiple star in the bright central part of the Orion nebula, a strikingly beautiful sight in such a large telescope. Keeler thought that he could glimpse the very faint star in the Trapezium that Clark himself had first discovered while testing the 36-inch lens with him and Captain Floyd that wintry night at Lick nearly eight years previously, but he could not be sure because the awkward mounting made it impossible to get a long steady look. Certainly the lens was acceptable and the requirements of the contract between Yerkes and Clark had been fulfilled.

Yet the effects of flexure, or slight bending of the glass itself, could be detected in some of the tests. Since these distorting effects increase rapidly with the diameter of the lens, Keeler's tests showed that forty inches was close to the upper limit of size to which a refracting telescope could be effectively built, as he immediately noted. In fact the 40-inch turned out to be the largest American refractor ever made, in no small part because of the efforts of Keeler and Hale with reflecting telescopes in the next few years. But for the moment it was sufficient to report to Yerkes that the lens was completed and ready for use.[145]

Hale had sent the street-railway tycoon his own preliminary version of the results of the acceptance tests immediately after they were made. As soon as Keeler filed his formal report, Hale forwarded it to Yerkes. The donor was very pleased and arranged for an exchange of letters between himself and Hale, to be published in the newspapers. Hale quoted liberally from Keeler's report in his letter, but omitted any mention of the wings or of the flexure in this public statement. The Allegheny astronomer agreed to this procedure, but insisted on mentioning the flexure problem in the short paper on the tests that he would publish in the *Astrophysical Journal*. It was an important fact that presumably would limit the size of future refractors and it could not be suppressed. Flexure resulted from the physical properties of glass and of lenses, and no one was to blame.

Furthermore, once the lens was permanently mounted and in use at Yerkes Observatory, the flexure effects would probably soon be discovered. If they had not been reported, questions would certainly be asked about why the so-called expert had not noticed them. No scientist could stand in the way of the truth.[146]

As Yerkes Observatory came closer to becoming a reality, Hale's relations with See, his colleague at Chicago, became worse and worse. Hale had first begun to have his suspicions years before when he had met two of See's fellow students from Berlin, who were surprised to learn that the University of Chicago had hired him. According to them, See was "very capable ... in certain directions, but ... very peculiar" and they "would say little more than to express a hope that he would turn out well." When Hale went to Germany, he found that Wilhelm Foerster, the director of the Berlin Observatory, under whom See had done his thesis, never mentioned his former student's name and, when questioned, declined to vouch for his work. Others at Berlin warned the young Chicagoan that See was very unhandy with instruments and would be a positive danger if allowed to use the Yerkes telescope. Hale, insisting he had *"absolutely no personal feeling"* against See, and that he only wanted to preserve the observatory, passed this information on to Harper. He regretted it very much, he said, but in order to save the telescope he felt the president must make it absolutely clear to See that his post would be in Chicago and that he would not be allowed to touch the instrument.[147]

Harper, hard pressed for money to operate the University and unwilling to add more observers to the Yerkes payroll, probably would have ignored Hale's suggestion, but he soon learned for himself that See was a very unpleasant character, with an extremely inflated opinion of his own importance. He was almost always engaged in involved, lengthy controversies with his detractors. Soon after coming to Chicago, See had written a long, legalistic document to the "Honorable Board of Curators of the University of Missouri" charging its former president, George D. Purinton, with issuing "malicious slanders" against him, but adding that in spite of these false accusations Harper had appointed him to the Chicago faculty and the Board of Trustees had confirmed his appointment. See grandly concluded that the Missouri educational establishment could use this information as the curators saw fit. This letter brought forth still more charges against See, but the combative astronomer discounted them as being the work of an accuser whom he called "the father of lies".[148]

See turned out to be a notably uncreative scientist, locked to the methods of the past, but trying to invest his own work, largely mathematical in nature, with an importance it did not have. He usually made a good initial impression, but anyone who worked closely with him, or ever studied his papers carefully, soon came to believe that he was arrogant, dishonest, and not to be trusted. Just at the time the 40-inch lens was being tested, See wrote a twenty-two page long, nearly hysterical letter to Harper, demanding that he be promoted to associate professor. He referred in extravagant terms to his own research, saying that he had sent reprints of his papers directly to the trustees so that they would know of his work – a blatant attack on Harper's integrity. See claimed that he had "saved" Yerkes Observatory by his advice "in saving you an annual wasteful expenditure of at least $15,000" that would have left the Department of Astronomy in "chaos and ruin." What he meant was simply that he had recommended that Harper disregard Hale's budget, forget about astrophysical research, and operate the Yerkes telescope, when completed, with a skeleton staff who, like himself, could only make old-fashioned visual observations. In a direct attack on Hale, See wrote "I decline absolutely to ask for advancement on any other ground than of scientific merit, and whatever others may do by way of social intrigue to obtain what they could not get on the merits of their work, I at least remain steadfast by this decision." Harper gently told See that the tone of his letter did him no credit, and that if his claims for his research had as little foundation in fact as his claim to have saved the university $15,000, "I am afraid that we should regard the amount of credit insufficient for a promotion." Harper concluded that he was "compelled to concede that there was some basis for the charges made in reference to your general bearing and spirit by those who were concerned in your recent trouble in the University of Missouri."[149]

This exchange finally convinced Harper that See could not be trusted. He was not promoted. He drifted off to Lowell Observatory, on leave from Chicago. He refused to resign his faculty position, but in 1898, when the university was undergoing a severe financial crisis, he bitterly informed Harper that he need no longer "labor under dread lest I desire to return to a position where I am not wanted." He contrasted the way in which he had fulfilled all of his own obligations to the university with Harper's "unmanly course." See concluded:

> I have such an abiding faith in the triumph of justice that I am confident the future will bring forth events in which you will long to be able to lay hands on trustworthy men such as I proved to you while at the University. In these emergencies you will find, like many unfortunate Roman Emperors of old, that it is not the two-faced sycophantic flatterers who will come to your relief.

Hoping you will continue to meet with all the good fortune in the administration of the University that you deserve,
I remain,

Very truly yours,

T.J.J. See[150]

Only a few months later See was fired from the Lowell Observatory post that he had thought would provide him a safe haven.[151] By that time he had lost whatever scientific credibility he had in the eyes of nearly every serious American astronomer. Rumors circulated that he was insane.[152] His papers were essentially barred from the pages of the *Astronomical Journal,* and its editor, the acid-penned Seth Chandler, wrote and privately circulated an anonymous parody of See's work, entitled "On the Fundamental Law of Increase of Gaseous Reputation." It was built around the equation

$$T = \frac{J}{J_1} C,$$

which every astronomer would immediately recognize as referring to the infamous T.J.J. See. All the factors in the equation were defined in derogatory terms, in a style that imitated See's. Thus T was "the scientific reputation that may be established by any means whatever, on the condition that it may be of a perfectly gaseous nature", and C was "the amount of his egotism, not to be regarded as a constant but as rapidly increasing at an enormous rate, approaching asymptotically to the limit of mental alienation."[153] This satire soon found its way into print in the *Observatory* in England, carrying See's reputation around the world. A year later the Council of the new Astronomical and Astrophysical Society of America, the original name of the present American Astronomical Society, rejected See's nomination for membership, an unprecedented action.[154] Nevertheless, he found a position as professor of astronomy in the Navy, stationed first at the Naval Observatory, then at Annapolis, and finally at the Mare Island Navy Yard, and lived on in increasing scientific isolation until 1962, long after Keeler, Hale, Chandler, Lowell, Campbell, Holden and a host of his other astronomical adversaries had passed from the scene.[155]

As this struggle worked itself out, construction continued at Yerkes Observatory. The progress was maddeningly slow, but in early 1896 Hale believed that the observatory could be completed and in operation by the fall of that year, if only President Harper would push the contractors into action. Hale immediately began thinking of plans for a spectacular dedication ceremony. He visualized a week-long astronomical "congress", with scientific guests from all over the world, special trains to Williams

Bay, and a banquet in Chicago, all to be hosted by Yerkes, the donor of the telescope. Hale asked Huggins, the pioneer English spectroscopist, to be principal speaker at the dedication, and when he learned that Keeler was going to the British Association meeting in Liverpool that September, he tried to get his friend to agree to bring the old man back with him.[156] By June, however, when Hale visited Keeler in Allegheny, it was clear that Yerkes Observatory would not be completed in time for a dedication in the fall. The young Chicagoan had gone east to attend his sister Martha's graduation at Smith College, and then had swung down to Pittsburgh on his way home. While he was staying with Keeler, he completed arrangements for borrowing a heliostat mirror from Allegheny Observatory to use for solar research at Yerkes. Hale stayed in Chicago most of the summer, however, for his wife was expecting, and in August they had their first child, a daughter named Margaret. Soon Huggins was joking that she and the Keelers' infant son Henry would some day be married and become the great astronomers of the twentieth century.[157]

As soon as it was certain the Observatory would not be completed until 1897 Hale decided that the dedication must be held in August of that year. The British Association would be meeting in Toronto that month, and he hoped to persuade a number of the prominent English astronomers and physicists who would attend it to come to the Yerkes dedication.[158] By the end of November Hale had moved to Williams Bay, and so had Barnard, who had left Lick to join the Yerkes faculty. He liked Hale much better than Holden, even though the 40-inch telescope was not completed, the work on it was proceeding rather slowly, and the bitter cold of wintry Wisconsin was "rather hard on us."[159]

The move from Chicago to Williams Bay required so much of Hale's attention that he had little time left for editorial matters, and none at all for writing papers. This caused an immediate publication crisis, as he wrote Keeler:

> I beseech you to send me anything or everything in your possession that could properly or improperly be printed in the January number of the Journal. I have nothing in hand but two articles, one of them very short, and the situation is far from a pleasant one ... Articles, notes, reviews, translations or anything you can manage to get hold of will be most gratefully accepted ... I am trying to get something out of Wadsworth, but he is so busy fitting up the shop that he maintains it will take him several months to get back to his normal condition as a writer of articles. Barnard is almost equally occupied with his new house, which has given him a world of trouble, and although I tried to get an article out of him today, I do not think it will be forthcoming for some weeks or perhaps months.

Keeler however could offer little help. His assistant, Henry Harrer, had been translating a German article, but it was a hard one, and he had made so many mistakes that Keeler decided to do it himself, starting afresh. He was leaving for Washington the next day, however, to spend Christmas with his mother and sister (his wife and children were in Louisiana visiting her relations on the old plantation) so he could not start until after the New Year. He piously hoped that Crew or Frost would send Hale some material.[160] The subscribers got a thin January issue of the *Astrophysical Journal* that year.

In January 1897 Hale visited Keeler in Allegheny again. He came after his lecture at Baltimore, in the Peabody Institute at Johns Hopkins. This was a weekly series of lectures, by recognized experts in a particular subject, and that winter the topic was "Recent Advances in Astronomical and Physical Sciences." The speakers included Young, Michelson, Rowland, Ames, Keeler and Hale, among others. Hale was originally asked by President Daniel Coit Gilman to speak on the planets, but changed his subject to the stars; Keeler was supposed to lecture on the moon, but undoubtedly filled the gap his friend had left with some results on Saturn's rings and the rotation of Jupiter as well. He certainly followed Gilman's admonition "Secure good slides!", the first rule of every successful astronomical lecture.[161]

Keeler, as president of the Academy of Sciences of Pittsburgh, arranged for Hale to give a popular lecture there immediately after his more formal effort at Baltimore. In spite of the name of the society the audience would be relatively unsophisticated scientifically, Keeler advised his friend, and he would be well advised to pitch his lecture accordingly. But it would pay his expenses, and enable the two of them to talk over many things. Cora and the children had returned from their trip to the South, so Hale saw the whole family. The lecture, the visit, and the discussions the two friends had were all great successes.[162]

No doubt they talked about Yerkes Observatory and the forthcoming dedication. Soon after Hale returned, he at last got his 12-inch telescope from Kenwood installed in one of the smaller domes at Williams Bay, and was able to observe the sun with it there for the first time on January 20, 1897. The dome was frozen solid, but Hale put the whole staff to chipping the ice off it, and they finally got it open. According to Hale the seeing was remarkably good for such a cold day.[163]

Hale was back in Allegheny again two months later. This time Keeler had arranged for the young director of Yerkes Observatory to receive an honorary Sc.D. degree from the Western University of Pennsylvania, of which Allegheny Observatory formed a part. It was Hale's first advanced degree, and was important to him, for many strangers would automatically address a scientist as "Doctor"; now he could silently and gracefully accept the title, while previously he would have been expected to deny it. Hale made a quick visit to receive the degree and stayed with the Keelers at their house again. He and Keeler, who presented him at the ceremony, were decked out in full academic regalia; Hale said afterward that he was only sorry that the universities' authorities did not "do the thing up brown and confer D.D.'s on two such dignitaries as ourselves."[164]

Just two months later, toward the end of May, the 40-inch telescope was at last completed. The lens was delivered at Yerkes Observatory in the custody of Clark, its maker, and was mounted in the telescope the next day. That night was cloudy, but the following day Harper and a delegation of trustees, faculty members and friends came out from Chicago to see the first light through the telescope. It was partly clear that evening, and they all marveled at the huge, bright image of Jupiter. After the guests had left, Hale and Barnard observed several other objects, giving the lens a more critical test. They found it fully up to specifications. Soon Hale was reporting to Harper and Yerkes that Barnard had discovered a previously unknown, very faint star near Vega with the 40-inch, which, although it had no astronomical significance, demonstrated "most conclusively that the Yerkes telescope is more powerful than the Lick telescope."[165]

However, just a few nights later, disaster struck. The financier Yerkes, fearful that some of his enemies or competitors would sabotage the telescope, had given strict orders that no one except the astronomers should ever be allowed anywhere near the telescope. He feared that "many persons – some of them high in the social scale ... would even be pleased to see an accident happen to the telescope." It must be guarded "against either accident or malicious acts of anyone who might feel disposed to injure it," he demanded, and every precaution should be taken to keep visitors away from it.[166]

His orders were fruitless, although it was a design failure rather than a saboteur that caused the crash. On the morning of May 29, just a few hours after Barnard and Ellerman left the dome at the end of the night's work, the giant rising wooden floor broke loose from some of the cables that supported it, came careening down, and smashed to pieces. The cables had not been securely fastened to the floor, and the weakened joints parted. As they broke, half the floor, unsupported, fell to the ground. If the accident had happened a few hours earlier, Barnard and Ellerman would probably have been killed; if it had happened a few nights earlier, Harper and half the University of Chicago high command might have died. As it was, the telescope was almost useless for months, while the floor was being rebuilt, this time with the cables securely fastened. No damage had occurred to the telescope or lens.[167]

As the work progressed on the floor, Hale began planning the dedication ceremonies. At first he had hoped to hold this event just after the British Association meeting in Toronto, but Harper consulted Yerkes and found that he would be on his usual summer trip to Europe at that time. The financier declared that he would be back in Chicago in October, so Hale and Harper chose the first of that month for the dedication, but they either misunderstood Yerkes or he ignored his promise, for at the last moment the ceremonies had to be postponed to the week of October 18.[168] The high point of the dedication was to be a major invited address on astrophysics, to be given by an outstanding practitioner of the subject. Hale's first choice for the speaker had been Huggins, and nearly two years before the dedication actually took place he had agreed to give the address. He was already in his seventies, though, and in failing health. As the date grew nearer, the old pioneer decided he was too frail for the trip and withdrew, in spite of Hale's pleas that he reconsider.[169] With Huggins out of the running, Harper and Hale decided to invite the famous English physicist, Lord Kelvin, to give the dedication address. He was to be in Toronto, but he replied that he could not stay over long enough for the ceremonies.[170]

Hale's next choice was Vogel, the director of the Potsdam Observatory and one of Campbell's most persistent critics, but he also declined. Then Hale shifted his attention from Europe to America, and asked Young, who had spoken at the dedication of his Kenwood Observatory six years before, to do the same at Yerkes. The Princeton professor was in his sixties and had been suffering from sciatica and diabetes; he declined the invitation also. By now the situation was getting desperate. The next candidate was Langley, Keeler's old employer and the inventor of the bolometer, but he turned out to be on a trip to Russia. He could not get back in time for the ceremony, and in addition felt out of touch with modern astrophysical research, so he declined also.[171]

At this point Hale played his last card and asked Keeler to give the principal address. He told Keeler that five eminent astrophysicists had already turned down the same invitation, but he had a convincing reason for each refusal, and insisted that actually Keeler was the best of the Americans. Young and Langley were no longer active researchers but had only been invited because of their past records, Hale maintained. He knew he could count on Keeler, and appealed to his friendship, writing, "I positively forbid you to refuse on pain of instant death." Actually, as

Keeler well knew, there was no reason for him to refuse, but he telegraphed Hale to try to get Pickering first, adding that if the Harvard director also refused, he himself would reluctantly accept.[172]

Very probably Hale then asked Pickering, but diplomatically let him know that Keeler would accept if he declined, and Pickering, equally diplomatically, did decline. At any rate, three days later Keeler firmly agreed to give the talk, and began thinking about what he was going to say. He chose as his title "The Importance of Astrophysical Research, and the Relation of Astrophysics to Other Physical Sciences," and asked Hale to be sure to invite Charles H. Rockwell, the wealthy New York amateur astronomer who had financed his education at Johns Hopkins, to the dedication. Brashear was ecstatic that Keeler had been chosen to deliver the speech, and assured Hale that the Allegheny director had improved very greatly in his public speaking, and was no longer "backward," as the peppery instrument maker evidently thought he had been when he was younger.[173]

Hale wanted to give Keeler a chance to use the new spectrograph on the 40-inch while he was at Yerkes, to try to measure the rotational velocities of the planets Venus and Uranus. Both of them had proved beyond the reach of the Allegheny 13-inch and the various high-dispersion instruments that Keeler and Brashear had improvised for it, but the 40-inch was a much larger telescope. Hale planned to observe with Keeler and share in the credit for whatever discoveries they made. He was providing the instruments; Keeler was providing the skill and experience. Keeler was not very hopeful of success with Venus, which has a similar size to the earth and Mars; presumably he expected its rotational velocity to be at most a fraction of a km/s, too small to detect. Uranus on the other hand is a giant planet like Jupiter and Saturn, and Keeler thought they had a real chance to measure its rotation. However after the rising floor was repaired and the new spectrograph had been adjusted, it turned out that the huge Yerkes Observatory dome was not quite round, so the wheels on which it rotated would continually jump off their rails. Further time-consuming repairs were necessary, and just as they were being completed, Keeler learned that Hale did not yet have a photographic corrector lens to use with the 40-inch for spectroscopy. It was obvious that they would not be able to make the observations that fall. Keeler suggested postponing them until the following summer, when he would bring along his family and combine astronomy with a summer vacation on Lake Geneva. The impatient Hale was disappointed, but in fact it was winter before the 40-inch was ready for serious spectroscopic work, and neither planet was accessible then.[174] By the next summer the situation had changed completely, as we shall see, and they never did get a chance to measure Uranus' rotational velocity.

Finally, the third week in October 1897 arrived. In Chicago on Sunday the popular Reverend Jenkin Lloyd Jones preached to his fashionable congregation at All Souls Unitarian Church, on the Yerkes dedication. The title of his sermon was "Astronomy – Its Struggles and Triumphs," and he was quoted as saying that astronomical research "should inspire us with new zeal for the quest, for such study releases us from the trammels of matter and carries us into the fellowship of the spirit ... The shackles of superstition fall off and the soul, unfettered, revels in the boundless universe of truth, beauty and love."[175]

The conferences connected with the dedication began at the observatory on Monday. The imposing buff-colored brick building stood on a large, level height overlooking Lake Geneva. A recent construction site, the ground was nearly bare, except for a few small, recently planted trees, and one of the three domes still had not been put on the building, giving it an unfinished look. The 40-inch and Hale's 12-inch were in complete working order in the other two domes. In the surrounding woods the maples were in brilliant autumn color, but the skies were gray and overcast. As the visiting astronomers arrived in Williams Bay, Hale met each train at the station with a string of horse-drawn hacks and buses, and accompanied them directly to the observatory. There were no hotels in the little town, and although the most important guests were put up in the faculty members' houses, the Pickerings and Keeler with Hale, and Simon Newcomb with Barnard, the rest had to stay in nearby farmhouses or at the observatory itself. Many of them slept in cots that were set up at night in the library, then taken down before the scientific conferences began the next day.[176]

No more than thirty astronomers were present for the first day, but the only scheduled events were the now regular annual meeting of the board of editors of the *Astrophysical Journal,* and informal scientific discussions. Nearly fifty more visitors arrived that evening, and had their first chance to look through the 40-inch, for the skies had temporarily cleared. Barnard and Burnham, who had come up from Chicago for the occasion, showed them close double stars, well-resolved by the giant telescope. For the next two days there was a busy round of conferences, with the astronomers presenting papers describing their recent research results. Keeler gave the most important paper Tuesday morning, telling of his photographic studies of the spectra of the "third-type" (cool) stars in the yellow spectral region. He showed enlarged slides of his spectrograms of many of these stars, arranged in order of spectral features, from α Bootis

Fig. 16. Group picture at the Scientific Conferences held at the dedication of Yerkes Observatory. Keeler is standing in the front row, to the right of the center of the picture, wearing a derby. Next to him is Edward C. Pickering wearing a soft, light hat, then (a little in front) Carl Runge. Henry Crew is at the far right, wearing a derby and a light-coloured overcoat, with his hands in its pockets. Charles H. Rockwell, Keeler's patron, is in the front, at the left, also with his hands in his pockets. Just behind Rockwell, at the far left of the picture, wearing a derby, is E. E. Barnard. John A. Brashear, wearing a light hat and with a beard, is near the middle of the second row from the rear, and Hale, wearing a soft, light hat, is at the right end of this same row (1897). (Reproduced by kind permission of Yerkes Observatory.)

(Aldebaran) to α Herculis, and explained how the various lines and bands strengthened or weakened along the sequence.[177] He and the other astrophysicists were searching for the meaning of these empirically discussed orderings, but too little was known of atomic physics, and success came only years later. The other important paper that day was by Runge, who had come from Hannover in Germany to visit Lick Observatory and to take part in the Yerkes dedication.[178] He spoke on his identification of oxygen in the solar spectrum; an identification that turned out later to be incorrect.

That afternoon Hale led off with an address of welcome, coupled with a description of the observatory, its instruments, and especially the 40-inch refractor. He described its light-gathering power and resolution in scientific terms to the visiting astronomers, but the *Chicago Tribune* translated his talk into terms its readers could understand by saying that the Yerkes telescope was one-fifth more powerful than the only other telescope with which it could be compared, the Lick 36-inch. The *Tribune* reporter had evidently been fed some background material by Hale or Burnham, for he added that the location of Yerkes Observatory was better than the Lick site on Mount Hamilton. This odd conclusion, attributed to "the opinion of practical men connected with both locations," was based on the statement that the seeing in the daytime (when solar work is done) was better at Yerkes, and that even if the nighttime seeing were on the average better at Lick, the really good nights at Yerkes were as good as those at Mount Hamilton. The first part of this statement is true, but the second is decidedly questionable, and the fact that there are many more clear nights at Lick than in southern Wisconsin was completely omitted. It was cloudy that night, so the visitors had to be content with a tour of the shops in place of the hoped-for observing session with Barnard, who had planned to show several nebulae, clusters and variable stars with the 40-inch.[179]

Wednesday was a full day of papers, the most interesting being Barnard's, on his work of photographing nebulae, and Pickering's, on the Harvard research program. Yerkes himself arrived on the train that evening, and Hale met him at the station and took him straight to the observatory to meet the visitors. Then after supper with Hale, Yerkes came back to observe stars with "his telescope," but it was cloudy again. Nevertheless, the big, florid magnate was pleased, and told the reporters, gesturing to Hale, that "the management could not be improved upon." He also predicted that before long a few comets and new stars would be discovered through "that tunnel," indicating the 40-inch refractor. He and Hale went into the library and listened for a few minutes to part of

Barnard's talk, illustrated with slides of his photographs, that was the substitute for the observing session that evening. Then Yerkes left for the night, no doubt spent as the guest of one of the trustees at a lakeshore mansion.

Thursday was the high point of the week's exercises. Two special trains, provided by the Northwestern Railroad, left Chicago early that morning, bringing about seven hundred trustees, faculty members, and guests to Williams Bay. The crowd was so large that several lake steamers were engaged to take most of them from the Williams Bay station, close to the shore, to the observatory pier. Only those too old or infirm to climb the steep hill from the landing to the observatory were conveyed directly to the door by horse-drawn buses. Entering the building, the spectators went straight to the 40-inch dome, where the floor was in its lowest position, down on concrete blocks. Even so, as they seated themselves on the wooden folding chairs that covered the newly rebuilt floor, some of them must have wondered just how strong it really was. The day was cold and gloomy, and the guests were advised to keep their hats and coats on in the unheated dome.

Soon after noon, the academic procession entered, everyone wearing caps and gowns, in the pageantry that President Harper loved. First came the faculty members from Chicago who made up the University Council and the University Senate. Following them, in pairs, were Hale and Newcomb, who was to be the speaker at the convocation in Chicago the following day, then Hulbert, by now dean of the Divinity School at Chicago and the Reverend James D. Butler of Madison, Wisconsin, the two chaplains for the occasion, then Ryerson, the chairman of the Board of Trustees and Keeler, and bringing up the rear, Harper and Yerkes. They all marched to a temporary rostrum that had been erected at one end of the floor and seated themselves.

The ceremonies opened with a prayer by Hulbert, and then the Spiering String Quartet, a Chicago group which had volunteered its services for the occasion, played a piece by Tchaikovsky. The acoustics in the cavernous dome left a good deal to be desired, but at least the audience could hear well enough to know when the selection was over, and applaud tumultuously. Then Keeler mounted the podium and delivered his address on astrophysics. He defined it as a subject closely allied to astronomy, chemistry, and physics, drawing material that could be profitably used from any science, and concerned with the nature of the heavenly bodies – what they are, rather than where they are, which had been the task of the old astronomy. Mostly it depended upon the analysis of light. Although astrophysics had little practical, money-making value, its subject, like that of astronomy, was nothing less than understanding the universe. Thus it was of the deepest interest to scientists and to the general population alike. Astrophysical research was difficult and demanding, requiring the highest mental discipline, training, and insight. It also required complicated apparatus. Some traditionalists might look back with nostalgia to the good old days when the human eye was the only instrument an astronomer needed at the telescope, but those days were gone forever.

Keeler went on to describe the importance of photography and of spectroscopy in astrophysics, and to sketch out some of the problems they might be expected to solve. He included the motions of the stars, the abundances of the elements, the physical conditions in the stars, the detailed structure of the sun, the surface of the moon, the atmospheres and surfaces of the planets, the intrinsic differences between stars, their evolutionary histories, the properties of binary stars and their orbits, and numerous other problems. It was a very modern, forward-looking speech. The topics Keeler described were in fact the main problems of astrophysics for the next fifty years, and more than one of them is still under active investigation today. He closed with graceful tributes to the munificence of Yerkes and to the skill of the builders of the observatory. Its staff, he said, were masters of both astronomical and astrophysical research, and their work would throw light on the dark places in nature and thus advance true scientific progress.[180] He sat down to more tumultuous applause, but as he had spoken for slightly over an hour, many of the non-astronomical guests were probably clapping more to warm up than to express their appreciation of the fine points of his paper.

Next the string group played the Largo from the Quartet in F Major by Dvořák, always a favorite composer in Chicago. Then Yerkes was introduced, and drew the most tumultuous applause of all. According to the *Tribune* reporter, "the street-car magnate blushed like a bashful maiden." To an appreciative audience like this, Yerkes could be surprisingly modest. His short speech paid tribute to Harper, the university, and the men who had built the telescope, but did not mention his own boyhood dreams or his adult fantasies. He gave a capsule history of five thousand years of astronomy, no doubt written for him by Hale, beginning with the Greeks and culminating in the spectroscope. Yerkes professed himself fully satisfied with the telescope, observatory and staff, and with a flourish turned the structure over to the University of Chicago. Ryerson expressed the gratitude of the Board of Trustees, and Harper then delivered an address describing the building of the observatory, and closed by thanking Yerkes fulsomely on behalf of the faculty and students.[181]

The final benediction was given by Butler, an eighty-three-year-old retired minister who was substituting for President Charles Kendall Adams of the University of Wisconsin, who was ill. The small, wiry Butler, a well-known Madison character who signed his letters, "Super-octogenarianically yours", in closing his prayer spoke directly to Hale, saying, "You are happier than Plato. When Plato was asked 'Where will your ideal heaven be realized?' he looked upward and answered, 'In Heaven.' But your ideal you behold becoming actual here and now in the midst of your best years."[182]

After this all the rest was anticlimax. A gigantic luncheon was served to the guests in the halls of the observatory. Then they walked through the building and inspected the telescopes, instruments and shops until the trains left for Chicago at 4 p.m. The whole day had been a great success.[183]

Friday was given over to the dedication at the University of Chicago campus. In the morning Michelson demonstrated some of his experimental work in light to the visiting scientists at Ryerson Laboratory, the physics building. Then they had lunch, hosted by Harper, and in the afternoon heard an address by Newcomb, the "rather grim dean of American astronomers."[184] Sixty-three years old, he had been in charge of the Nautical Almanac Office from 1877 until his retirement a few months before the dedication, and he represented the apotheosis of the old-time astronomy. Newcomb was a master of gravitational theory, and had made many contributions to understanding and predicting in detail the motions of the planets in their orbits. His article on "Abstract Science in America," published in the *North American Review* in 1876, was one of the first appeals for support for pure science, as opposed to applied science, in the United States, and according to one admiring biographer he had done more than any other person since Benjamin Franklin to make American science respected and honored abroad. In addition to his duties at the Nautical Almanac Office, Newcomb had been a part-time lecturer on astronomy and mathematics at Johns Hopkins, where he had taught Keeler, and he had also written and lectured on economics. He had been the principal astronomical adviser to the Lick Trust when they were planning and building the 36-inch refractor, although his main expertise was in theoretical, rather than observational work. Over the years he had given many invited lectures at commencements, dedications, and scientific congresses.[185]

When Newcomb rose and launched into his address at the Kent Lecture

Theater in the chemistry building, the audience saw a very distinguished, determined-looking, patriarchal figure. He walked to the rostrum with a slight limp, but spoke slowly and forcefully. His address, "Aspects of American Astronomy," was a graceful, somewhat florid, but backward-looking speech. He paid many pleasant compliments to Chicago, and even cracked a joke at the expense of the reporters, who had "made known everything that occurred, and, in an emergency, requiring a heroic measure, what did not occur." The conferences on every aspect of astronomy had been inspiring, and the younger generation of astronomers would "reap the reward which nature always bestows upon those who seek her acquaintance from unselfish motives." Newcomb then gave his version of the history of astronomy in the United States, beginning "in the middle of the last century" (just before the American Revolution). He tied it adroitly to the history of "the great metropolis of the West," Chicago, "just pride of its people and the wonder of the world." Nearly all the astronomical history he related had to do with celestial mechanics and astronomy of position; he said not a word about Langley or Young, and referred to astrophysics only very briefly, mentioning spectra taken at Harvard and "discoveries too varied and numerous to be even mentioned [made] at Lick." Newcomb paid very graceful compliments to Burnham, Barnard and Hale, without giving their names, but describing them by their deeds in astronomy and emphasizing their fame in the scientific world. He ended up with some extremely vague speculations about the future of astronomy – "it is important to us to keep in touch with the traditions of our race" – and closed with the words, "The public spirit of which this city is the focus has made the desert bloom as the rose, and benefited humanity by the diffusion of the material products of the earth. Should you ask me how it is in the future to use its influence for the benefit of humanity at large, I would say, look at the work now going on in these precincts, and study its spirit."[186] At the time, delivered in Newcomb's ringing orator's voice, his address probably made far more of an impression than Keeler's had the day before, but to us today it seems shallow and unprophetic.

That evening Yerkes gave a banquet for the visiting scientists and the university officials at Kinsley's, a fashionable downtown restaurant. Over two hundred guests were present. A long series of toasts were proposed by the locals and the visitors, and at the end Yerkes made a speech expressing his gratitude to all concerned. He said that the visiting astronomers should regard Yerkes Observatory as a sister observatory, not a rival. He hoped that all rivalry would cease, and that the 40-inch could be put at the use of others. On this high note the banquet and the dedication ceremonies ended. A few details remained to be settled, as when Hale had to insist to the University comptroller that the bill submitted by Harley Williams of Williams Bay should be paid. "Mr. Williams," Hale wrote, "is a real estate agent and notary public, it is true, but he adds to these public functions that of teamster, coal purveyor, etc., etc. The teaming was done by him."

Keeler's old friend Rockwell, who had been very pleased with all he had seen and heard at the dedication, went back to Tarrytown on the same train with Yerkes, who was returning to his new home in New York City. Keeler himself spent a few more days with Hale at Williams Bay before heading back to Allegheny.[187]

Keeler's address at the Yerkes dedication marked him, as Hale said, as the leading working astrophysicist in the United States. He had been asked to give the main address on astrophysics at the ceremonies inaugurating the new largest refractor in the world, and had succeeded admirably. His research was known and respected by astronomers everywhere. At forty, a long career and many high honors seemed to lie before him.

Hale had been tireless in promoting the Yerkes dedication. He had tried to persuade as many research astronomers and physicists to come as he could. He was particularly anxious to have a representative from faraway Lick Observatory at the ceremonies, and had done his best to get Campbell to attend. However, the active young spectroscopist was preparing to leave on a long voyage from San Francisco to observe the solar eclipse on January 22, 1898 in India, and when the date of the dedication slipped back to the third week in October, it was too late for him. Hale also urged Charles D. Perrine, the Lick secretary who had become a hard-working researcher and had discovered several comets, to come to Williams Bay for the dedication, but he could not make it either. The one person from Mount Hamilton who did arrive for the conferences, but only because they had been postponed, was Allen L. Colton, former photographer and assistant astronomer at Lick.[188]

This shy, quiet, self-effacing young man must have provided one of the most titillating topics of conversation among the assembled astronomers, for he had just struck the final blow that brought down the mighty Edward S. Holden, and had been the means of forcing him out of the directorship of Lick Observatory that he had held for the past nine years.

8

The best man for the place

"The Devil", "the Czar", "an unmitigated blackguard", "the Dictator", "Prince Holden", "the great I am", "the Great Mahatma", "that humbug", "that contemptible brute", "that immoral and incompetent man", "our former colleague and fake" – these are some of the terms in which men who worked under Edward S. Holden, or were close confidantes of others who did, referred to him in their private letters.[1] Yet, for nine eventful years he retained the confidence of several of the most influential American astronomers, including Simon Newcomb, and of a majority of the regents of the University of California.[2] With their support, Holden continued as director of Lick Observatory through the summer of 1897. Then came the fall.

Holden was a graduate of West Point and a convinced believer in the "Duty-Honor-Country" system inculcated at the Military Academy. After a few years in the Army, he was appointed a professor at the Naval Observatory in Washington in 1873. There he worked closely with Newcomb, the most influential American astronomer of the time.

Although Holden was highly intelligent and a hard worker, full of interest in astronomy, he was not a particularly good research scientist. Most of his papers had very little importance. Nevertheless, on the recommendation of Newcomb he was chosen by the the Lick trustees as the prospective director of their observatory in its very early planning stages. In 1881 Holden was appointed director of the Washburn Observatory of the University of Wisconsin in Madison.[3] He notified Captain Richard S. Floyd, who was by then the president of the Lick Trust, that he had taken the position in Madison, but assured him that he regarded it only as a preparation for his future work at Lick Observatory, at which construction had barely started.[4] During his five years at Madison, Holden continued as a very active adviser of the Lick Trust, exchanging literally hundreds of letters with Floyd.

Four years later, when the University of California needed a president, Holden was named to the post, as described in an earlier chapter. He accepted it with the understanding that he would give it up when Lick Observatory was completed, and that he would then be appointed as its director.

While Holden sat in the president's office in Berkeley, Captain Floyd worked on the summit of Mount Hamilton, pushing the observatory through to completion. Holden persuaded him that he needed a knowledgeable assistant, familiar with astronomy, to help at the site, and James E. Keeler was hired, as we have seen. He worked directly for Floyd, but acted, with the captain's approval, as Holden's eyes and ears on the mountain, sending him frequent written reports on the progress of the work.[5]

Holden sent hundreds of suggestions, ideas, plans, schemes and questions to the captain. As the observatory neared completion, Holden's relationships with Floyd and the rest of the trustees deteriorated badly. Just before the University of California took over the observatory, Holden submitted a bill to the Lick Trust for $6,000 for his personal services since May 1876, when the third Board of Trustees, with Floyd as its president, had been appointed. In the nine-page, legalistic document, Holden stated in considerable detail the advice he had tendered, the letters he had written, the instruments he had ordered, the specifications he had drafted, the plans he had supervised, and the trips he had taken on behalf of the trust.[6] He had in fact done nearly all the deeds he specified, but he completely ignored the fact that he had begun, and until he left for Madison, done all the work under the supervision of Newcomb, whose name he never mentioned in the bill. Floyd had always considered Newcomb as the trust's chief astronomical adviser, and Holden merely as his protégé and assistant.[7] The trustees never considered paying his claim, and were prepared to fight it in court if he pressed it. They already resented his continued assertions that the plans for Lick Observatory had originated almost entirely with him; the bill was the last straw, and they never trusted him again.[8]

As director of Lick Observatory Holden worked very hard, but he soon managed to antagonize nearly all his staff members. He became locked in bitter quarrels with S.W. Burnham, E.E. Barnard, and after Keeler left for Allegheny Observatory in 1891, with Henry Crew, appointed that year as the first Lick staff member with an earned Ph.D. degree. Armin O. Leuschner, who had been the first graduate student at Lick, but went to Berkeley as a part-time instructor in mathematics in the fall of 1890, became another bitter enemy of Holden. All of them considered him inadequate as a scientist, and arrogant and insensitive in his dealings with them as human beings. Burnham and Crew both resigned from the Lick staff in the summer of 1892, with public blasts, widely reported in the newspapers, at Holden's regime.[9]

Keeler agreed with W.W. Campbell, who came to Holden's defense. "Your position in L[ick] O[bservatory] matters is a just one", the Allegheny director wrote his young friend. "Burnham is a poor ship-of-the-line, but a splendid independent cruiser. I had a good time with him & Hale in Chicago lately. We must let other people fight their own battles, – to a large extent. I intend to send a note to Astr[onomy] and Ast[ro-] Phy[sics] explaining that I did not leave the L[ick] O[bservatory] on account of unfair treatment."[10] This letter exactly sums up Keeler's attitude throughout his life. He was tolerant, amused and unwilling to take sides. He always sought to put the best construction he could on anyone's activities, to emphasize the positive, and never to criticize unless absolutely necessary. It was perhaps not the most courageous philosophy in the world, but it had taken him far, from Florida to Johns Hopkins, Mount Whitney, Allegheny, Germany, Lick, and now Allegheny again, and it had avoided a lot of stress and strain along the way. He was never to change it.

Keeler had in fact already written to Holden, soon after an anonymous letter titled "Misery on Mount Hamilton" had appeared in the *San Francisco Examiner,* saying that he had heard nothing concerning Burnham's resignation since he had left Lick, and was sorry to see his name linked with it in the newspapers. He himself had never criticized Holden publicly, and would always say that he could not have asked for a fairer chance than the director had given him at Mount Hamilton. Comparing it with his previous experience as Langley's research assistant, he said, he could appreciate his scientific freedom at Lick at full value. Holden, who was embroiled in a struggle with Barnard at the time, immediately sent Keeler's letter to the regents to prove that the newspaper attacks were overdrawn.[11]

In his letter, Keeler had said he was willing to repeat that he had always been well treated at Mount Hamilton to anyone Holden wished. The director now wrote his former staff member a warm, friendly letter, in which he asked him to repeat the statement to Charles A. Young, whose support he badly needed. Furthermore, Holden delicately suggested, "I am not sure it is not worth saying publicly – & very briefly – but that is for you to say."[12]

Just a week after receiving this letter, and the letter from Campbell defending Holden, which had been sent from Mount Hamilton almost simultaneously, no doubt not by coincidence, Keeler sat down and composed a diplomatic statement for publication. He wrote Payne and asked him to insert it in the next issue of *Astronomy and Astro-Physics*. It read "As my name has been mentioned in various articles commenting on recent changes in the Lick Observatory staff, I desire to say that I resigned more than a year ago for private reasons which were in no way connected with the administration of observatory affairs, and that I left Mount Hamilton with the good will of the Regents and on the best of terms with the Director and all my associates." As he first wrote it, the end of the statement had simply said "on the best of terms with all my associates", but Keeler evidently realized that this might be misinterpreted as excluding Holden, and he reworded it slightly. The statement was published, exactly in the form Keeler wrote it, in the November 1892 issue of *Astronomy and Astro-Physics.*[13]

It was a far from ringing cry, but without attacking anyone, or even defending any of Holden's actions, Keeler had clearly indicated that he was not one of the director's enemies. Holden naturally greatly appreciated it. He thanked Keeler immediately, and kept up for some months a flattering stream of letters about publishing *all* of Keeler's Jupiter drawings, his popular article on the planets intended for *Century* magazine, and similar subjects.[14]

To Barnard, on the other hand, it was a betrayal, but one that was more to be understood and pitied than abhorred. "I am not so much surprised at the note from Mr. Keeler, after knowing as I do the power that was brought to bear on him when he left and since he left," he wrote his friend George Davidson.

> It is pretty hard to withstand a severe attack of flattery and adulation. This was piled on for political reasons – and no one knows how to carry a point that way better than the Director of the L[ick] O[bservatory]. I regret that Mr. Keeler has done it because I well knew his feelings and his estimate of the Director during the greater part of the time he was here, what ever may be his feelings and opinions now. I suppose the Director will make full use of the note. Well there are two men in this world who will go down to the grave without writing such a note and one is S.W. B[urnham] and the other can be no one else but E.E. B[arnard]. I have a great regard for Mr. Keeler and I cannot believe he has carefully weighed the manner in which his note will be used by one who never had any love for him and who has flattered him only for political reasons. Ah! Me. So goes the world. But they will all be even, and all rights will

be straightened out and all wrongs will be distorted and twisted and finally come out on top when we are all dead!

In this letter, as in most of his others written when he was under stress, Barnard added that he still felt "under the weather".[15]

More realistically, John A. Brashear also wrote Davidson from far-off Pennsylvania. Keeler and his wife had just been over for dinner, he said, and the Allegheny director had "split his sides laughing" when Brashear read him a parody Davidson had written about how the "blooded horses" (Keeler, Burnham, Crew, and presumably Barnard also, in the near future) all had gone, and only "the Jack[ass]" (Holden) was left in the end. But, he added, as Davidson knew, "Keeler is sort of non-Committal – Simply because he is one of those independent fellows who went right ahead & did his duty independent of his Environments."[16]

At any rate, the following May Keeler received a telegram from San Jose. It read "Regents gave you degree Doctor of Science yesterday[.] My nomination[.] Congratulations[.] Holden". He had first recommended Keeler for the honorary degree the previous year, but as nothing had happened, he had renewed the suggestion just a few months after Keeler's statement was published in *Astronomy and Astro-Physics*. Holden recommended both Keeler and Ladislaus Weinek, a Czech astronomer who had done the most actual research with the moon photographs taken at Lick Observatory, for honorary degrees. According to the director, Weinek deserved a degree as a graceful acknowledgment of his very important services to Lick Observatory, while Keeler's research would have earned him an advanced degree "a dozen times over" if he had done it as a student. They were awarded the first two Sc.D. degrees ever given by the University of California, in May 1893. Certainly no one could seriously maintain that they did not deserve them, but neither could anyone imagine that Holden would ever recommend Burnham or Barnard for a similar honor.[17]

Keeler was very pleased to receive the doctor's degree from the University of California. He knew it was valuable currency for a person in academic life. He immediately notified Chancellor William J. Holland, his superior at the Western University of Pennsylvania, and dashed off letters to President Daniel Coit Gilman at Johns Hopkins and to the secretary of the Royal Astronomical Society in London, advising them to include D.Sc. after his name in their respective directories.[18]

Barnard hung on a few years longer at Lick, constantly quarreling with Holden. Finally in 1895 he departed for Chicago and a position at Yerkes Observatory, two years before it was completed, just as he had arrived at

Fig. 17. E.E. Barnard, Keeler's friend and "anti-twin," standing at the eyepiece end of the 36-inch refractor of Lick Observatory, soon after his discovery of the fifth satellite of Jupiter (1892). (Reproduced by kind permission of the Mary Lea Shane Archives of Lick Observatory.)

San Francisco back in 1887, before Lick Observatory was ready for operation.[9]

The man who ultimately brought Holden down was William J. Hussey, whom he hired as Barnard's replacement. Within little more than a week of the date on which the regents confirmed Hussey's appointment, the mirror and other optics of the Crossley reflector were delivered on Mount Hamilton, undoubtedly linking the man and the telescope in Holden's mind.[19] The acquisition of this telescope was the short-lived high point of triumph of Holden's administration of Lick Observatory. He had learned in early 1893 that Edward Crossley, a wealthy English amateur, wanted to sell his "Three-foot [diameter] Reflector". It had been originally built by Andrew A. Common, a Newcastle engineer and astronomer, who used it for several years, and then sold it to Crossley and went on to build a new "five-foot reflector" for himself. The skies in Halifax, where Crossley lived, were notoriously cloudy for astronomical observations, and after ten years Crossley's interests had shifted from science to religion. He decided to dispose of the telescope.[20] When Holden heard this news, he wrote to ask Crossley's price. To his surprise, he learned that the North Country businessman would sell the telescope, completely equipped and with its dome, for £1,150 which, as Holden carefully noted on the telegram, was only $5,750 in American money. This figure was less than half what Crossley had paid for the telescope, only three percent of Brashear's estimate of the cost of a new refractor of the same light-gathering power, or one percent of the sum the Lick trustees had spent to build their complete observatory. Holden dashed off a letter asking Crossley to hold the telescope for Lick Observatory, and started trying to get the needed funds from wealthy Americans whom he had marked out as potential donors. Crossley was extremely receptive and not only gave Lick Observatory the right of "first refusal" (to buy the telescope at the asking price), but when he did receive an actual offer from another would-be purchaser, notified Holden and gave him time to try to match the bid. The Englishman, convinced of the advantages of the Mount Hamilton skies for astronomical research, wanted his telescope to find its new home there.[21]

However, the panic of 1893, one of the worst depressions the United States had ever suffered, had extended into 1894, and rich men were hoarding their capital, not giving it to equip observatories. Neither Andrew Carnegie nor D.O. Mills came forth with the few thousand dollars needed to buy the Crossley reflector for Lick Observatory. Holden was forced to give up his hopes.[22]

A year later, the situation unexpectedly changed. Joseph Gledhill, Crossley's house astronomer, advised Holden confidentially that the sale had not gone through, and that his master might be prepared to let Lick have the telescope free. Summoning up all his promotional skills, Holden composed a letter to Crossley, artfully designed to convince him to give the telescope to the University of California, for the good of science.[23] Evidently the spinning mills of Halifax were doing better business than the steel mills of Pittsburgh, or Crossley was more sympathetic to the research ideal than Carnegie, for Holden's plea achieved its purpose. Crossley agreed to donate the telescope, provided only that the University of California would pay the costs of disassembling the instrument and its dome and shipping them from England to Mount Hamilton, which he estimated as $1,000. This was the kind of challenge grant that the wealthy people of San Francisco found attractive, and Holden had no trouble in raising the sum from a galaxy of names that included Charles F. Crocker, William H. Crocker, Collis P. Huntington and Levi Strauss. The regents adopted a high-blown resolution thanking Crossley for his gift, and voted to name the telescope for him. In the end it cost a little more to take down and ship the instrument and dome than he had estimated, but Crossley absorbed the additional expense, and they were soon on their way to the New World.[24] The University of California paid nothing; Holden even had to go begging to the Wells Fargo Express Company and the Southern Pacific Railroad to ship the telescope and dome free from New York to San Jose. The telescope arrived in July 1895, just as Hussey was appointed to the Lick faculty. The dome came by a more roundabout route and did not arrive until the fall. Holden was already pressing to get the Crossley reflector into operation by the following spring.[25]

Holden placed Hussey in charge of erecting and using the Crossley reflector, but the new staff member was not at all interested in this telescope. It was a notoriously awkward instrument. Barnard, always contemptuous of anything associated with Holden, gibed that it was "no good" and that he would not pay $5 for it. A year later Hussey referred to the telescope as "a pile of junk". Their testimony must be discounted, but Keeler's description of all the changes he had to make still later to render the Crossley usable shows clearly that it was a very poor instrument indeed to begin with.[25]

Hussey resisted his new assignment, which he regarded as an infringement of his scientific freedom. Charles D. Perrine, the observatory secretary, Allen L. Colton, Holden's young assistant, and Campbell, by now the director's enemy, all joined in his revolt. The climax came when Colton resigned and the case went public. The regents lost confidence in Holden, and he realized that he would have to resign. He left Mount Hamilton on September 18, 1897, never to return. A few weeks later he submitted his resignation, to become effective January 1, 1898; until then he would remain on leave. John M. Schaeberle was placed in temporary charge of the observatory as acting director.[9]

As the train carrying Keeler back to his home in Allegheny rolled across northern Ohio in late October 1897, a month after Holden's departure from Mount Hamilton, he had time to reflect on his career in

Fig. 18. The Crossley reflecting telescope at Lick Observatory, as Keeler used it (1900). (Reproduced by kind permission of the Mary Lea Shane Archives of Lick Observatory.)

astronomy.[26] Just over forty years old, he was considered by his astronomical colleagues to be the outstanding active astrophysicist in America. Only a few days before he had given the main address at the dedication of the recently built observatory of the University of Chicago, summarizing the new science of astrophysics, its relationship to the other sciences, and its future as he saw it – a forecast which in the event proved to be amazingly correct. Keeler's younger friend George Ellery Hale, the director of Yerkes Observatory, counted on him as his closest collaborator in editing the *Astrophysical Journal,* the new professional journal the two of them had founded, and wanted him to come and use the new 40-inch refractor, the largest in the world, for joint observational research programs.

Yet professionally, Keeler was isolated in a little observatory with a small telescope at a poor site. He was grandly titled the director of Allegheny Observatory, but his entire staff consisted of one assistant, Henry Harrer, whom he was training to be an astronomer, much as Samuel P. Langley had trained him years before. His telescope was a 13-inch refractor, originally made for an amateur society forty years previously. With a highly efficient photographic spectrograph, built to his own design by his close friend Brashear, Keeler had been able to get important new research results on the spectra of gaseous nebulae, of the "helium stars" (which we know today to be hot stars), and of the planets, especially of the rings of Saturn. He had made a very good start on analyzing the spectra of the "third-type stars" (cool stars), in the longer-wavelength spectral region, which was especially well suited both to them and to his telescope.

All Keeler's research at Allegheny had been accomplished in spite of his small telescope, on the basis of his own superior training, insight, knowledge and skill. He was a pioneer, and the field of observational astrophysics was ripe for exploration. But the skies at Allegheny were rapidly becoming worse for astronomical work. It was never a good climate, for many nights were naturally cloudy. But since 1891, when Keeler had come to Allegheny, the iron and steel mills of that city, and of Pittsburgh across the river, had grown ever larger, belching black soft-coal smoke into the sky. Now it was a rare night indeed when the winds were in the right direction to blow the smoke away, and in moments of desperation Keeler was inclined to think that the Allegheny skies were the worst in America.[27]

Furthermore, he knew that Campbell, his former young assistant at Lick Observatory, who had taken charge of the spectroscopic program there when Keeler left, had regular access to its 36-inch refractor, equipped with a good new spectrograph, designed specifically for photographic work. Mount Hamilton was a much better site than Allegheny, and Campbell, with his sheer instrumental power, had already overtaken Keeler in getting research results. The Allegheny director realized this all too well, for he had watched the new spectrograph being built in Brashear's nearby shop, and had tested it for Campbell before it was sent to California and then afterward helped him troubleshoot it by correspondence. Keeler knew that the results the Lick astronomer could obtain with this instrument on the big telescope, with nearly eight times the light-gathering power of his Allegheny refractor, would be far superior to anything that he himself could do.[28]

Furthermore, Hale with the new 40-inch refractor and its even greater light-collecting ability, would soon be competing with Keeler as well. Previously the young Chicagoan had only been equipped for solar work at his Kenwood Observatory, which his father had built for him. Now, at Yerkes he would be starting stellar spectroscopy, and as he had no doubt told Keeler, he was to begin by assigning his assistant, Ferdinand Ellerman, to observe the very same cool stars that the Allegheny director had studied.[29] Keeler could not complain; science is an open subject and no stars are reserved for anyone. He had led the way, and the fact that Hale was following showed that the trail he had blazed was a good one. Still Keeler would have been less than human if he had not felt that he himself, with his superior training, skill and experience, could use the big telescope to better effect than Ellerman and Hale.

The Allegheny director also had every right to feel dissatisfied about his salary. He was earning $2,000 a year, the same amount he had been getting ever since he came to the Western University of Pennsylvania, only $200 more than the salary he had been making when he left Lick. Campbell was by now earning $2,400 at Mount Hamilton, with a house rent-free to boot, and Keeler knew that his own salary would have been at least that high if he had stayed. Barnard, Keeler's friend from Lick days, almost exactly his age, was making $3,000 a year at Yerkes. Newcomb, the patriarch of American astronomy and a very knowledgeable observer on money matters (he was an extremely conservative economics analyst on the side), believed $2,500 to $3,000 to be a suitable salary for an average director of a one-man college observatory. Yet Keeler, whom everyone considered the outstanding practitioner of the new field of astrophysics, was getting less than even the minimum figure. He was supposed to have the director's house, attached to Allegheny Observatory, rent-free, but it was old and dilapidated, and he considered it quite unsuitable for his family. So the observatory housekeeper, who had worked for Langley when he was at Allegheny and Keeler was his young assistant, occupied the house with her husband while the new director paid rent from his own pocket for a more appropriate residence nearby. His wife received only a very small income as an inheritance from her wealthy aunt, Cora Lyons Floyd. Keeler was not rich, like his friend Hale, and he and his wife needed every cent they could get.[30]

Keeler had hoped for better things when he came to Allegheny. With the strong encouragement of Brashear, the most active member of the Board of Trustees' Observatory Committee, the new director had drawn up plans for a larger telescope and a building to house it. The only possible sources of funds were the millionaires who had profited so greatly from the development of Pittsburgh as the great center of the coal, iron and steel industries in America. Brashear had targeted Carnegie, Henry Clay Frick and Henry Phipps as potential donors. Keeler was hopeful at first, but soon after he came to Allegheny, the economy of the nation began to slide. Naturally, the heavily industrialized "forge of the nation" at the junction of the Allegheny and Monongahela Rivers was one of the first areas to suffer. The Homestead strike at the Carnegie Steel plant, in which Frick was shot and wounded by Alexander Berkman, a union leader, was one of the early symptoms of this economic decline. It grew into the "Panic of 1893", but lasted almost until the Spanish-American War began.[31]

In such a climate, the wealthy men of Pittsburgh wanted to hang onto their money, not find new ways to give it away. Carnegie advised Brashear that he was sympathetic to the idea of a new Allegheny Observatory, but that he could not consider putting any money into it "until coke sold for $1.50 a ton." Keeler did not want the largest telescope in the world – that would be a bigger project than Pittsburgh could handle, he knew – but he hoped for a 30-inch, which would keep him competitive with Lick. Phipps pledged a sizeable contribution, and Frick promised to at least think it over.[32]

Keeler was basically optimistic about the project, but at times he grew disheartened and considered giving up the idea of a bigger telescope, and settling instead for a new lens for his 13-inch refractor. The glass in that old telescope was very yellowish, absorbing all the ultraviolet and most of the blue light that was so important for photographic spectroscopy. Keeler made a virtue out of it and worked in the yellow-red region, but he knew it restricted his research. With the railroads threatening to stop subscribing for the Allegheny time service, the problem was more one of finding the money to keep the observatory running, than of financing a new telescope that was coming increasingly to seem an impossible dream.[33] However, the city of Allegheny deeded two acres of land in Riverview Park, on a hill west of the foundries and steel mills and out of the smoke, as the site for a new observatory whenever the money could be raised.[34]

Frick backed out of the observatory project when his infant daughter died, and he and his wife decided to build a children's hospital in Pittsburgh in her memory instead. Keeler went on trying to raise funds from time to time, in a desultory way, but never with the all-out drive that paid off so well for his young friend Hale in Chicago. Keeler was a better research scientist, but a far less successful money raiser. He had a complete plan for an observatory ready and waiting, however, just in case the funds turned up.

Keeler was deeply interested in reflecting telescopes, and recognized their great potentialities for spectroscopic work. He eagerly sought information from Campbell on how the newly acquired Crossley reflector was performing at Lick (it wasn't), and although he had little personal experience with reflectors, he was convinced that the general prejudice against them was unfounded. A large reflector in a well-built mounting, rather than in a cheap, under-engineered one of the kind used for all reflecting telescopes that American astronomers had then seen, would be the best possible telescope for spectroscopy, he believed.[35] Nevertheless, the observatory Keeler had planned was based on a 30-inch refractor, not a reflector, apparently as a concession to the potential donors. In addition, the old 13-inch Keeler had been using as a research instrument was to be moved to the new site too, "forever free for the use of the people" – a favorite idea of Brashear's.[36]

He had always pushed for the new observatory, and took a more active part in the fund-raising activities than Keeler. In his business, Brashear had a wide acquaintance with the wealthy industrialists of the Pittsburgh area, but he could see no hope of getting the money needed to build a 30-inch telescope and a new observatory at the Riverview site until better times returned.[37]

Hale was well aware that Keeler was an outstanding scientist who was working far below his potentialities at Allegheny. From the moment that he could see that Yerkes Observatory was going to become a reality, "the largest and best in the world," he began scheming to get together a staff that, if not the largest, would at least be the best in the world too. Keeler, America's leading active astrophysicist, figured in his plans from the beginning. He was not only a great scientist, but a warm and personable human being who had become Hale's close friend and associate. The young Chicagoan wanted Keeler on the Yerkes staff.

Hale broached this idea for the first time in a letter from Germany, in the spring of 1894, the year he spent in Europe, never quite getting down to studying, while Keeler ran the *Astrophysical Journal*. Hale wrote that if Keeler did not succeed in getting a new large telescope at Allegheny, he wanted him to consider moving to Yerkes Observatory. "There is not an astronomer I would so much like to have as yourself, and the position – when it materializes – will be at your disposal. The salary will probably be

$3,000, the same as I expect to get then – at present, as Associate Professor, I get $500 less. You would have charge of the stellar spectroscopic work, with the 40-inch two nights a week."

Keeler replied that if conditions at Allegheny were not likely to improve, he would accept such an offer at once. "It is well enough to be Director of an observatory, but if the observatory is not in a condition to accomplish results, the honor is a rather empty one." Keeler was still hopeful that the wealthy Pittsburghers he was cultivating would put up the money to build a new observatory at Allegheny, so he did not want to come to a decision at once, but as Hale had given two years as the time it would take to complete Yerkes, he would be glad to discuss the subject further after the younger man had returned to America. Hale, who had by now moved on to Florence, wrote back that he was "very glad . . . that you might perhaps consider favorably an appointment to the Yerkes Obsy." and assured Keeler that "I would rather have you than anyone else for the position."[38]

That fall, true to his word, Hale included a plea to hire Keeler in his budget request to President William Rainey Harper of the University of Chicago. According to the director it was only a modest request, for little more than one-third the Harvard College Observatory budget, one half those of Paris or Pulkowa (in Russia), or about equal that of Lick Observatory. Keeler, whom Hale flatly called "the ablest stellar spectroscopist in this country," would come if the Allegheny expansion fell through, he predicted. Hale also wished to add Barnard and Victor Schumann, the young German astrophysicist he had met in Leipzig, to the Yerkes staff. Keeler and Barnard would not come for less than $3,000 each, the young director stated.[39]

The president was less than enthusiastic. His idea was always to get one of each kind of professor for his university. He already had an astrophysicist in Hale, he probably reasoned, why get another? The greatest economy must be practiced at the observatory, he advised the director. When he counted up the teaching faculty on the campus, plus Burnham, the double-star observer who as a court reporter in Chicago would work as a volunteer (free), Harper could only see the need for *one* more observer, "namely Barnard." As we have seen, Barnard did join the Yerkes faculty the following year. Soon after he had come, Hale sat down to prepare the next budget request. Again he emphasized the need for an enlarged staff, "worthy of the observatory." Barnard's appointment had been very favorably received by the astronomical community, he told Harper, but now it would be waiting to see who came next. The most important field of research for the 40-inch telescope would be stellar spectroscopy. Charles T. Yerkes, the donor of the observatory, had ordered (from Brashear) "the best stellar spectroscope ever built," and it only remained for the University of Chicago to appoint the right astronomer to use it. Fortunately, Hale went on, "the ablest investigator in this line in the United States" would probably be available – Keeler. He urged that he be appointed in the summer of 1896, and miraculously enough, according to Hale, there would be money left over from the 1895-96 budget to pay his $3,000 salary for the first year.

Probably it was clear to Harper, at least, that Yerkes Observatory would be nowhere near ready for operation by the summer of 1896. Hale's budget proposals were always long, involved documents, with many more requests than he could possibly expect would be granted. This year again, Harper did not give him Keeler's salary, but did let him hire Frank Wadsworth, the young instrumentation specialist, as an assistant professor. Hale was suitably grateful, but emphasized that he and his father hoped that Keeler would be appointed the following year.[40] Since Hale's father, William E. Hale, supported the observatory very substantially from his personal fortune, this "hope" had more force behind it than is at first apparent.

By budget-preparation time in 1896, Hale, Barnard, Wadsworth and Ellerman had actually moved to Williams Bay. The director again asked for the appointment of Keeler, "of whose qualifications it is unnecessary to speak." His salary was to be $3,000, the same as Barnard's, while Hale's was $2,500, Wadsworth's $2,000, and Ellerman's $900. Still Harper dragged his feet.[41]

Keeler was willing to bide his time. Unlike the impetuous Barnard, he saw no reason to charge off to a new observatory before he was certain that it was really built, and ready for operation. He could continue to work to the limit of his instrumental capabilities at Allegheny. So matters drifted on until Yerkes Observatory neared completion. Even then he need not take a job to use the big new telescope; Hale was urging him to stay at Williams Bay as his guest and try to measure the rotational velocity of Venus spectroscopically. Keeler might succeed because the long focal length of the 40-inch telescope would give an image of the planet three times larger than his little Allegheny refractor, and therefore a correspondingly wider spectrogram. Actually, the 40-inch telescope and its spectrograph were not yet ready for use, and Venus was poorly placed in its orbit for observations that autumn anyhow, so this scheme fell through. When Hale asked Keeler to make the main address at the Yerkes dedication his chief purpose was to get the best working astrophysicist in the country, after all the grand old figures of the past had declined. But he also wanted, as he admitted to Keeler, to show him off to Harper and the University of Chicago trustees.[42] They would see their prospective new faculty member at the center of the stage, and hear the assembled scientists applaud him. Keeler succeeded admirably, as we have seen, and the trustees were no doubt suitably impressed. That was the situation as Keeler returned home from the dedication.

He had been invited by President Gilman of Johns Hopkins University, his alma mater, to repeat his address on the importance of astrophysics and its relationships to the other sciences there, but after careful consideration he had politely declined. Newcomb, who had given the other main address, on the campus in Chicago, was the high priest of the older positional astronomy and had for some years been a part-time professor at Hopkins. Keeler knew that the touchy old man would resent an upstart astrophysicist being invited to speak at an important occasion on his own turf, rather than himself or some other representative of classical astronomy. The young spectroscopist preferred to avoid the situation; he begged off and returned home instead.[43]

Back at Allegheny, Keeler went on with his regular cycle of activities. The winter was always the worst season for observing, because of clouds and smoke, as well as the cold weather that was all too likely to ice up the dome. This winter was no exception, and Keeler did little work at the telescope. The previous year, he had been able to get some good spectrograms of Mars during the winter season. Keeler had taken them in an effort to solve a scientific controversy concerning the red planet, always interesting to the general public as the earth's near neighbor in the solar system, and doubly so in those days when Percival Lowell was touting it as a possible or even probable abode of life. To support life, Mars must clearly have an atmosphere something like the earth's. Whether or not it did was the question that was dividing astrophysicists at the time, and into which Keeler plunged.

Spectroscopy could settle the question, for nearly every gas has its own characteristic absorption lines or bands, closely packed groups of lines, that signal its presence. Oxygen, O_2, the gas on which animal life on earth depends for survival, has only weak bands, and those only in the infrared spectral region. It was essentially undetectable to the astrophysicists of Keeler's generation. Water vapor, H_2O, is another story. It has strong bands in the infrared region, and weak features in the yellow and red. Several of the giants of the earliest days of astronomical spectroscopy, Jules Janssen in France, William Huggins in England, Angelo Secchi in Rome, and Hermann Vogel in Germany, had reported observing these features in the spectrum of Mars, and thus had apparently proved the presence of water in its atmosphere.

However, at Lick Observatory, the irreverent young Campbell always felt it was worthwhile to check up on the statements of his scientific elders. He had much superior equipment to any of the earlier investigators – a large telescope, a modern, well-designed spectroscope, and a much better site. The latter is quite important, for Mars is observed through the earth's atmosphere, which itself contains H_2O. The bands of this molecule are therefore bound to be present in the spectrum of Mars, just as they are in the spectrum of every celestial object observed from the earth. The question is, whether the bands are stronger in the spectrum of Mars, as a result of additional H_2O along the line of sight in its atmosphere, and not simply the same strength as in other objects, because of the terrestrial H_2O alone. Mount Hamilton is a high, and, particularly in summer, very dry site. Thus there is much less terrestrial H_2O above it than above any of the observatories in Europe where the previous work had been done, all of them essentially at sea level and in normally humid climates. Finally, since central California is further south than Europe, Mars at favorable oppositions is nearly at the zenith and can be observed along a nearly vertical ray which passes through a much shorter path in the earth's atmosphere than along a slant ray close to the horizon, as in England, France or Germany.

For all these reasons Campbell came much closer to eliminating the effects of the earth's atmosphere than any of the earlier observers, as he stated no doubt with great pleasure. He carefully planned his observations, looking in the Martian spectrum at the wavelengths where the laboratory measurements showed the H_2O lines were most densely clumped, and comparing them with the clear regions between the lines. Yet, observing visually as all the others had, he saw no sign of H_2O in Mars. This, as Campbell carefully reported, did not mean there was no water vapor there, but rather that there was less water vapor than about one quarter the amount in the earth's atmosphere above Mount Hamilton on a dry summer night.[44]

To understand this upper limit, it is important to realize that the measurements are made by a comparative method. Mars, like the other planets and the moon, shines by reflected sunlight. All the observers, including Campbell, had compared the spectrum of Mars with that of the moon. In both objects they saw the spectrum of the sun, with its own absorption features that arise in its atmosphere, plus the weak H_2O features of the earth's atmosphere. What they were looking for were slight strengthenings of these features in the Martian spectrum, which, if present, would indicate the presence of H_2O in the red planet's atmosphere. Campbell did not see such strengthenings, and was forced to conclude that the previous observers had been mistaken. Furthermore, he believed he could see such strengthening of the H_2O features in the spectrum of the moon when he observed it at about 45 degrees zenith distance, that is through about 1.5 times as much H_2O in the terrestrial atmosphere as along a nearly vertical ray. Since sunlight goes through Mars' atmosphere twice on its way to the earth, once on its way down to the surface, then back up again after it is reflected, Campbell concluded that 0.25 as much H_2O on Mars as in the earth's atmosphere would have been detectable. He had not seen it. Hence this was the upper limit to the amount of water vapor there could be in the Martian atmosphere.

The older men did not take Campbell's corrections of their work lying down. Both Vogel and Huggins repeated their observations, Vogel visually and Huggins photographically, using yellow-sensitive orthochromatic plates. Both of them insisted that they saw the H_2O bands in Mars' spectrum again. Vogel even cited two witnesses, who observed with him, and said that they saw the bands too. He politely said Campbell's negative result deserved consideration, but claimed that since the telescope he himself had used to confirm his earlier observations had a shorter focal ratio than the Lick refractor, and hence produced a brighter image of Mars at the slit of the spectroscope, his positive result deserved more weight.[45] In fact, Vogel's analysis was mistaken, and he clearly did not at all understand how much drier Lick really was than Potsdam.

Even worse, from Campbell's point of view, the American Lewis E. Jewell, Henry A. Rowland's assistant at Johns Hopkins, also jumped into the fight. He had been studying the terrestrial lines in the solar spectrum, using very high-dispersion gratings and prisms. This he could do because the sun is so bright; its spectrum can be spread out much more than any star's or planet's, and still be easily observable. Jewell correctly stated that the best way to observe the H_2O lines is with very high dispersion, so they are resolved individually, and not blended with the neighboring continuous spectrum. From this he drew the incorrect conclusion that even though Mars was too faint to observe with the Johns Hopkins super-high dispersion instruments, Campbell should have used higher dispersion than he did at Lick Observatory. Furthermore, he criticized Campbell's remarks about the dryness at Mount Hamilton, using as evidence weather and climatic records and concepts that applied in Baltimore![46]

Campbell, always a fighter, countered immediately. He politely corrected Vogel's error; what really mattered, as Keeler had pointed out on many previous occasions, was the brightness in the focal plane of the spectroscope, not at the slit. Using dimensions of his own and Vogel's instruments, he proved that they were on equal terms in this respect, but because of the larger size of the Lick refractor, he himself had a wider Martian spectrum than Vogel, making it correspondingly easier for him to detect faint spectral features if they were in fact present. Huggins he simply ignored in his published paper, but he tore into Jewell with abandon, gleefully pointing out how wrong he was about the California weather pattern and the water-vapor distribution. As to the dispersion question, since Mars was not bright enough for anyone to be able to resolve the lines, fairly low dispersion, under which the individual H_2O lines blended together and produced a broadened band that was easier to see, was best, Campbell insisted. He had made the trial on the moon, and could see the H_2O lines better with low dispersion than with high. His reasoning was practical and based on experience, not theoretical and extended to faulty conclusions. Jewell shot back a paper in which he answered these objections to his own satisfaction, Campbell fired back another answering Jewell's, and it was at this point that Keeler entered the fray.[47]

At Allegheny, he had done some experiments in which he followed the sun and watched its spectrum, as it sank toward the horizon. Observing with a small, low-dispersion spectroscope, he could easily see the H_2O lines in the yellow spectral region strengthen as the sunlight penetrated progressively longer paths through the earth's atmosphere, and thus through more water vapor. At the same time he took a series of high-dispersion photographic spectrograms of the sun, no doubt using one of the spectrographs he had borrowed or built to try to measure the rotation of Venus. On the high-dispersion spectrograms, where the individual H_2O lines were partly but not completely resolved, he could not see them increase nearly so clearly. This demonstrated, according to him, that Campbell was right and that lower dispersion was better than high for this problem. It was not quite a fair test, as Keeler himself realized, because it compared visual observing of weak spectral features with photographic, in the yellow spectral region where the eye is especially sensitive and the photographic plates of that day were not, but it was the best comparison he could make at the time.[48]

The main problem with the orthochromatic plates was not their low sensitivity, for Mars is bright, but rather their rapid variation of sensitivity with wavelength, precisely in the yellow spectral region where the H_2O bands lie. However, the great advantages of the photographic method are that it removes the tremendous difference in brightness between Mars and the moon, and that it provides a permanent record that can be studied and compared objectively with other spectrograms. During the winter of 1896-97 therefore, Keeler worked very hard to get the exposure times exactly right in the region of the H_2O bands, and succeeded in obtaining several very good spectrograms of the red planet and the moon. Just as in Campbell's visual observations, he could find no differences whatever in the spectra of the two planets. From tests on the sun and moon at various altitudes, he concluded that he would have been able to detect the effect of half as much water vapor as in the earth's atmosphere, this time in the winter above Allegheny. Such an effect would have been produced by one fourth as much H_2O in the Martian atmosphere, because of the double passage of the reflected sunlight through it, as we have seen. Keeler, who had suffered through several parched summers on Mount Hamilton, ended his paper by saying he agreed with Campbell that the altitude and dryness at Lick Observatory were very important advantages for such observations.[49]

Although the European astronomers were slow to accept the fact, Campbell and Keeler were right and they had been wrong. The amount of water vapor in the atmosphere of Mars is very small. Over the intervening years many astronomers (including Campbell himself on several occasions) returned to this problem and succeeded in pushing the limit lower and lower, but never detected H_2O until 1963, using the strong bands in the near infrared spectral region to do so. Space probes since have confirmed their result, and today we know the amount of H_2O in Mars is only about one three-hundredth that on the earth, about one hundred times less than the early upper limit set by Campbell and Keeler.[50]

As president of the Pittsburgh Academy of Science and Art, Keeler was heavily involved in the early dissemination of knowledge of X-rays. This penetrating form of radiation was first discovered by Wilhelm Roentgen in Germany in November 1895. Only three months later Keeler was writing to his friend Joseph S. Ames, the physicist who worked under Rowland at Johns Hopkins, asking him to review Roentgen's experiments for the readers of the *Astrophysical Journal.* "The subject is so much out of my line that I am afraid to tackle it," Keeler confessed.[51] However, he read everything he could find in the physics and engineering journals on the new rays. It was evident that they would be a great help in medical diagnosis, particularly of broken bones or for locating foreign objects lodged in the human body. Soon the Allegheny director was the local expert on the subject, and was collecting slides for a lecture before the Pittsburgh Academy. He got an especially good set from Michael Pupin, of Columbia University, who was later to make so many important contributions to improving long-distance telephone and telegraph lines. Pupin was the originator of the idea of letting the X-rays fall on a phosphorescent screen, and then photographing the screen, rather than letting the X-rays strike the photographic plate directly. This scheme greatly increased the sensitivity of the process, and made it possible to get X-ray pictures in a minute or two, rather than a half hour. His slides ranged from the simple "Hand" or "Hand with dislocated thumb" through the gruesome "Scull (of a living person) showing a ten cent piece which was photographed through it. This photogr[aph] showed that the man did not have a bullet in his head as he supposed owing to his delirious condition produced by a shot wound in the side of the head" to the complex "Thumb of my assistant, showing buckles and trouser's buttons, also collar button and shirt stud."[52]

The talk Keeler gave, using these slides, was a great success. A large crowd attended and saw not only the projected pictures, but some actual demonstrations as well, for by that time Keeler had put together an apparatus for generating X-rays himself.[53] The lecture was entertaining and educational, but it was also a fund-raising event. The physicians of the Pittsburgh area wanted an X-ray apparatus for diagnostic purposes. The Academy of Science and Art provided the technical know-how "for the good of the public" through a committee consisting of Keeler, his close friend Reginald A. Fessenden, professor of electrical engineering at the Western University of Pennsylvania, and Brashear. The Academy of Science and Art solicited financial contributions for the apparatus, and much of it was constructed locally, the electrical coils in Brashear's shop. His wife, a large and heavy woman, broke her ankle in a fall down a flight of stairs, and was one of the first patients in the Pittsburgh area to have a bone set with the aid of X-rays.[54] Keeler and Fessenden continued for several years as the local experts on X-rays and the methods and apparatus used to generate them.[55]

The years at the end of the last century were the start of the rapidly developing era of modern physics. Less than twelve months after Roentgen's discovery of X-rays, Henri Becquerel in Paris discovered another new kind of radiation. Keeler soon heard of it and again asked Ames to review the subject for the readers of the *Astrophysical Journal*.[56] These rays were what we now know as the energetic particles and γ-rays resulting from natural radioactivity in uranium and thorium.

In addition to such practical technical matters as X-rays, Keeler was constantly called on for more theoretical advice. One question that interested him very greatly, but which he could not answer, was from a correspondent who asked whether gravitation is a force that propagates instantaneously or whether it travels at a finite (if very large) velocity, like light. This is a question closely related to the general theory of relativity, and could not be satisfactorily answered in Keeler's time. His answer was very physical, but as he said, incomplete. He urged the questioner, if he could find a satisfactory explanation from some "more competent person," to let him know it as well.

To a Unitarian minister who asked his opinion as a scientist of Swedenborg's writings, Keeler could give a more definite reply. He found the Swedish theologian's philosophy "somewhat repellant," and his scriptural interpretations "frequently arbitrary and sometimes even fantastic." Swedenborg had written that all the planets must be inhabited because they were created for that purpose, which was to Keeler "a very unsafe argument, and one that cannot be recognized as having a scientific basis." The Allegheny director very much doubted that any planets except the earth were inhabited at the present time. Unfortunately it seemed impossible to test Swedenborg's ideas by scientific research "for while he describes very minutely the supposed inhabitants he says almost nothing about the physical characteristics of the planets themselves, or of such matters as come within the province of observation."[57] Keeler was always the rationalist and skeptical observer.

A month and a half after returning to Allegheny from the Yerkes dedication, Keeler received a letter from Holden, who was now living in New York and supporting himself by his pen. He was collecting material for an article on the teaching of astronomy in American universities and colleges and asked Keeler about the courses and students at the Western University of Pennsylvania. Also, of course, he was letting his former staff member know where he was, and testing how he felt about him now. Keeler's reply was polite, cool, and distant. He furnished the requested information, including the fact that he himself gave a series of ten or twelve lectures on practical astronomy to the engineering students each year, but he did not make any advances or send any suggestions to his old chief. Holden was finished as an astronomer, and Keeler knew it. Hale, at Yerkes Observatory, who had received a similar letter from the ousted Lick director, dispatched an even colder reply than Keeler's. Neither of them wanted to be associated with a failure, whom they had not really respected even in his heyday.[58]

After Holden's dismissal and final departure, the regents of the University of California had placed John M. Schaeberle, the senior staff member, temporarily in charge of Lick Observatory. Keeler knew him well. At Mount Hamilton they had been the two principal members of the bachelor "mess," the group of single men who hired a cook and took their meals together. Schaeberle, five years older than Keeler, had been born in Germany but brought to America by his parents as a baby. He grew up in Ann Arbor, Michigan, was trained as a machinist in Chicago, and then returned to school and finally graduated from the University of Michigan in 1876. He became an assistant at its observatory, and later a faculty member. He left Michigan to become a member of the original Lick staff in 1888. At first he was in charge of the meridian-circle, or fundamental positional-measurement work, but he was never highly successful at it. Schaeberle had very good mechanical skills, and made several mirrors for reflecting telescopes, with which he experimented at Mount Hamilton. He became interested in eclipses, and developed the long focal-length camera, later called the Schaeberle camera, with which he obtained the best large-scale photographs of the corona taken in those days, beginning with the 1893 eclipse at Mina Bronces, Chile. Schaeberle had spent a fair amount of time observing with the 36-inch refractor at Lick, but had not made any very large impression as a scientist, nor carved out any field as his own.[59]

At the University of Michigan he had taught Campbell and Hussey, as undergraduates, as well as Leuschner, who was now in charge of astronomical work on the Berkeley campus. All three of them came to dislike Holden intensely, but Schaeberle oscillated back and forth right up to the end of the struggle, sometimes supporting the director, sometimes the rebels. As a research worker he had the reputation of being unsystematic, and he was highly quarrelsome if his own work was questioned. He took life easily, usually only observing one night a week and doing a good deal of reading. Schaeberle seldom left the mountain, sometimes staying on Mount Hamilton for months at a time between short visits to San Jose or San Francisco. Personally, his friends considered him rather tactless, and very absent minded.[60] Yet he was one of the main contenders to be appointed the permanent director of Lick Observatory.

At the Yerkes dedication, a month after Holden's departure, there was naturally much excited speculation about who would be chosen to succeed him. Some of the assembled astronomers supported Schaeberle, others Davidson, others Keeler, and still others George C. Comstock, who had been at Mount Hamilton with Keeler in the summer of 1886, in the Lick Trust days. Other possibilities suggested in California newspaper stories were Barnard, Burnham and Leuschner.[61]

George Davidson, the pioneer West Coast scientist whom Keeler had first met in San Francisco in 1881, when he had aided Langley's Mount Whitney expedition, was now seventy-two years old. He had been rewarded for a lifetime of service in the Coast and Geodetic Survey by being abruptly sacked two years previously, when he had reached the age of seventy. Such was not the standard procedure in those days before government pensions, but an arbitrary act of the Democratic administration of President Grover Cleveland, unsympathetic to science and determined to reduce the staff of the Survey. Davidson's friends had all rallied around him and denounced the politicians who had dismissed him, but it was to no avail and he had opened an office as a consulting engineer in San Francisco. By 1897 the Republicans had returned to power and removed from office General William W. Duffield, the head of the Survey who had dismissed Davidson. President David Starr Jordan of Stanford University and Senator George C. Perkins of California campaigned strongly to have the old man appointed to succeed Duffield, but neither he nor Holden, who had also tried hard for the job, was chosen. Instead, Henry S. Pritchett got the post. Davidson was free, and what was more, very much wanted the Lick directorship.[62]

Davidson always believed that, as the man who had convinced James Lick to provide the funds to build the observatory, he had a special relationship to it. He considered himself as good an astronomer as any other man, and now he wanted to cap his career with the Lick directorship. Neither Burnham nor Barnard had any interest in coming back to Lick and said so when Davidson delicately hinted that they were being considered as rival candidates; Burnham supported Davidson while Barnard remained neutral. Leuschner had no serious chance for the job; he was far too young and inexperienced and the only possible reason for mentioning his name was that his father-in-law, Ernst A. Denicke, was a powerful member of the Board of Regents.[63]

Keeler, on the other hand, was a highly viable candidate, although the California newspapers hardly realized it, and the regents only did so much later in the game. He was an outstanding research astronomer, recognized throughout the world for his astrophysical results on nebulae, stars, and planets. He had been the first astronomer on the staff of the Lick Trust, and after five years at Mount Hamilton had departed, not because of a fight, but for a better position and with the good will of all of his colleagues and of the university administration. Keeler had always kept up his interest in Lick Observatory and California; he had a bank account in San Jose and his wife hung on to some property there. Her cousin, Harry Floyd, who often visited them for long periods of time, had inherited Kono Tayee, her father's estate on Clear Lake, one hundred miles north of San Francisco. Keeler always identified with Lick, even years after he had left it. And, through Brashear, he continued to receive frequent reports on the rows going on at Mount Hamilton between

Holden and his staff. Keeler usually laughed at them, and avoided expressing an opinion on who was right, but toward the end he realized that
Holden would have to go, and no doubt started thinking about who would
succeed him.[64]

The Lick staff members were united in not wanting Davidson as their
next director. He was too old, too fussy, too combative, and his scientific
ideas were too old fashioned. Hussey and probably several others would
have favored Campbell, a vigorous, active, successful research worker,
over the colorless Schaeberle, but the younger man would not hear of it.
He was about to leave on a long trip to India, to observe the solar eclipse
of January 22, 1898, but before going he emphatically rejected any notion
that his name should be considered, and led the staff in signing a petition
to the regents that Schaeberle be appointed as Holden's successor. The
last thing any of them wanted was a contested selection that would open a
new round of bitterness on the mountain.[65]

Perrine, the observatory secretary, wrote to Keeler at the same time
this petition was submitted, and asked him to write to the regents in
support of Schaeberle. We shall never know what Keeler thought when
he read this letter, but we do know that he waited a long time before he
replied. In his note he said he had little doubt that Schaeberle would be
selected as director, but that he hesitated to send his own unsolicited
opinion to the regents. If they wanted it, they would ask for it, he said; if
they did not, it would do little good, and might even cause "unnecessary
enmity." Then he blandly went on to describe the pleasant time he had
had at the Yerkes dedication, and the regrets of all that no one from Lick
had been able to attend.[66] In this way he avoided expressing an opinion,
which would certainly have been used against his candidacy, without
giving offense to anyone.

After the Yerkes Observatory dedication, Hale renewed his attempts
to add Keeler to its staff. He emphasized to President Harper the strong
consensus of the astronomers who had taken part in the dedication that
the most important research that could be done with the new 40-inch
telescope was spectroscopic measurements of stellar radial velocities. No
stellar spectroscopist could do this work as well as Keeler. Furthermore,
Hale pointed out, now was the moment to strike, for times were getting
better and the wealthy men of Pittsburgh might soon loosen their purse
strings and build a new Allegheny Observatory for him; if that once got
started he would never leave. Finally, Hale advised Harper, Keeler was
being mentioned frequently as the most likely successor to Holden as the
director of Lick Observatory. "Should [Keeler] be appointed to one of
the positions named it would be impossible for us to obtain any other

observer who could be considered in the same class with him," Hale concluded.[67]

Aware that the president, who often treated him as a child rather than as an important department head, would be skeptical of his unsupported enthusiasm, Hale backed it up with glowing testimonials from some of the most important senior astronomers in the English-speaking world.[68] Edward C. Pickering, director of the Harvard College Observatory, wrote that he considered stellar spectroscopy "one of the most important [fields] to which the telescope can be applied," and Keeler as "one of the foremost representatives of this line of astronomical work, ... most admirably fitted to take charge of such a department." Huggins, the English pioneer who had discovered the gaseous nature of planetary nebulae when he first observed their emission-line spectra, called Keeler "preeminently the man to carry on successfully stellar spectroscopic work at your observatory." And Charles A. Young of Princeton University, Huggins' contemporary and one of the first American solar spectroscopists, said that it was of the utmost importance that the 40-inch be used for stellar spectroscopy, and that he knew of no one better fitted for the work than Keeler. Even the great Newcomb sent an equally strong recommendation.[69] Harper surely realized that Hale had solicited these letters, but he also knew that these famous figures would not write anything they did not themselves believe.

Just at that time, Hale received a letter from Martin Kellogg, the president of the University of California, asking him whom he would suggest to the regents as possible candidates for the Lick directorship. In a postscript Kellogg asked Hale directly "Would Prof. Keeler consent to be a candidate (to guess at the question)? And is he the best man?" Hale dashed off a letter to Keeler and informed him of Kellogg's questions. The Allegheny director modestly and no doubt disingenuously wrote back that he had no idea that his name was being seriously considered for the Lick position. He said he would do nothing to seek its directorship, but that if it were offered to him, he would probably accept it and do his best to run Lick Observatory successfully. Keeler told Hale that although he felt confident he could do a good job if he were appointed to the Yerkes staff, he was not at all sure that he would make a good director of Lick. He could see all too many difficulties there. But "one ought not to be afraid to tackle a job because it was hard," so Hale could tell Kellogg that he would probably accept if the position were offered to him. As to his qualifications, Hale should say what he believed, but what Keeler prized above all was his younger friend's "disinterested friendship."[70]

Hale really was extremely supportive, for he immediately sat down and wrote Kellogg a very strong letter recommending Keeler as "the best man for the place." He was not only an outstandingly successful scientist, Hale wrote, but also "possesses in an eminent degree the tact and executive ability which will be so necessary to the future Director of the Lick Observatory." Furthermore, Hale added, he had reason to believe that although Keeler would not seek the directorship, he would probably accept it if it were offered to him. However, he warned Kellogg, he was still trying very hard to get Keeler for the Yerkes staff, and all that stood in his way was finding the necessary funds to do so.[71]

The University of California regents were supposed to meet and choose a director on December 14, as Holden's resignation did not actually become effective until the first of the year. As the moment approached, the Lick staff members reiterated their support of Schaeberle. Davidson believed that he himself had enough support to be elected. However, at the regents' meeting the decision was postponed. No reason was stated, but it was generally believed that Phoebe A. Hearst, newly appointed as the first woman member of the Board of Regents, had requested the postponement. She was the widow of Senator George R. Hearst, and the mother of William Randolph Hearst, and she had resolved to move back to California from the East and apply her life to supporting the University of California. As she had a very considerable fortune, it was politic for the board to wait if that was what she wanted.

Very probably, however, the postponement of the decision represented the reluctance of the more progressive members of the board, including Mrs. Hearst, to limit their choice to Davidson and Schaeberle, who were considered the two viable candidates at the time. The one was too old, the other too colorless and too little-known, except among the Lick staff. Behind the scenes, Keeler's name surfaced. Arthur Rodgers, an active, interventionist regent, ascertained through a mutual friend, who telegraphed Keeler, that he was interested in the directorship. Another new name thrown into the hopper at the last moment was that of Newcomb himself, who previously had been considered unavailable. Six months before, his wife had asked Joseph LeConte, the elderly professor of geology at Berkeley, to propose him for the directorship if Holden ever resigned. Newcomb had turned down the position years before, but now that their daughters were married and he had been retired from the Nautical Almanac Office, he would take the job, she said. He needed to get out of Washington for his health, she added, and in spite of chronic problems with his leg he was enjoying his work more than ever. LeConte, who no doubt knew that President Kellogg was not happy with the thought of either Davidson or Schaeberle as the next director, chose this moment to pass her suggestion on to him. The geology professor, himself seventy-four years old, added on his own that Newcomb was "only 62" and had been making excellent addresses lately at observatory dedications.[72]

Kellogg, as soon as he received this suggestion from LeConte, wrote to Gilman at Johns Hopkins, who had been the president of the University of California years before, to ask his opinion of both Davidson and Newcomb. No doubt Kellogg himself considered "only 62" far too old for a man to start such a demanding job as director of the astronomical outpost on Mount Hamilton, but everyone knew that a reply could not get back from the East in less than two weeks from the day he sent his letter. Very probably this was the convenient pretext he used for recommending a postponement of the vote on the directorship.

As Kellogg no doubt expected, Gilman wrote that both Davidson and Newcomb were distinguished, Newcomb far more so, and that either of them would be "a creditable appointment" if he were forty or fifty years old, but he questioned whether they were young enough for the job now. There were perfectly qualified men twenty years younger, Gilman wrote, but the University of California should get professional help in identifying them. He recommended setting up a search committee of astronomers from around the country. Kellogg no doubt quickly circulated this note to the regents, and although they did not set up an expert committee, the note probably shook the confidence of some of them in Davidson, and made them resolve to take the letters from outside astronomers more seriously than they otherwise would have done.[73]

The delay gave Hale a little more time, but he knew that he would have to act quickly. President Harper still refused to earmark regular University of Chicago funds for another faculty appointment in astronomy; there were too many needs in the other departments. Hale therefore had no choice but to try to raise the money for Keeler's salary from private donors. He proposed to try to get the funds to endow a professorship from Yerkes himself, who had provided the money to build the observatory originally. Lately he had proved recalcitrant, however, and had sarcastically advised Harper to seek funds for the further needs of astronomy at the University of Chicago not from himself but from "the great and good few who represent the great and good part of our city, ... ,who uphold all the charities of our city, who are always fairly throwing away their wealth so that others may be benefited thereby, ... while such people as myself, according to the theory of your friends, are doing their best to pull it to pieces and destroy what little honor and integrity and worth is left in our community." The ebullient Hale, however, never gave up, and continued to pepper Yerkes with letters flatteringly designed to convince him that the 40-inch was the greatest telescope in the world, that stellar spectroscopy was the most important research that could be done with it, and that Keeler was the person who should be hired to do it. Hale's father, ever practical suggested that instead of trying to raise an endowment fund for Keeler's salary, which would require $60,000 (a sum that would earn five percent in those good old days), he simply ask Yerkes to promise $3,000 a year, "a mere bagatelle." Then they could concentrate on persuading the robber-baron financier to put up some really significant money to endow the whole observatory![74]

Hale acted on this suggestion, but decided to hold Yerkes in reserve and try to get the money for Keeler's salary from Catherine W. Bruce first. This woman, then eighty-two years old, was one of the great benefactors of astronomy. She had inherited a fortune from her father, George Bruce, an immigrant from Scotland who had invented a highly successful type-casting system. She lived as a recluse in her mansion on Fifth Avenue in New York and devoted her life to supporting various charities, education, and science. By January 1898 she had given over $150,000 for astronomy, and those were dollars that were worth about ten times as much as our present-day dollars. Edward C. Pickering, the Harvard College Observatory director, was her chief astronomical adviser, and over $50,000 of her money had gone to his institution, but she had given many smaller amounts to others. Lick Observatory had been the beneficiary of several $500 and $1,000 contributions from her, one of them to help bring the Crossley reflector from England to Mount Hamilton, and another for a spectrograph to use with it. Her gift of $1,000 had paid off the first year's deficit of the *Astrophysical Journal* in 1895, and had made its continued existence possible.[75] Now Hale was going to try to get the money from her to pay Keeler's salary.

One of the most difficult aspects of securing a contribution from Catherine Bruce was getting to see her. Old and feeble, she lived as an invalid, and apparently spent most of her time in bed in her room, seeing visitors only very occasionally. Her younger sister, Matilda W. Bruce, served as her doorkeeper. The previous year the famous and personable Barnard had managed to penetrate the screen, during a visit to the East during which he had given a popular lecture at Pittsburgh, under Keeler's sponsorship. But the main purpose of his trip had been to pay a call on the two sisters. He later reported to President Harper that it had made his heart ache to see how feeble Catherine Bruce was, and he had thought it would be heartless to bring up money in her presence. But he had told her all about his wonderful photographs of the Milky Way and comets, taken with a wide-field camera at Lick Observatory, and "intimated only very indirectly" of his need for a similar instrument at his new post at Yerkes. He had charmed both sisters with his warm personality and expert knowledge, and when he followed up the next summer with a letter with a specific plan and proposal, Catherine had rewarded him with a check for $7,000, enough to build the Bruce Photographic Telescope and its dome at Yerkes Observatory.[76] No doubt hearing the story of this triumph from Barnard whetted Hale's appetite, and he resolved to try himself to make Keeler's appointment a reality.

He began with a long letter to Catherine Bruce, setting forth the now familiar story of the great potential of Yerkes Observatory for frontier astrophysical research if only Keeler, "without a doubt the ablest stellar spectroscopist in the United States and without any superior in the world" could be added to its faculty. The University of Chicago trustees wanted desperately to bring Keeler to Williams Bay, but unfortunately did not have the necessary funds to do so. Ordinarily, he would not dream of imposing on her generosity, Hale wrote, but this was an extraordinary opportunity. He did not know if she would want to endow a Bruce professorship of astrophysics for Keeler – that might perhaps be more than she would wish to undertake – but if she could pledge $3,000 a year for the next five years, the university might possibly be able to offer him a permanent position. Hale was going east in January on a trip which would include lectures at Brooklyn, Vassar and Pittsburgh, and visits with Pickering and Keeler, and he asked if he might call and see her. On the train headed for New York, Hale met, apparently by chance, Yerkes, and had a long talk with him, laying the groundwork for an approach if Catherine Bruce did not come through.[77]

Unknown to the young Yerkes director, even before he left Chicago by train, the old spinster promised to provide $7,500, half the necessary amount to guarantee Keeler's salary for five years, provided that "the men and women of the great city of Chicago" would provide matching funds to assist in "securing the services of one of the most distinguished living spectroscopists." Hale thought the chances of raising the money in Chicago or through Harper were slim, and arranged to call on the Bruce sisters and present his case for the full amount. Pickering supported him with a letter of recommendation sent directly to the wealthy spinster. Catherine was too "ill" to receive any but her closest relatives, but Hale presented such a convincing case to Matilda, her channel of communication with the outside world, that after a few days hesitation the invalid consented to give the full $15,000. This time her condition was that the University of Chicago also promise to set aside an equal amount for

Keeler's salary, for future years. Hale, pleased with his success, immediately informed Harper. Since time was of the essence if a possible Lick offer were to be forestalled, the director urged the president to earmark $15,000 of the funds received from John D. Rockefeller, the founder and chief benefactor of the University of Chicago, for this purpose. Even this was impossible, Harper maintained. Rockefeller was in principle opposed to any long-term designation of his funds. Furthermore, Harper had managed to see Yerkes, and he also had refused to make any additional contributions. Would it not suffice to tell Catherine Bruce that $15,000 of Rockefeller funds were already going into Yerkes Observatory annually (it was the entire university appropriation) and would undoubtedly continue to do so? Hale had to go back to her again, through her sister, with this proposition. She accepted, subject at last only to the condition that Keeler accept the appointment. Now Hale had his money, if he could get his man.[78]

Keeler expected and wanted a permanent appointment if he were to go to Yerkes Observatory. This, Harper said, the trustees would be unwilling to grant. At first the president suggested asking Keeler to accept a five-year appointment; that is, one with no commitment from the University of Chicago beyond Catherine Bruce's $15,000. "The ablest stellar spectroscopist in the United States" was not at all willing to do so, as he told Hale when he came to Allegheny and stayed with the Keelers before and after his Pittsburgh lecture. Just a few days before, Hale had managed to see Frederick T. Gates, the former Baptist minister who was Rockefeller's personal adviser on University of Chicago matters, and explained to him that what Keeler wanted was not a "permanent" appointment for all time. What he did want was to be on the same footing as all the other faculty members, who knew they would be kept on the staff so long as they did their jobs and the University had money to pay them. Keeler did not want a special appointment, unlike anyone else's, that was only for a limited term.[79]

Even though Harper knew all this, when he finally wrote Keeler and offered him the position, as professor in astrophysics at the University of Chicago at $3,000 a year, he said that he only had money in hand to pay his salary for six years, one year beyond the period for which Catherine Bruce had provided. The president went on to promise that the university would make every effort to secure the funds to continue the appointment, and that in all probability they would come in, so that he felt personally that Keeler should not hesitate to accept. Hale considered this the equivalent of a permanent appointment and thought that the Allegheny director would probably take it if he were not offered the Lick job. Still, it was an extraordinarily grudgingly made offer, no doubt partly because of internal competition from other departments, but mostly as a result of pressure from Gates and Rockefeller to keep their options open. It was hardly an offer calculated to sweep Keeler off his feet. He at once wrote to Hale, describing the terms as Harper had stated them, and asked if he could not wait until the Lick directorship was decided before answering the president.[80]

It is impossible to know now whether Keeler would actually have joined the Yerkes Observatory staff, if nothing else had happened. Probably he did not really know himself. It is easy for an astronomer, frustrated by his equipment and observing conditions, to say he would take another job if it were offered; to actually uproot himself and his family and go is another story. The 40-inch was the largest refracting telescope in the world, but Keeler would have to give up the independence he had enjoyed for seven years at Allegheny if he moved to the University of Chicago. And, although he would be getting away from the Pittsburgh smoke, the skies were no clearer in southern Wisconsin, and the winters were even colder. Very probably during his visit Hale had shown Keeler the report that he had received from Barnard, who had been left in charge at Yerkes while the director was away. Four days before Christmas the shutter on the big dome had frozen shut, and Barnard and the night assistant had hiked down from the observatory, through the snow, to a nearby camp to borrow a ladder. They brought it back, climbed up on top of the dome in the dark, and after five hours work, starting at midnight, succeeded in getting the shutter open for one hour's observing before dawn. On Christmas Day it was clear again. Barnard and his wife had invited the other staff members and their families for dinner, but as Burnham, who was assigned the telescope that night, did not come up from Chicago, Barnard "had to leave the company to observe."[81] Hale and Keeler probably had a good chuckle together at the ever eager Barnard's expense, but the prospective new staff member must have wondered just how often the sky really was clear at Williams Bay.

Brashear, the chairman of the Allegheny Observatory Committee, soon learned of the Chicago offer, as Keeler knew he would. The astronomical instrument-shop owner wanted desperately to keep his friend at Allegheny. He asked Keeler what it would take, and the director told him he did not want to leave if he could be assured of a competitively sized telescope and a decent salary. He mentioned the figure of $3,000 that the University of Chicago was offering, and Brashear thought that the Western University of Pennsylvania could match it. But that alone was not enough. What Keeler really wanted was a bigger telescope, a 30-inch if he could get it, at the new site in Riverview Park, out of the smoke and grime of the city. Brashear had the plan Keeler had drawn up two years before, and he sent it off to his friends Worcester Warner and Ambrose Swasey, by now the leading telescope-making firm in the country, for a rough cost estimate. It soon came back; the total cost of the telescope and the building would be $160,000. Brashear launched a whirlwind fund-raising campaign, aimed at the wealthy industrialists of Pittsburgh, beginning with Carnegie. Speed was of the essence, he emphasized, if they were to keep Keeler at Allegheny. The director himself kicked off the public part of the fund drive for the new observatory in the sixth and last of the series of lectures on astronomy he was giving at the Carnegie Library.[82]

However, events in the larger world overtook the project. Spain's American empire was coming to pieces, and jingoists in the United States, led by the yellow press, were doing their best to help it along. On January 25, 1898, four days before Harper offered Keeler the Yerkes job, the battleship *Maine,* the pride of the United States Navy, arrived in Havana harbor, where an insurrection was in progress. On February 15, a week after Brashear approached Carnegie for a contribution for the new observatory, an explosion tore through the ship. It sank with the loss of 260 American lives. The rest of Brashear's and Keeler's negotiations were carried out amid the tension of a nation moving ever closer to war with Spain. Appeals for funds had to compete for attention with cries of "Remember the Maine." It was not an atmosphere conducive to astronomical entrepreneurship.[83]

Keeler had very clear ideas of how his own, or any other university, observatory should be built and operated to maximize research results, and at the same time allow selected students to learn astronomy. The most important requirement was to have a large telescope, devoted to research. One instrument – for instance a spectrograph – might be used on the telescope for months, if that was what the research program demanded. Students on the other hand should have the experience of seeing and using many different instruments, but these should be small and simple, so they could clearly understand what they were doing. They did not need a large telescope – a small one would do. Preparing lectures and hearing recitations were time-consuming tasks that interfered seriously with research; a very limited amount of teaching however, in classes strictly confined to a few graduate students who were themselves training to go on as professional astronomers, was beneficial for the research scientist. It sounded ideal from the faculty members' point of view, and was in fact very similar to the program that Keeler was to set in motion in the next two years at Lick Observatory.

In a personal letter to a friend, Keeler outlined his hopes for a new Allegheny Observatory. The new 30-inch telescope was to have a special "orthochromatic" lens, designed by Keeler's good friend Charles S. Hastings, his former physics professor at Johns Hopkins, now on the faculty at Yale. Such a lens would have a very flat "color curve," so that it brought light of all wavelengths to very nearly the same focus. Thus it could be used for spectroscopy over a wide spectral range, without the loss of any light at the slit. He estimated the cost of the whole new observatory, including the telescope, at $150,000, but hoped for a total endowment of $200,000. Such a sum would not only guarantee his own salary, but would ensure sufficient operating funds for the observatory. Even after the *Maine* was sunk, Keeler believed the prospects for raising the money were very good, although by no means certain.[84]

Meanwhile, in California, the decision on the directorship was postponed at the January and February regents' meetings. The Lick staff believed that Davidson was losing ground, and that Schaeberle was practically certain to be elected. Keeler was mentioned as a possible dark horse, but he was not thought to have much support. Also Newcomb was rumored to be a willing candidate. Although he was a great figure in American astronomy, he was considered by many who knew him "to be an extremely arbitrary man, and personally disagreeable in temper," and at least one Lick faculty member hoped that he would not be chosen. Keeler could not afford to show his interest publicly, but behind the scenes he kept a close watch on what was going on in California, as did Hale.[85]

Finally, in March, the regents decided they could put off their decision no longer. Their meeting was scheduled for Tuesday, March 6. The weekend before it occurred, Jacob Reinstein, one of the newer regents, proposed a deal. Reinstein, a short, vigorous attorney, was one of the strongest pushers on the board for building up the University of California. He wanted to make it a great university, not just continue the status quo. Reinstein had been appointed to the regents by Governor James H. Budd, whose views on the future of Berkeley were far less advanced. Budd and the other less academically oriented members of the Board supported Davidson for the Lick directorship, chiefly for sentimental reasons. He was a long-time Californian, he had known James Lick, he needed a job. Reinstein and the other activist members favored Keeler. Although she was the most recently appointed regent, Mrs. Hearst was particularly potent in their group because of her vast fortune which she was willing to spend on the university.[86]

Undoubtedly their faction favored Keeler largely because of the extremely strong letters of support for him they had received from leading American astronomers. In addition to Hale's glowing letter, Pickering had written strongly endorsing Keeler, as had Langley, and Newcomb himself had recommended either Keeler or Comstock. The Harvard director, in his carefully chosen prose, said that he had formed a very high opinion of Keeler and of his research work, and that "[h]e would doubtless be my own first choice." Davidson he damned with faint praise, saying that he "has a high reputation in geodetic work. I have not met him for some years, but should suppose that he now would be rather old for a position requiring so much enthusiasm and active work. His own work has been rather in the direction of the older astronomy than in the modern application of physics to astronomy which I suppose opens the whole field for usefulness for a great telescope like that on Mt. Hamilton. Of Professor Schaeberle I know less than of the others, but believe him to be a rising and active young astronomer of whose qualifications you can doubtless judge better than I." Only M.W. Elkin of Yale, a far less well known figure than Pickering, Langley or Newcomb, did not directly recommend Keeler. A classical astronomer himself, Elkin came out strongly for Newcomb, Lewis Boss and S.C. Chandler, three exponents of the "older astronomy" that Pickering had criticized by implication, but added that "[a]s regards the two names you mention, I think I can say that I imagine the majority of astronomers would consider Keeler the sounder and more painstaking scientist of the two, and I also think he would prove the more efficient director."[87]

Faced with such overwhelmingly united support for Keeler from the outside astronomical community, the regents who wanted to build a great university at Berkeley could hardly support a seventy-two-year-old folk hero whose whole ideas on astronomy were based on looking through telescopes and investigating the rotation and shape of the earth. But, to preserve the harmony of the board, Regent Reinstein put forward a compromise. He and his friends would vote for Davidson, assuring his election as the sentimental favorite by a large majority, if the old man would guarantee to resign the directorship after a few months. Then they would elect Keeler, again presumably by a nearly unanimous vote. Davidson would be guaranteed a professorship, "an easy berth," in exchange for his resignation as director of Lick. He could name his subject; they would found a chair for him and pay for it. According to Davidson, Reinstein even told him that if he chose to become the professor of commercial fisheries, they would give him a yacht. But they wanted his promise to resign.

Davidson was repelled by the idea. He though that he had a majority of the votes, but a narrow one. He wanted to be the director of Lick Observatory, but if he were opposed by Reinstein, Mrs. Hearst, and the other Regents who took such an active interest in the university, how could he be effective? By his own account he did not refuse the offer immediately, but promised to discuss it with his wife and grown children. He left the interview with Reinstein with a cold chill and shivers. All of his family urged him to repudiate the proposition. He went to bed with a raging headache. The next day he woke up determined not to give in, and went out to call on his supporters on the board and tell them he would fight to the end.[88]

On Tuesday came the climactic meeting. It was open to public and press, and was carried on in an atmosphere close to that of a modern political convention. Judge William T. Wallace, a seventy-year-old politician who had been a member of the Board of Regents since 1875 made an eloquent speech nominating Davidson. Timothy Guy Phelps, three years older than Wallace and the long-time chairman of the Lick Observatory Committee, read out a recommendation for Schaeberle, the acting director, whom the staff supported for the permanent appointment. Wallace made a sarcastic speech in reply, and Phelps answered him back in kind. Then Rodgers, a forty-nine-year-old attorney and long-time regent, nominated Keeler. He made no speech, but instead President Kellogg read aloud the letters strongly recommending Keeler for his astronomical ability, and for his tact and executive ability as well, from Hale, Langley and Pickering. They were identified as the directors of the "Chicago," Smithsonian and Harvard observatories, and apparently no one at the meeting attacked their appraisals.

Then the ballot was taken, on a roll-call vote. Twenty-two of the twenty-three regents were present. It was immediately apparent that Schaeberle had very little real support, while Keeler, the supposed dark horse, had a solid bloc of votes. The final tally showed a tie, ten votes for Davidson, ten for Keeler, and only two for Schaeberle. By the standing rules of the regents, Schaeberle's name was eliminated and they proceeded immediately to a second roll call. Both the regents who had voted for Schaeberle, Phelps and Samuel T. Black, switched to Keeler. William T. Jeter, another long time member of the board, who had voted for Davidson on the first round, abstained on the second round. Thus Keeler was chosen as the director of Lick Observatory, on a split vote, twelve to nine. Immediately afterward Davidson was voted an appointment as professor of geography in the College of Commerce at a salary of $3,000 a year, so he got his "soft berth" in the end anyhow, without compromising his ideals.[89]

The selection of Keeler was a triumph of the progressives. Although he was supported by Phelps, Kellogg and James W. Martin, the oldest man on the board and a regent since 1871, most of the rest of his support came from younger members like Reinstein, Rodgers, and Hearst, who wanted to change the University of California into a modern, research-oriented institution. The result was a surprise to the Lick staff. They had expected that Schaeberle would be selected, and were initially disappointed that he was not, but they were happy with the thought that at least they had been saved from Davidson. And they realized that Keeler was "a first-rate man" who had done good research work in his own line. He had not sought the post, and had wide national recognition. In the long run, they came to realize, he would be better for the observatory than Schaeberle.[90]

However, all was far from settled. Keeler began to have second thoughts as soon as he received President Kellogg's telegram offering him the Lick directorship. Either taking the Yerkes position or staying at Allegheny, if the money could be found for the new observatory, would mean less administrative responsibilities, and more time for his own research. Keeler's friends in the Pittsburgh area, especially Brashear, had been working so actively to raise funds to build a new telescope that he felt it would be unfair to them to leave now, although he had made no promises to stay. He had already delayed his decision for a week; now he asked the University of California for two more weeks. If in that time Brashear and his committee could complete their drive and get enough money to build the observatory, "to provide the director with a suitable residence," and to guarantee him "the increased salary properly pertaining to such a position" for a period of ten years, Keeler promised to stay. If they could not do it, he would have to accept the Lick appointment, in fairness to himself.

In seven years Keeler had sunk his roots deep in Allegheny. He had made many friends, and was an important member of the community. Sentimentally, he wanted to stay. But his wife had never reconciled herself to living in the Pittsburgh area. She thought its wintry climate was positively dangerous to their son's health, and she wanted to return to California, which she now considered her home. She would accept whatever decision her husband made, but she desperately hoped that the fund drive would fail, and that they could then go back to Mount Hamilton. All of Keeler's professional friends considered the Yerkes offer and then the

Lick directorship as scientific opportunities that he had fully earned and deserved. They could not imagine that he could stay in Allegheny, which even with a 30-inch telescope at a new location would still be far less than the best.[91]

Still Keeler waited to see what would happen. Brashear stepped up the campaign, in a last desperate attempt to raise the funds. Although a few large pledges came in, including one from Carnegie and one from George Westinghouse, in the end the nearly $150,000 that was promised was only about three quarters of the necessary total.[92] It was none the less a wrenching personal decision for Keeler, but finally Brashear himself told him that he should take the Lick job, "and tho' it will be a sad parting, you go with my best wishes and with the best wishes of those who have done so much to keep you here."[93] Keeler telegraphed his acceptance on April 2, 1898.[94]

Thus Keeler had the position he had wanted from the start, and yet he retained the friendship and keen support of all his friends. Not only Brashear, but Hale too, understood and accepted his decision. He had hoped up to the end that Keeler would come to Yerkes, but now at least the future of Lick Observatory was safe. The young Yerkes director insisted that his friend must not give up his position as co-editor of the *Astrophysical Journal*. Even though he were in far-off California, he could support the *Journal* in many ways, not least by the prestige of his name and of the institution of which he would be director.[95] Charles Burckhalter, the Oakland astronomer who knew all the Lick people, including Keeler, very well, happened to pass through Pittsburgh just at this time. He was on his way back to California after observing the January 22, 1898 eclipse in India. Burckhalter was overjoyed to learn that Keeler had been chosen, and that he had accepted the post. Brashear reported to a wealthy patron in Pittsburgh that "Burckhalter says No Other man in the world will smooth the troubled Waters at Lick Observatory as Keeler will – and I believe he is right."[93] The Oakland man congratulated the new director, and gave him a detailed personal description of the situation on Mount Hamilton. Keeler was sure that he would be able to patch up all the quarrels on the mountain, and get everybody working together peacefully on astronomy. As his first move in this direction he assured Burckhalter that he would be glad indeed to have Barnard back on the Lick staff, if only he would agree to come.[96] Undoubtedly he knew that it was a harmless gesture for Barnard would never return to the scene he associated with so much unhappiness. Keeler had already told him that he would be glad to have Burnham back at Lick too, and that if he came, California and astronomy itself would be the gainers, "but if [he] is unwilling of course the matter must be dropped." Initially Barnard felt sympathetic to Schaeberle, who had not gotten the directorship, and was noncommittal about Keeler's appointment. Before very long, however, as we shall see, his old friend's research accomplishments at Lick, together with his sympathetic diplomacy, won him over.[97]

Holden sent his congratulations from New York; probably he would have preferred Schaeberle to get the appointment but he undoubtedly realized Keeler was a better choice.[98] Even Davidson was reconciled to Keeler's appointment. The old man understood nothing about astrophysics, but he had always admired Keeler's planetary work, and recognized his drawings of Jupiter and Saturn as outstanding. They had always been on relaxed, friendly terms, and nothing was changed now. Davidson could not help but believe that he himself deserved the directorship, and that he would have done a better job at it than anyone else, but he reserved all his scathing criticism for President Kellogg and the regents, and did not blame Keeler in any way.[99]

Schaeberle could not take the decision so philosophically. He had expected to be appointed director and had announced, long before the final selection was made, that he would resign and leave Lick if he were not chosen. Most of his colleagues thought that this was a threat intended only as a bargaining ploy, and not a very good one at that. They did not believe that Schaeberle would follow through. However, he was adamant, and the day after Keeler's election was announced, he sent in his resignation. To him it was inconceivable that he could serve under his successor if the regents, who had placed him in the position of acting director, did not have enough confidence in him to give him the post on a permanent basis. The regents voted their appreciation of his services, and urged him to reconsider and stay at Lick as an astronomer. Keeler sent him a warm, friendly letter urging him not to resign. But Schaeberle would not consider it. He would remain on the job until Keeler arrived, but then he would turn over the directorship to him, and leave the mountain forever. He planned to travel all over the world, for two years at least, not doing any astronomy. The regents offered him a year's leave, at full salary, but even that he rejected. Then they gave him an honorary doctor's degree, which he gratefully accepted, but that was the only thing he would take from them.[100]

Even before he accepted the Lick directorship, Keeler began to be bombarded with job recommendations and applications. Hale put in a strong pitch for William H. Wright, who had worked with him for a year at Chicago and at Yerkes. Wright was a product of the University of California, who had done his undergraduate work at Berkeley in civil

engineering, and had spent more than two years, including one summer at
Lick, as a graduate student in mathematics and astronomy. He had
worked with Leuschner and with Campbell, and had made a very good
impression on both of them, as well as on Barnard and Holden. They had
recommended him to Hale for a fellowship at the University of Chicago,
and after his year there, they had brought him back to Lick in the summer
of 1897 on a temporary one-year appointment. Wright's job was to fill in
for Campbell, who was to be gone for more than eight months to observe
the solar eclipse in India. This expedition, including the salary of the
substitute for Campbell, was financed by Charles F. Crocker, the regent
who had provided support for several earlier similar eclipse trips. Young
Wright had proved an excellent spectroscopist and Hale would have liked
to have him back on his staff at Yerkes, but he knew that he preferred to
remain in California.[101]

Hale also recommended Robert G. Aitken, the young man from the
University of the Pacific, for a permanent position. Like Wright, Aitken
had been hired for a temporary, one-year appointment at Lick, mostly to
assist Holden. Although Wright considered him poorly trained (he was a
graduate of Williams College, in Massachusetts) and little more than a
toady of the director, even the critical Burnham thought him an excellent
double-star observer. Keeler had heard many good reports on both
Aitken and Wright, and he saw that both of them were kept on the Lick
staff with permanent appointments.[102]

The jobs Keeler controlled as the newly appointed Lick director were
not the only ones he was involved in filling; the positions that he himself
had not kept and had not taken were equally in demand. As soon as Hale
realized that Keeler was not going to accept the Yerkes position, he
cabled Carl Runge to see if he would take it, but the young German
spectroscopist was unwilling to leave Europe.[103] Then Hale asked Edwin
B. Frost, the "young friend" of Young who had translated Julius Scheiner's book on spectroscopy, and was now a professor at Dartmouth, if he
would come to Yerkes instead of Keeler. Frost indicated he would, and
Hale turned his considerable talents to convincing Catherine Bruce that
he would make an acceptable substitute. No one could possibly match
"the ablest stellar spectroscopist in the United States," but he wrote that
"[n]ext to Professor Keeler I consider Professor Frost better qualified
than any one else we could secure for the place." Keeler cooperated by
sending the generous old spinster "as nice a letter ... as I know how to
write" in support of Frost's appointment. She then agreed to using her gift
to pay Frost's salary, instead of Keeler's. Frost was eager to come to
Yerkes for the astronomical opportunity, but hesitant to leave the New

England he knew and loved so well for the wilds of southern Wisconsin. A personal visit from the ebullient Hale convinced him to take the job. A year later Hale persuaded Catherine Bruce to put up an additional $2,300 for a spectrograph for Frost to use on the 40-inch telescope. It was specifically designed for radial-velocity work, based on Campbell's experience with the D.O. Mills Spectrograph at Lick.[104] The Catherine W. Bruce Spectrograph was one of her last gifts to astronomy before she died.[75]

Keeler's own job at Allegheny Observatory was another valuable prize. Frank Wadsworth, the assistant professor who helped Hale with instrumentation and drawings, wanted it badly. Before coming to Yerkes, he had worked as an assistant for Langley at the Smithsonian Institution, and then for Albert A. Michelson on the University of Chicago campus. Both of them he had left after quarrels about authority, responsibility, and salary. Now he was repeating the pattern. Hale could get along with anybody he could use, and Wadsworth, whose skills were basically those of an engineer and designer, was very useful indeed. But Wadsworth wanted to be promoted immediately to full professor. He put forward a complicated "proposition", according to which he would receive this rank, but be paid only $2,000 as a half-time salary. He proposed to earn it by working through the weekend, from Friday afternoon through Monday night, thus including four nights observing and three days in the Yerkes shop. This schedule would leave him three full days, Tuesday, Wednesday and Thursday, for his own private engineering and consulting practice, from which he intended to make at least $2,000 more per year. Evidently he did not intend to rest ever.

President Harper would not hear of this arrangement and rejected it out of hand. Hale had recommended it, but in such an uncharacteristically mild way that there is little doubt but that he was already resigned to his assistant's imminent departure. Wadsworth was a master of tactlessness, and raged at Harper, accusing him of bad faith and almost of lying. One of the last of his letters, pious old Thomas W. Goodspeed, Harper's right-hand man, endorsed with the words "This man we can get on without very well indeed."[105]

To Wadsworth, Allegheny Observatory seemed a good place for his talents. There would be plenty of consulting opportunities for him in the busy Pittsburgh industrial area. At the observatory he would be in charge, with no supervisor to thwart his plans, he thought. Hale, always willing to help a friend in need, asked Keeler to recommend Wadsworth as his successor when he resigned. He begged Keeler not to recommend Frost, whom he needed himself so much at Yerkes. Keeler was less convinced of

Wadsworth's talents than Hale appeared to be, and was well aware of his quarrelsome nature, but he promised to put in a good word for him, and did so ultimately. Frost, he thought, was better qualified for the job, but Keeler agreed not to "interfere with your plans," and left his name off his list of recommendations. However, when Keeler departed for California he recommended that Brashear be put in temporary charge of Allegheny Observatory until the fund drive was completed, and that no permanent director be named until the trustees were ready to begin building at the new site. This is what they did. A year later they chose Wadsworth as the new director, and he took over from Brashear on January 1, 1900.[106]

Keeler spent his last few weeks at Allegheny in a frenzy of packing, devoting whatever spare moments he had to measuring the spectrograms of the third-type stars he had been taking ever since he had first put his spectrograph into operation on the 13-inch refractor. Just a few days before his departure, Keeler learned that he had been awarded the Rumford Medal of the American Academy of Arts and Sciences. This medal, given only annually or less frequently, originated in a gift from Benjamin Thompson, Count Rumford, to the Academy just over a century before Keeler received the award. Rumford, a soldier of fortune by profession, was one of the first experimental physicists, and he specified that the medal should be given for important discoveries in the subjects of light or heat. It is one of the highest marks of recognition to which an American physicist or astrophysicist can aspire. The year before Keeler the inventor Thomas Alva Edison had received the Rumford Medal, and other previous winners included Michelson, Rowland, Langley and Pickering. Keeler was given the medal "for his application of the spectroscope to astronomical problems, and especially for his investigations of the proper motions of the nebulae, and the physical constitution of the rings of the planet Saturn, by the use of that instrument."

Hale congratulated his friend warmly for the richly deserved prize, but could not resist twitting him that it would be only the first in a long series of honors, "which may perhaps end up with the Nawabship of Cawnpore, or something of equal dignity". Keeler, he suggested, as the successor of Holden should expect in time to equal the Great Mahatma himself in pride and ostentatious display. The new director took the chaffing in good spirits.[107] Then, on May 20, 1898, he, his wife Cora, their five-year-old son Henry, and their three-year-old daughter Cora boarded the train for the west. They had to change railroads in Chicago, and spent nearly a whole day there, much of it with Hale at his parents' house in Kenwood. He had come in from Williams Bay to entertain them and see them off. That evening it was on to California on the Overland Limited.[108]

Three days later they arrived in San Francisco. The Keelers stayed in the city for a few days at the California Hotel, and on May 27 Regent Denicke and his wife hosted a large reception for them at their home at Mason and Sacramento Streets. Leuschner, Denicke's son-in-law, had been the first graduate student at Lick, had worked as Keeler's assistant on Mount Hamilton, and had accompanied him on the eclipse expedition to Bartlett Springs in 1889. Leuschner had also spent part of the summer of 1893 at Allegheny with Keeler. During a year's sabbatical at the University of Berlin in 1896-1897 he had earned his Ph.D. with high honors, for his thesis on orbit determination, his field of specialization. Many of the Regents were at the Denickes' reception with their wives, including Governor Budd, Lieutenant Governor Jeter, Lick Observatory Committee Chairman Phelps, Andrew S. Hallidie, Charles W. Slack and Reinstein. Over half of them had voted against Keeler on the first ballot, including Phelps and Denicke himself, but now the wounds were healed, and everyone congratulated the new director. President Kellogg and several senior faculty members were present, among them LeConte and Davidson, the newly appointed professor of geography in the College of Commerce, Keeler's rival in the contest a few months before. No doubt he now congratulated the victor and wished him every success. Three days later there was a second reception for the Keelers, this one across the Bay in Berkeley. The hosts this time were the Leuschners, at their home on Bancroft Way near the university campus. Many of the faculty members were present, along with the few advanced astronomy students.[109]

On June 1 Keeler went up to Mount Hamilton, and took over as director. Schaeberle left the next day, after handing the observatory over to him. Keeler said it was hard to see "dear old Schaeberle" go away; it seemed almost as if he himself were crowding his friend out of the place, but he had done his best to persuade the older man to stay and did not feel that he could be blamed.

On the same stage coming up to Mount Hamilton on June 1, W.W. Campbell returned to Lick along with Keeler.[110] He was just getting back from observing the solar eclipse of January 22, 1898 in India, a trip that had taken him around the world and that had lasted the better part of a year. Accompanied by his wife, Campbell had left Mount Hamilton on October 16, 1897, just after Holden had finally resigned the directorship. They had sailed from San Francisco on October 21, with several tons of instruments and scientific equipment. After stops at Honolulu, Tokyo, Kobe, Nagasaki, Shanghai, Hong Kong (where they changed ships), Singapore (where they stayed at the Raffles Hotel), Penang, Ceylon and Bombay, they had taken a train and ultimately a bullock cart to the site at

Jeur, in the path of totality. Campbell originally had planned to go to a larger town, Karad, but changed his plans when he learned that a bubonic plague epidemic was raging there. They arrived at Jeur over two months after leaving San Francisco.

The scientific program at the eclipse involved photography of the corona with several different cameras, including the 40-foot focal-length instrument developed by Schaeberle, high-dispersion spectroscopy to measure the rotational velocity of the corona by the Doppler effect, and low-dispersion spectroscopy to measure the wavelengths of the coronal lines. In preparation Campbell drilled a corps of English governmental officials and army officers who volunteered to help operate the various instruments. On the date of the eclipse the sky was perfectly clear and all the instruments and observers worked as planned. It was so hot at Jeur that Campbell could only develop the photographic plates between midnight and dawn. The spectrograms would only be measured the following winter, months after he had returned to Lick. After the eclipse the Campbells traveled through India, visiting Jaipur, Delhi, Agra, Benares and Calcutta before sailing from Bombay. On their ship they met the famous English physicist Lord Rayleigh, who discussed with Campbell the still-recent spectroscopic discovery of the element helium. After stops at Suez and Cairo they traveled through Italy, France and England, visiting astronomers and observatories along the way. Finally they sailed from Liverpool on May 4, and after visits with relatives in Ohio and Michigan, Campbell got back to California in time to ride up Mount Hamilton with Keeler on his first day as director.[111]

Campbell had come to Lick as Keeler's assistant at Lick Observatory back in 1890. As his successor on the Lick faculty he had made many important spectroscopic discoveries, and now with the Mills spectrograph that he had designed, was in the first stages of a large radial-velocity program. With Leuschner in charge of astronomy at Berkeley, and Campbell as his right-hand man on the mountain, Keeler was truly home at Lick Observatory. Both his former assistants were touchy characters, and he was to have problems with each of them initially, but he won them over and finally made Lick Observatory the outstanding astronomical institution of which James Lick, Richard S. Floyd and Edward S. Holden had all, in their highly individualistic ways, dreamed.

9

An ideal director and investigator

When James E. Keeler returned to Mount Hamilton as director of Lick Observatory on June 1, 1898, he found it little changed in the seven years since he had departed for Allegheny Observatory. As the stage from San Jose wound up "Lick Avenue" to the 4,200-foot high summit, Keeler could see that the sparse vegetation on the mountain was already beginning to dry out and turn brown as it did each year after the rainy season. The huge dome of the 36-inch telescope, at one end of the Main Building, dominated the top of the mountain; at the other end stood the smaller 12-inch. Keeler moved his books into the director's office, next to the big dome, where Edward S. Holden had officiated until the previous fall. John M. Schaeberle, who had served as the acting director following Holden's forced resignation, turned the observatory over to his old friend and colleague and left the mountain forever on the very next day.[1]

The new director, his wife Cora, Henry, five years old, and little Cora, soon to be four, moved into one side of the Brick House, the large, three-story residence, just down the slope from the Main Building. With their cook, whom Mrs. Keeler had hired in San Francisco, they occupied the same apartment that Holden had lived in for so many years. The Keelers had not been able to bring much of their furniture with them; the big pieces were too expensive to transport across the country. But Holden had left much of his furniture on Mount Hamilton for the same reason, selling it to the observatory in one of his last acts before he gave up the directorship. The Keelers used it, supplemented by a large shipment of additional pieces sent up by Harry Floyd, Cora's young cousin and the heiress to the fortune, the San Francisco house, and the Lake County estate of Captain Richard S. Floyd, the builder of Lick Observatory. They were huge pieces, "quite too gorgeous for the rest of our belongings," according to Keeler, "though somewhat faded, 'vergangener Pracht' [departed splendor]."[2]

Although the buildings at the summit were little changed, there had been almost a complete turnover in the staff during the years Keeler had been in Allegheny. The three senior astronomers were W.W. Campbell, who had been Keeler's volunteer assistant his last summer at Lick, and was now in charge of stellar spectroscopy; William J. Hussey, who specialized in double-star measurements; and Richard H. Tucker, a meridian-circle observer. Tucker, a bachelor two years younger than Keeler, lived in a few rooms in the other side of the Brick House, while Campbell and Hussey, both five years younger than the new director, were with their families in two recently built houses on a side ridge named Ptolemy. On it also stood the Crossley 36-inch reflecting telescope, the famous gift that Holden had acquired for the observatory in 1895, but from which absolutely no research results had flowed to date.

There were also three younger staff members, designated assistant astronomers in contrast to the more senior astronomers. They were Charles D. Perrine, the observatory secretary, also a dedicated observer who had discovered several comets; William H. Wright, who worked as Campbell's assistant in spectroscopy; and Robert G. Aitken, another double-star observer. Keeler, with his excellent training in physics at Johns Hopkins and a year of graduate study in Germany, had by far the best scientific education in the group. None of the others had any formal graduate training except Wright, who had earned an M.S. degree at Berkeley in mathematics and physics, with a minor in astronomy for which he had worked with Campbell at Mount Hamilton. Both Hussey and Campbell had been "special students" at Lick in summer vacations before joining the staff, learning by doing, and Tucker had received thorough on-the-job training, first as an assistant to Lewis Boss at the Dudley Observatory in Albany, New York, and then with Benjamin A. Gould at the Cordoba Observatory in Argentina. It was a distinctly practical group, but not at all unusual for its time. At Yerkes Observatory, the only astronomical research institution in the country with a larger telescope than the Lick 36-inch refractor, none of the staff had any real graduate training, not even the director, Keeler's friend George Ellery Hale.

The astronomers at Mount Hamilton were assisted by a small group of workmen who maintained the observatory, the telescope and the instruments. Their foreman was John McDonald, a machinist who could fix anything. He had first come to the mountain in the old Lick Trust days, when he had worked under Thomas Fraser, and had been chief of the work crew since June 1, 1888, the day the University of California took over the observatory.[3] There were two other highly skilled workers, J.I. Bane, a carpenter, and Emil Zengeler, an instrument maker. W.C. Pauli, the janitor who doubled as a photographic technician, and the brothers Paul and N.D. Soto, two general laborers, completed the staff.[4] With families, relatives and servants, the little mountain community numbered about fifty people. There were sixteen children, and their parents banded together to hire a teacher, who boarded with each of the families in turn. She taught her seven students in a little one-room schoolhouse, provided by the university. As in Keeler's earlier days on the mountain, the housing situation was definitely crowded. The junior astronomers lived in various apartments on the same side of the Brick House as Tucker, while the workmen had the old small, wooden cottages that had housed the original construction crew.[5]

Mount Hamilton was an isolated astronomical village. The inhabitants were thrown pretty much on their own resources. There were few cultural diversions except reading and music, although occasionally Tucker would organize a community play. Outdoor activities included hiking, riding, hunting everything from pigeons and quail to deer, and fishing in nearby Smith Creek. There was even a tennis court, and a rough-and-ready eight-hole golf course, laid out by Tucker at the Brickyard, a fairly flat area not far below the top of the mountain. Keeler could occasionally be persuaded to try his hand at golf, but never went hunting or fishing; he spent nearly all his time either working or with his family. He and his wife soon established themselves as friendly, gracious hosts, inviting the other astronomers and their families to frequent dinners. Visiting scientific dignitaries, or important faculty members and officials from Berkeley, were always invited to stay with the Keelers as their guests.[6]

One of the new director's first moves quietly demonstrated his loyalty to the traditions of Lick Observatory, at the same time that it distanced him from his predecessor's policies. Keeler had a large portrait of Floyd brought up from San Francisco and hung in a conspicuous place in the Main Building.[7] The captain, more than anyone else, had been responsible for building the observatory, but Holden had tried to arrogate the credit to himself. They had become enemies at the end of Floyd's life, and Holden had consistently downgraded the captain's role after his death. Most of the Lick staff hated their former director for having, as they saw it, tried to pass off their research as his own; no doubt they enjoyed seeing their new director's symbolic gesture in reestablishing Floyd.

Keeler's first important decision was one that would have a long and beneficial influence on Lick Observatory. As a result of Schaeberle's resignation, the new director had a vacant faculty position to fill. President Martin Kellogg of the University of California had advised him of this even before he left Allegheny, so that he could make plans for hiring a new staff member. Keeler's old friend E.E. Barnard recommended Sidney D. Townley, who had been an early graduate student at Lick, while Armin O. Leuschner, once Keeler's assistant and now associate professor of astronomy on the Berkeley campus, put forward the claims of the graduates of his department. But Keeler had more far-reaching ideas than simply filling the one vacancy that then existed.[8]

Committed as he was to graduate training, Keeler decided to leave the faculty position unfilled, and to use the salary money thus saved to support several students at Lick Observatory. Except for Townley one year (he had held the first, last and only Phoebe A. Hearst Fellowship at Mount Hamilton in 1892-93), and Ernest F. Coddington, whom Holden had managed to include in the budget his last year, there never had been money for graduate students at Lick Observatory.[9] As a result, there had been very few students, except the "specials", mostly faculty members at other colleges and universities, who would come to Mount Hamilton in the summers.

Yet as Keeler knew from his Johns Hopkins days and his year in Germany, graduate students were the lifeblood of a research institution. They were young, eager to learn, hard working, and often even creative. They could help the faculty members with their research, particularly in the time-consuming observing and reduction of the data, at the same time they were being trained themselves for future faculty positions, either at Lick or elsewhere. Hence Keeler decided to convert the one vacant faculty position into three Lick Observatory fellowships. The students who received them would be required to spend the summer and fall, the good observing seasons, at Mount Hamilton, assisting the astronomers with their research, and thus learning by doing. The winter and spring, the cloudy seasons in central California, they were to spend on the campus in Berkeley, taking courses in astronomy and related subjects.

This last provision was very sensible from the educational point of view. In addition, it made explicit one of Keeler's chief goals, to bring about cooperation, rather than conflict, between Berkeley and Mount Hamilton. All previous attempts had foundered on Holden's intense suspicion of Leuschner. Holden had seen the need for graduate students, but he had insisted on keeping complete control of their programs in his own hands. He would not agree to any plan that took them to the Berkeley campus, even for only a few days at a time. In a long, emotional report he made in 1895, he had rejected any thought of cooperation.[10] As a result, under his regime there was practically no money for graduate-student support, very few graduate students, no formal courses at all, and not a single graduate degree granted.

Keeler wanted to have students, and he wanted them to get a good education. He wanted Leuschner's help, not his opposition, and he wanted the help of the Berkeley professor's father-in-law, Regent Ernst A. Denicke. Leuschner, for his part, admired and respected his former mentor, and was eager to hitch his little campus operation to the Lick Observatory star. As soon as Keeler explained his plan for the fellowships to him, at the Denickes' reception just after his return to California, Leuschner became an eager exponent of it. Keeler sold his young former assistant on the idea so well, in fact, that after the director's sudden death Leuschner published an account giving himself credit for the idea.[11] The most positive interpretation is that a true diplomat is one who can make the person he is working with believe he has originated all the best ideas.[8]

Thus, soon after his arrival on Mount Hamilton, Keeler's very first letter to President Kellogg expressed his concerns regarding graduate students and teaching. Kellogg sent Keeler copies of some of the papers that had been generated in the previous debates between Holden and Leuschner on the relationships between Mount Hamilton and Berkeley, and emphasized his own desire for close cooperation between the observatory and the campus. He said that Leuschner would come to Lick to discuss the problem with the director. Keeler invited Leuschner to bring his wife and the Denickes along, and asked them all to stay at Mount Hamilton as his guests for the weekend. With all this preparation, the conference was a complete success. Leuschner agreed fully with Keeler's recommendations for the Lick Observatory fellowship program.[12] So did Kellogg, and the next step was to get the regents to approve the plan.

Here there was a breakdown in communication. Keeler had no difficulty in selling the plan to Timothy Guy Phelps, the powerful old politician who had spat tobacco juice on the marble floors of the Main Building on his first inspection visit ten years before, and who had been chairman of the regents' Lick Observatory Committee ever since. Phelps was always extremely supportive of the Lick director, whether it was Holden, Schaeberle or Keeler. He had invited Keeler to his palatial home at San Carlos, south of San Francisco, and then taken him to the June regents' meeting, where he must have been pleased to hear the new director report the discovery of a comet by Coddington, and describe Lick as second only to Harvard as an astronomical research institution.[13] Now Phelps was doubly certain that the fellowship plan was a good one for not only had the new director proposed it, but as the old man wrote, "Dr

Holden always thought that three or four Fellows Could render a great deal of Service while pursuing their Studies, or rather in the line of their Studies, and in work in which it would be necessary that they Should perfect themselves. The pay of the present Astronomers I always favor advancing the Men we have according to their Merits, and as fast as our Means will Allow."

His last sentence referred to the other part of Keeler's plan, which involved using the rest of the money originally budgeted for Schaeberle's salary to provide raises for Tucker and Hussey, from $2,000 to $2,200, for Perrine from $1,320 to $1,500, for Coddington from $480 to $600 (which would bring him close to the three proposed new fellows' salaries), and for N.D. Soto from $330 to $720 (this involved his going from part-time to full-time).[14] All these raises were quite reasonable, but Phelps neglected to drum up support for them among the other regents before the meeting. They got the idea that Keeler was trying to put over some kind of back-door raid on the university budget. As a result, although they approved the fellowships, they turned down the raises.

Then Phelps swung into action, and Leuschner, who had been on a vacation with his family in northern California, returned to Berkeley. He briefed Denicke, and soon all the other regents were brought to realize that they had misunderstood their new director, and that in actuality he, like them, was working for the best interests of Lick Observatory. At their next meeting, in August, they approved all the faculty raises without a murmur.[15]

The candidates for the new fellowships had all been picked long before by Leuschner, with Keeler's approval. They were Russell Tracy Crawford, Frank E. Ross, and Harold K. Palmer, all three of them Berkeley graduates. In accordance with the custom of the time, Crawford, who was married, originally was to be paid $800 a year, while Ross and Palmer, who were single, were to get $648, but the regents reduced them all to $600, the same figure they had approved for Coddington. Crawford and Palmer accepted at once and were at work at Mount Hamilton by August, while Ross, who had taken a summer job with the Coast Survey, came a month later.[16] Crawford and Ross worked chiefly with Tucker and Hussey; Palmer, as we shall see, assisted Keeler, and Coddington continued to work mostly with Hussey, but to help Keeler as well. So began the Lick Observatory fellowship program, which continued down to the 1960's, and produced many outstanding astronomers in the seventy years of its existence.

The courses that the original fellows took in Berkeley were almost entirely mathematical in nature. For instance, in the second semester of

1898-99, Palmer studied Elliptic Functions, Linear Differential Equations, Mechanical Quadratures, and Perturbations, while Crawford, who had been a graduate student in Berkeley before getting the Lick Fellowship, over a two-year interval took Elliptic Functions, Linear Differential Equations, Partial Differential Equations, Transformation Groups and Differential Equations, Theoretical Astronomy, Perturbations, and one so-called physics course, Dynamics of Rotation. The Theoretical Astronomy course, taught by Leuschner, was actually on orbit theory, and every one of these courses, whether called mathematics, astronomy, or physics, was in fact almost entirely concerned with mathematical manipulation, involving very little in the way of physical concepts. In part this was the result of Leuschner's strong bias toward traditional gravitational astronomy, his own field of research, but even more it was because Berkeley was very weak in physics. For many years John LeConte, a gentlemanly teacher of the old school, had been the professor of physics. After LeConte's death, Holden, who nearly always had superb ideas (though his execution often left much to be desired), had recommended Albert A. Michelson, then at Clark University, for the vacancy. According to Holden, he was "a *genius,* [who is] young – ambitious – an admirable teacher – & rich." But the University of Chicago got the future Nobel Prize winner, and the University of California hired Professor Frederick Slate, who did not believe in dirtying his hands in the laboratory.[17]

Nevertheless, although their graduate courses did little to prepare them to become astrophysicists, these men had received a good deal of practical training from Leuschner as undergraduates, and were highly useful assistants at Lick Observatory. As Campbell later wrote of Crawford when he had finished his Ph.D. thesis and was looking for a job,

> His knowledge of Physics is that of an undergraduate, but he was well taught. His knowledge of Astronomy along the old lines and of Mathematics is entirely satisfactory ... *All* my experience goes to show that for night work in an Observatory an astronomical training is a *sine qua non.* One whose training has been confined to the physical laboratory is at a great disadvantage for many months. Reese [a Johns Hopkins, laboratory-trained physicist] is a good man, but it has taken him nine months of *very hard work* to train himself so that I can tell him to "get a spectrogram of β Herculis and measure it" and feel reasonably sure that he will be right.[18]

Keeler did not simply hand over control of the graduate students to Leuschner. He made sure that his approval as director of the observatory was required, as well as Leuschner's as head of the campus department,

Fig. 19. Group picture at Lick Observatory. Back row, left to right, Cora M. Keeler, W. H. Wright, Keeler, Ernest F. Coddington, R. Tracy Crawford, Frank Ross, William J. Hussey, Robert G. Aitken. Front row, the astronomers with umbrellas (to cut down the contrast of the brilliant Mount Hamilton sunshine) are, from left to right, W. W. Campbell, Charles D. Perrine and Harold K. Palmer. The two children in the front row are Cora and Henry Keeler; Harry Floyd is seated to the left of Cora (1898). (Reproduced by kind permission of the Mary Lea Shane Archives of Lick Observatory.)

for any graduate degree in astronomy, whether the student came to Mount Hamilton or not. And he thought that although a purely theoretical student need not be *required* to spend a period in residence at Mount Hamilton, still the student might well *benefit* from being exposed to the observational work going on there. (Frederick H. Seares, another Berkeley product, was at that time Leuschner's one graduate student on the campus. He did not register for a Ph.D. degree, but went abroad in 1899 and continued his studies in Munich and Paris.) Keeler held the power of assigning the work of the students at Mount Hamilton firmly in his own hands.[19]

In addition to supervising the Lick graduate students, Keeler kept in close touch with the progress of Henry Harrer, his former assistant at Allegheny, whom he had been training on the job, just as he himself had been trained by Samuel P. Langley fifteen years before. John A. Brashear was serving as the caretaker acting director of Allegheny Observatory, but he had little time to spare for it, especially as his optical shop was swamped with government orders for range finders for the Spanish-American War. Brashear was completely untrained scientifically and could only hope to preserve the observatory, and let the public view the stars. Harrer was thoroughly familiar with the practice of astronomical spectroscopy, and Keeler encouraged him to keep using the Allegheny telescope for research. At his former director's suggestion, Harrer got a good series of spectrograms of the bright irregular variable star α Orionis (Betelgeuse) as it grew fainter, reached minimum light, and then brightened again. Although its brightness had varied considerably, the spectrum had hardly changed at all. This was a new and interesting result, and Keeler advised Harrer on how to write up the results in a paper, and then shepherded it through to publication.[20]

Besides the graduate-student fellowships, the other issue on which Keeler and Leuschner achieved instant agreement was undergraduate teaching on the Berkeley campus. The regents had from the first wanted to involve the Lick faculty in the educational program, but Holden always balked at it. Berkeley was too far away, the astronomers had to work at the telescopes every night, and Leuschner could do all the elementary teaching anyhow, if only he did not offer the graduate courses he preferred, were the first director's constant refrains.[10] But Keeler knew it was essential to strengthen Lick's contacts with the campus. Hence he proposed that during the winter, he, Campbell, Hussey and Tucker would each come down to Berkeley and give one week's lectures in the second term of the Modern Astronomy survey course taught by Leuschner. Each of the Lick astronomers would pick his own subjects, within his own field of research. Keeler would count on President Kellogg to schedule the class at times that were convenient for the students, but that the Mount Hamilton faculty members could meet. Again, his plan was simple and straightforward, and by putting his trust in Kellogg and Leuschner, rather than trying to set up complicated conditions as Holden had, Keeler gained the good will of the campus faculty, the university administration, and the regents.[12]

By August the list of subjects to be taught the following winter was complete. Keeler was to lead off himself with lectures on astrophysics and on the spectral classification of stars, then Campbell would give two on his spectroscopic measurements of stellar velocities. He would be followed by Tucker, giving two lectures on "Astronomy of Precision – Aims and Methods", which must have been among the duller lectures the students heard that year. However, Tucker would be followed by Hussey, lecturing on Mars and on double stars, the first at least, a very exciting subject at the time. Each pair of lectures was given on a Tuesday and the following Thursday, and Leuschner gave his own lectures on alternate weeks, so that he could prepare the class on the Lick faculty members' subjects. The university paid the traveling and living expenses of the astronomers when they went to Berkeley to lecture, which made the trip attractive to most of them. Typical reimbursements, including fare on the stage between Mount Hamilton and San Jose, the train between San Jose and San Francisco, the ferry between San Francisco and Berkeley, and the street car in Berkeley, together with three nights in a hotel in San Francisco, one in San Jose, and a few meals thrown in, totalled just under $20. Keeler and Campbell both enjoyed teaching, and liked the opportunities to go to Berkeley and solidify their contacts there. Keeler would usually give a popular lecture somewhere in the Bay area, between his Tuesday and Thursday classes. Only Tucker, who dreaded teaching and disliked Leuschner intensely, resented having to make the trips.[21]

In 1899, the first year the Lick astronomers taught, the astronomy class was scheduled early in the morning, and this proved very inconvenient. They had to come down from Mount Hamilton the day before the lecture, stay in a hotel in San Jose, and leave "in the middle of the night" to get to Berkeley in time for their first lectures. The next year Keeler got the class moved to the hour just before lunch. An additional advantage of the change was that no class was scheduled for the next hour, so the astronomers, who were not used to finishing their lectures at an exact time, could go on longer if necessary. No doubt they frequently did, to the students' distress. The subjects were updated for the 1900 lectures, Keeler lecturing on reflecting telescopes, and on his results with the Crossley reflector, which, as we shall see, had become his main field of research. Campbell gave a new lecture on his results from the eclipse expedition to India. Tucker again lectured on the meridian circle and on star catalogues, and still found it "a very disagreeable task." A New England Puritan, he insisted on paying his own hotel bills, to demonstrate that he would not, like the others, cash in on a "nice thing."[22]

Keeler did not object to collecting what was due him, but made certain that he never was even suspected of profiting from the directorship. For his first business trip to San Francisco, to attend the June regents' meeting, he had listed the round-trip stage fare from Mount Hamilton to San Jose as $3.00, but when he later received the bill and discovered that it cost only $2.00 for mountain residents, he immediately notified the university secretary in Berkeley of his mistake and returned $1.00. One of the many charges against Holden had been that he secretly rode free on the Mount Hamilton stages, but publicly declared that he was being billed for the rides.[23]

One immediate dividend from the much improved relationship with the Berkeley campus was that Keeler met the engineering professors, and saw the well-equipped machine shop they had. He arranged for Zengeler, the Lick instrument maker, to go and work there on special jobs such as cutting gears, which required tools he did not have on Mount Hamilton. For his part, Keeler was glad to let a friend of Leuschner's come up and make a set of slides and transparencies of Lick astronomical photographs for the Berkeley department. He welcomed Leuschner's classes to Mount Hamilton, and invited their professor to stay overnight at the mountain as his guest.[24] Such simple, friendly interactions would have been unthinkable at the end of Holden's reign.

Keeler's first year back at Lick was marred by one blow-up, which taught him that he must be even more careful than he had been with his staff members, particularly the somewhat touchy Campbell. As director, Keeler followed a policy of publicizing the new findings made at the observatory, giving full credit to their authors. One frequent recipient of such recognition was Campbell, beginning with a press release recounting his "discovery of great scientific and considerable popular interest", that Polaris (α Ursae Minoris, the pole star) is a triple-star system. It appeared as a single, unresolved star even with the 36-inch refractor, but Campbell had measured the radial velocity by the Doppler method, and found that it varied in a four-day period, which he interpreted as resulting from the orbital motion of a close pair of stars about each other. He had also found a much slower variation of velocity, with a period of at least several years, which could only result from the orbital motion of this pair about a third, much more distant companion, much as the earth and the moon orbit about each other, as a close pair, and together orbit around the more distant sun.[25] (Today we realize that the short-period variation Campbell had found is not orbital motion but rather results from the intrinsic pulsation of a single star; Polaris is in fact a double, not a triple star.)

However, Keeler came to grief when he did not take into account his staff's tender feelings on their official titles. Since the founding of Lick Observatory, the senior men had been appointed as astronomers, while the more junior researchers were called assistant astronomers. These ranks did not fit, however, with the standard professor titles used on the campus. Therefore, the question came up of what the equivalent ranks of the Lick faculty should be. President Kellogg thought the astronomers should rank as professors, and the assistant astronomers as assistant professors, but said he would accept the director's judgment. Keeler, without much thought, concluded from the comparative salary scales that he, as director, should rank as a professor, the astronomers as associate professors, and the assistant astronomers as instructors. The comparison was true, but neglected the fact that the Lick staff members were badly underpaid, but nevertheless stayed on at Mount Hamilton because of the research opportunities. Campbell and Tucker had both turned down full professorships elsewhere.[26]

Their feelings can be imagined when the annual catalogue appeared in March 1899, and they found themselves listed as the newest associate professors in the University of California, outranked by Leuschner, whom both of them disliked and considered far beneath themselves scientifically. Campbell complained vigorously. Intensely suspicious, he was sure that Leuschner was behind it all. When Campbell learned that the recommendation had come from Keeler, he must have felt that his director had betrayed him. Keeler immediately realized that it was a much more serious problem than he had originally understood. He apologized to all concerned, accepted all the blame himself, withdrew his recommendation, and said that the president's original plan was better. Best of all, he said, would be to return to the old terms, and do away with the professor titles for the Lick faculty altogether. President Kellogg agreed and restored the astronomer and assistant astronomer titles.[27] Such a retreat in the face of opposition by his staff would never have been considered by Holden, but it seemed reasonable to Keeler, and everyone was happy again, except perhaps Leuschner.

The new director kept on polite, but cool and impersonal, terms with his predecessor, and never criticized him. Keeler did, however, quietly drop the photographic moon atlas, and the huge catalogue of supposedly accurate positions of stars, derived from old observations made at the Naval Observatory, that had been two of Holden's pet projects. He had often mentioned them in glowing terms, but not much had come of them. A few sheets of the moon atlas had been issued, and although they were spectacular and drew much praise, to the critical eyes of Keeler and other professionals they were technically not of the highest quality, and scientifically represented a waste of valuable telescope time. The star catalogue had never had any hope of success, as Holden probably realized himself by 1897. Both of these programs had been severely criticized by Holden's enemies, and formed the bases of many of the charges against him.[28] Keeler simply let them die.

He never tried to tell his staff members how to do their research. As he saw his task, it was to determine the general policy of the observatory, and in what general field each of the astronomers on the faculty would work. This, however, was largely based on their skills and experience; obviously Campbell would do spectroscopic observations of radial velocity, and Hussey and Aitken would measure double stars. The details he left entirely to them. "The Director does not interfere without cause," was his motto, and he seldom seemed to find cause to do so. In his opinion no elaborate rules were necessary. The observatory thrived under his benign leadership. The telescopes were in use every clear night, and the results flowed out. As Keeler assured the director of the new Nice Observatory in France,

> Except for the students, there are no regulations fixing the number of hours spent at the telescope, or in computations. Everything is safely left to the judgment and zeal of the individuals. It has usually been necessary here to restrain the inclination for work, rather than to encourage it.[29]

Keeler considered astronomical research the reason for existence of Lick Observatory, and not only its by far most important task, but the most important task of every astronomer in it, including himself. When he came to Mount Hamilton he brought with him his spectrograms of "third-type" stars, or as we would call them today, G, K and M stars, taken at Allegheny Observatory. He had not finished analyzing them and writing up the results for publication, and he intended to do so at Lick. Keeler also brought with him, on loan from Allegheny, the spectrocomparator he used to measure accurately the wavelengths of the absorption lines in the stars' spectra on the tiny glass photographic plates. It was a precision optical device, made by Carl Zeiss in Germany, in which two spectra could be viewed simultaneously, and either one could by measured by turning a precision screw, centering each line in turn on a cross-hair in the high-power eyepiece. Keeler had long used it at

Allegheny, and Brashear was glad to let him borrow it. Keeler ordered another one for Lick Observatory. It cost just under 500 marks (about $100 at the time), and was made and delivered from Jena within eight months.[30]

By that time Keeler and his assistant, Palmer, one of the fellows, had finished all the measurements of the spectrograms, but he was never to complete the discussion and publish this work. He was too busy with his new research program at Lick. His preliminary results on the third-type stars, which he had presented at the Yerkes dedication in 1897, were published only in a brief abstract. Keeler sent his further measurements to Hale for oral presentation at the Harvard meeting in August 1898, at which the first steps were taken to form the Astronomical and Astrophysical Society of America, now the American Astronomical Society.[31] These measurements were not published, nor did Keeler ever work up a complete analysis of the spectra. This unfinished program represents his greatest failure as a scientist.

The reason he did not complete this research is that he could never find time for it. As director, he had many administrative tasks he simply had to do. There were thousands of vouchers, checks, and orders that he had to sign personally, even though Perrine, the observatory secretary, wrote most of them out for him. And there was a torrent of correspondence with the administration in Berkeley, with suppliers of everything from optical parts to toilet seats for the observatory, with prospective students and established scientists, and always with the public, interested in astronomy and countless other subjects on which astronomers were imagined to be experts. Typewriters were not in general use in universities in 1898, and Keeler wrote nearly every letter by hand himself. Some of the letters that came to Lick Observatory, the outstanding research institution in the West at the time, appealed to Keeler's sense of humor, and brought out the best of his restrained, tongue-in-cheek manner. For instance, to a benefactor in Phoenix, Arizona, who sent an unsolicited gift to the observatory, he wrote,

> I beg to acknowledge, with thanks, the receipt of your "Gila Monster" – alive and in good biting order. Such specimens are a little out of our line. When we have derived sufficient pleasure from the company of your pet, perhaps you will allow us to present him, in your name, to the California Academy of Sciences in San Francisco, where I think he is likely to be more useful to science. In the mean time we will take good care of him.[32]

And, to a cowpuncher in Whitehorn, Colorado, who wrote on a

friend's recommendation, no doubt given as a prank, to order a pair of glasses, Keeler replied,

> I have yours of Nov. 22 – but as the Lick Observatory sells nothing whatsoever, your "pard" must have been misinformed. In general it is better to get spectacles specially fitted to one's eyes than to buy them at random.[33]

To other correspondents he gave advice that he could see no advantage to dividing the year into thirteen months instead of twelve, detailed instructions on how to make and set up a sundial, and the firm statement that "there are no astronomical phenomena expected to accompany, or precede, the second coming of Christ." Another, more factual, question he answered by saying that approximately one hundred million stars are bright enough to be seen with the 36-inch Lick refractor, but that the estimate was a very rough one, because many of them would be just at the limit of visibility.[34] All of these are answers to typical questions that still flow in to observatories today.

On a more unusual tack, an officer of a chemical company in Wilmington, Delaware, asked if there were any seismograph records of explosions. Lick Observatory in fact had an operating seismograph from its earliest days and had recorded hundreds of earthquakes, but this was a touchy question. Years before, Holden had been widely accused of claiming to have detected with the seismograph a planned dynamite explosion in San Francisco, but then when he learned that the explosion had been postponed, of suppressing the news release of the purported detection. Keeler did not refer to this episode, but merely told the correspondent that "none of the numerous explosions in the vicinity of San Francisco has certainly been detected or has left a record on these instruments".[35]

One of the director's responsibilities was to preserve the property and security of the observatory. The horses and sheep that strayed onto the reservation, eating brush and young saplings, and generally annoying everyone who lived on the mountain, were a constant problem. Keeler believed that the sheep herders drove their flocks onto the university property at every opportunity for free grazing, knowing that the observatory was shorthanded and could not keep them off. He waged a ceaseless war against the sheep.[36]

Another of the director's duties was to publicize the observatory by giving astronomical lectures to college classes, clubs, and other groups. This he did frequently, especially in San Jose, Berkeley and San Francisco. Keeler received several invitations to speak in Los Angeles, but it was too far away, and he always put them off for the future. In the end, he

never did get to Southern California to lecture. Nevertheless, he succeeded in recruiting John D. Hooker, the very wealthy Los Angeles oil
and steel magnate who later provided funds to build the 100-inch Mount
Wilson telescope, as a member of the Astronomical Society of the
Pacific.[37]

To friends who were astronomers or scientists, Keeler was always
available as a source of help. For Professor J.H. Montgomery of little
Allegheny College in Meadville, Pennsylvania, not far from Pittsburgh,
the Lick director sent detailed advice on how to plan and build an
observatory for the college's 7-inch Clark refractor. To Professor Henry
C. Lord of Ohio State University he sent a reasoned letter, skillfully
designed to persuade the administration there to build a house for Lord
close to their recently erected McMillin Observatory, so that he could
"take advantage of every favorable turn of the weather" by getting to the
telescope quickly. And to help Father Jerome S. Ricard of Santa Clara
College, Keeler was glad to come to the campus, built around the picturesque old mission not far from San Jose, and with him pick out the site
for its new observatory. Afterward Keeler studied its plans in detail, and
suggested a few improvements based on his own experience.[38]

Keeler was even willing to engage in a little polite philosophical disput-
ing with his friends on the faculty of the Jesuit institution. To Father
Anthony Cichi he defended the attitude of astrophysicists such as himself.
The priest had imagined that they believed they knew indirectly measured quantities, such as the temperature of the sun, with absolute
certainty. Not so, Keeler assured him. Scientists push their studies as far
as they can, and the derived temperature was the logical result of the laws
of physics as they knew them, and the measurements of the solar radiation that they had made. But, as scientists, they were in the best position
to realize that "imperfections" in both the method and the data make the
result uncertain, and this always has to be kept in mind.[39] The result is
always subject to improvement by better physical knowledge, or better
measurements, but it is better than no result at all, or than a guess based
on no knowledge or facts. And to a Los Angeles physician, who had
written to ask if he did not agree that the "Divine Purpose" would not let
the earth and the solar system and the sun be destroyed in a collision with
another star, Keeler replied that he found his argument "rather shaky in
some respects." He explained that "[w]e cannot say that a thing is not so,
because we think it not to be so. If the evidence favors a final catastrophe
such as you imagine, ... it is not discredited by our opinion that such a
catastrophe could not form a part of the Creator's design, for of this we
know nothing whatsoever." Everything we understand about the motions

in the solar system, Keeler said, can be understood as resulting from the action of the force of gravitation, and he saw no necessity to assume the existence of any other force outside the solar system. As the stars are moving in random directions, there is nothing to prevent their colliding, except their almost infinitesimal dimensions in comparison with the distances between them. The chance of a collision is almost infinitesimally small. Keeler closed with the thought that "[t]he phenomena of 'new' stars show that catastrophes of some kind do occasionally occur in space, but their nature is at present quite uncertain." Today we know that novae (or "new stars") are not the results of stellar collisions, but rather are episodes in which a star flares up suddenly to a high luminosity, and then becomes faint again, but the rest of his letter is strikingly modern in tone and content.[40]

Along with all the letters he had to write, Keeler managed to find time for a correspondence with David T. Day, his Johns Hopkins classmate, friend, and brother-in-law, who had become the chief of the Department of Mining Statistics in the U.S. Geological Survey in Washington. The two, both by now important scientific administrators, amused themselves by exchanging communications in exaggeratedly bureaucratic officialese. An example is Keeler's note in reply to his brother-in-law's advice about making a relief map:

> We note what you say with reference to Mr. Horrell's ability to make the best possible map of the L[ick] O[bservatory] Reservation; but while admitting the facts you mention, we beg to observe that we have some engineering and artistic, as well as astronomical talent among us, and hope to gain not only a valuable map, but some valuable experience. We will inform you later how our experiments turn out.

No doubt it was side-splittingly funny when received from a boyhood friend at the end of a long, tiring day, if less so to us today. Only a few of these mock-official letters survive as a tiny reminder of Keeler's personal life; almost all of his family letters and papers were lost in the disastrous 1923 Berkeley fire.[41]

The director's many administrative duties were necessary for the smooth running of the observatory, and even its continued survival, but he had taken the job to do science, and he made time for it. He began almost as soon as he arrived at Mount Hamilton. His first published paper from Lick (after his return as director) was completed on July 6, 1898, and came out in print the next month. It reported his spectroscopic confirmation, with the 36-inch refractor, of Campbell's earlier discovery of the "hydrogen envelope" surrounding the Wolf-Rayet star BD +30°3639. This object is actually, as we know today, a minute planetary nebula, with relatively high density and low ionization. As a result, the characteristic green nebular lines that are so strong in the spectra of most planetary nebulae, and that Keeler had measured so carefully in his earlier days at Lick, are suppressed, and the main features visible in the spectrum are the hydrogen lines. The German potentates of spectroscopy, Hermann Vogel and Julius Scheiner, could not see the spectral lines of the tiny nebula with their small telescopes, nor even record them photographically. They believed that what Campbell had seen was simply the optical illusion of spreading of light, or "irradiation" effects, from the emission lines in the star at the center of the nebula, a member of the unusual class of stars with broad emission "bands" or blends of wide lines in their spectra. The spectra of these Wolf-Rayet stars had also been surveyed by Keeler soon after the 36-inch went into operation on Mount Hamilton, and after he had departed for Allegheny they had been studied more systematically by Campbell.

Keeler, as Campbell had before him, simply used the visual spectroscope of the huge 36-inch refractor as a monochromator. That is, he opened the slit of the spectroscope to a width of a little over 0.5 mm, wide enough so that all the light of the star and of the envelope went into it. Looking into the eyepiece of the spectroscope, he could then see the continuous spectrum of the star with its wide emission bands, and also the round image of the nebula itself at the wavelength of Hβ, at which most of its visible light is concentrated. The image at Hβ was clearly not stellar, or point-like, but had a finite size and thus was nebular, although like Campbell, Keeler persisted in referring to it as an "envelope" about the star. When he narrowed the slit of the spectroscope, to see the true width of the spectral lines, Keeler could observe that the Hβ line in the nebula was narrow, just like the lines in other nebulae, not broad, like the Wolf-Rayet bands he could see in the star's spectrum. Clearly the "envelope" was real, and since it emitted Hβ radiation it contained hydrogen. Campbell had already pointed out that the existence of such an extensive envelope around a Wolf-Rayet star must have an important bearing on theories of stellar evolution and of emission-line stars, and Keeler reemphasized his statement.[42] Today we know that planetary nebulae are directly observable cases of the return of material from stars to interstellar matter. Many planetary-nebula stars, in their earliest stages, just after the shell has been thrown off and is at its densest, have Wolf-Rayet type spectra, indicating continued mass loss.

Keeler's paper was important chiefly for its conciliatory tone, so characteristic of him, and so unlike Campbell's polemical outbursts, particularly where controversies with the Germans were concerned. They could not grasp, and would not admit, that their small telescopes had been rendered technologically outmoded by the great Lick refractor on its superior site, and that even youngsters like Campbell, with only a little training and experience, could make discoveries that were impossible for them. Keeler, because of his education at Johns Hopkins, his period of study in their country, and his immense reputation as a spectroscopist, was at least a little more credible in their eyes. It was hard for them to ignore his publications.

Another very similar situation prompted the other paper that Keeler wrote soon after his return. Campbell, as we have seen, had discovered that the relative strengths of the two green nebular lines changed greatly with respect to the strength of Hβ from one part of the Orion nebula to another. In the central, brightest, regions, Hβ was approximately the same strength as λ4959, the weaker of the two green lines, and definitely fainter than λ5007, the stronger. On the other hand, in the faintest, outermost parts of the nebula, Hβ was as much as five times stronger than λ5007, and λ4959 could not be seen at all. Campbell had interpreted these observations as indicating that the substance, whatever it might be, that emitted the two green lines (Sir William Huggins and Agnes Clerke, the historian of astronomy, had independently suggested the name "nebulium" for it)[43] was concentrated in the central parts of the nebula, and that hydrogen was concentrated in its outer parts.

Again, the German observers had not been able to confirm the effect with their small telescopes, and had doubted its reality. Several of the other Lick astronomers had confirmed it with the 36-inch, but Vogel and Scheiner had still been skeptical. Then, in the fall of 1897, just before the Yerkes dedication, the German spectroscopist Carl Runge had traveled through the United States and had visited Mount Hamilton. At Campbell's invitation, he had looked at the Orion nebula with the spectroscope on the big refractor, and had confirmed the variations the young American had seen.[44] Now Vogel and Scheiner could no longer claim to doubt it. Instead, as many other scientists who have missed a discovery have done before and since, they began to belittle its significance. Scheiner now asserted that the spectrum was not really different in different parts of the nebula, but looked different to Campbell as a result of the "Purkinje effect." This is the name of a well-known physiological phenomenon. It is that the color sensitivity of the human eye changes slightly with the level of illumination. For bright light, the maximum sensitivity of the eye is at somewhat longer wavelengths than for faint light. Scheiner believed, on the basis of some photometric experiments he had made in the laboratory, that this Purkinje effect explained what Campbell and Runge had seen.

Keeler, in his paper, agreed with Scheiner that the effect was in the right direction, for Hβ at a wavelength of λ4861 is relatively enhanced slightly with respect to λ4959 and λ5007 at faint light levels, as in the outer part of the Orion nebula. But the amount of the physiological effect is quantitatively far too small, he correctly stated, to explain the very large differences between the spectra of the inner and outer parts of the Orion nebula that Campbell had discovered and Runge had confirmed. Keeler went on to describe how the reality of the observation could be confirmed, either visually by artificially reducing the brightness of the central parts, or better by the objective method of photography. (He was writing in the summer, when the Orion nebula is in the daytime sky; the following winter Keeler himself made both the visual and the photographic confirmations, as we shall see.)

Scheiner had not actually experimented in the laboratory on the slight differences in visual sensitivity between the blue λ4861 spectral line, and the green λ5007, which would have been far too small for him to measure. He had experimented on the differences between Hβ and the red Hα line, and he believed that his experiments had also shown why it is that Hα is relatively faint in the spectra of the Orion nebula and of planetary nebulae, and could be observed visually in only a very few of them, while Hβ could be seen much more readily. In contrast, in a laboratory hydrogen tube (or lamp, as we would say today), Hα appeared relatively brighter with respect to Hβ than in nebulae. The explanation, Scheiner believed again, was the Purkinje effect. Further, he had cooled his hydrogen tube to −200 °C with liquid air, and found that its spectrum was no different from the spectrum at room temperature. Hence, he concluded, there was no reason to suppose that the spectrum of the hydrogen in nebulae was influenced by the cold of surrounding space, and the apparent weakness of Hα could only be due to the Purkinje effect.

This was a most interesting subject to Keeler. At Allegheny he had often puzzled over the obvious differences between the relative strengths of the hydrogen lines in the spectra of the nebulae and in laboratory sources.[45] Again he was sure, from his own observations of nebulae, and his own comparisons of their spectra with the spectra of laboratory hydrogen tubes, that the Purkinje effect was not the whole story. The hydrogen-line spectra of nebulae were different from the hydrogen-line spectrum of a hydrogen tube. "This difference, if it is real, as I believe it to be, may be a key which will finally unlock some of the many mysteries by which the nature and constitution of the nebulae are still surrounded" were the words with which Keeler concluded his paper.[46] Very perceptive words they proved to be, for a generation later astronomers came to realize that under the very low-density conditions in nebulae, the hydrogen lines are emitted in the radiative processes following recombination of electrons with positive hydrogen ions, or protons, in contrast to the situation in the much denser laboratory hydrogen tubes, in which collisional excitation at high temperature leads to the emission of the same lines with very different relative intensities. No one could understand this in Keeler's day, because the relevant physics was unknown, but his accurate observations and careful analysis were pointing in the right direction.

However, Keeler's principal research at Lick, after his return as director, was not based on the 36-inch refractor, nor was it further analysis and interpretation of previously published observations. Instead, as soon as he reached the observatory, he announced that he intended to take over the Crossley reflector himself, and use it for his own research. It was a completely unexpected decision to the Mount Hamilton staff.[47] No doubt they had wondered how Keeler, who had done all the spectroscopic work with the 36-inch refractor before he left Lick Observatory back in 1891, would arrange matters with Campbell, who had succeeded to his position, made his mark as a spectroscopist, and was now embarked on a mammoth program of measuring the radial velocities of the stars. Keeler himself regarded it as the most important program to which a "great telescope" like the 36-inch refractor could be devoted.[48] Campbell was notoriously quick to take offense and would certainly have resented Keeler's taking over the program from him.

The Crossley reflector had sat untouched since Holden left the mountain. Hussey, whom Holden had assigned to put it into working order and use it for research, instead had led a revolt that resulted in the director's forced resignation. In the course of the struggle, Hussey had drawn up a long memorandum summarizing his opinion of the research that had been done with the Crossley, and that could be done with it in the future, as "No Work of Importance."[49] He would have absolutely nothing to do with the poorly mounted reflector, and had gone back to full-time visual observing with the 36-inch and 12-inch refractors, devoting most of his time to double stars.[50] Schaeberle, as acting director, had assigned no one to observe with the Crossley, nor had he done anything with it himself, although he was the one person on the staff who had some experience with reflecting telescopes.[51]

Thus, when Keeler let it be known that he personally would take over the reflector, and devote his own research time to it, both Campbell and

Hussey must have breathed long sighs of relief. Always the diplomat, Keeler certainly knew what he was doing. Almost immediately after accepting the directorship, he had written to President David Starr Jordan at Stanford University to thank him for his congratulatory message on the appointment. "It is very pleasant to know that I shall meet with a cordial reception in California," Keeler told the man who behind the scenes had vigorously supported the movement that led to Holden's downfall, "and I can assure you that in carrying on the work of the observatory I shall aim to secure the friendly cooperation of all my associates."[52]

Yet in taking over the Crossley, Keeler was interested in far more than securing the friendly cooperation of Campbell and Hussey. As we have seen, he had long been aware that reflectors were in principle superior to refractors as light collectors for astrophysical instruments in general, and for spectrographs in particular. They were achromatic, did not absorb the photographically active blue, violet and ultraviolet light, and could be made with much larger mirrors than the lenses on refractors, which at forty inches diameter he had seen showing the first signs of flexure. Keeler's friend Hale had recently published a paper listing the advantages of reflectors over refractors. Although he gave them in more quantitative form, they were essentially the same as those David Gill and Howard Grubb had stated twenty years earlier, before the construction of the Lick refractor was begun.[53] At the dedication of the Yerkes Observatory, with its largest refractor in the world, the assembled scientists had seen George W. Ritchey, working away in the optical shop in the basement on a 60-inch mirror for a reflecting telescope that Hale hoped to erect next.[54] Keeler knew that in the Crossley 36-inch reflector, he had a telescope that would collect large amounts of light, especially photographically active light, from faint objects. He wanted to use it to do spectroscopic work on the nature of these objects – stars and nebulae that had been too faint for him to observe with his Allegheny refractor with its little 13-inch lens that was so nearly opaque to all but yellow light.[55]

When Keeler arrived at Mount Hamilton, there was no working spectrograph for the Crossley reflector. Actually, Holden had secured a contribution of $500 from Catherine W. Bruce to build one very soon after Edward Crossley had announced his gift of the telescope.[56] Campbell had then designed the Bruce spectrograph and Holden had sent his drawings and specifications to Brashear for a cost estimate. His price turned out to be too high, as Holden and Campbell had feared; the complete instrument would cost $1,160, over twice the amount they had. Holden decided to order the optics for the spectrograph (the prisms and lenses) from Brashear, but to hire an instrument maker to build it in-house at Mount Hamilton. He thought he would save money that way.[57] This was the origin of the first spectrograph actually built at Lick Observatory, and it turned out to be a disaster.

The man Holden hired, Percy Urmy, was a skilled watchmaker, but he had no experience whatsoever with precision optical instruments.[58] He did as good a job as he could, but it took an immense amount of Campbell's time to supervise him and keep him on the right track. Urmy left for a better job before he could finish the Bruce spectrograph, which remained in limbo for months. Then Holden was able to hire Zengeler, a regular instrument maker, primarily to work on the telescope itself, but also on the spectrograph, which he had supposedly "essentially com-plete[d]" by March 1897. Holden wrote a boastful letter to Brashear, claiming that by his own accounting the spectrograph had cost the observatory only $850 (he did not count anything for Campbell's time) rather than the $1,160 price that the optical-shop owner had quoted. Brashear was naturally deeply hurt. He knew that he did good work and never overcharged for it. He liked nearly all the other astronomers he did business with, but this experience only confirmed his strong antipathy for Holden. According to his calculation, including Campbell's time and an allowance for overhead, the real cost of the Bruce spectrograph to Lick Observatory was at least $200 more than what he would have charged.[59] And in fact it was still not ready for use by the summer of 1898.

There were some very serious design flaws in the design that Campbell had provided. An astronomical spectrograph must be matched to the telescope with which it will be used, and Campbell had not paid enough attention to the peculiarities of the unfamiliar reflector. Perhaps the problems it presented were insoluble with the techniques of his day; certainly the Bruce spectrograph, built to his design, was unusable. The spectrograph had to be mounted at the prime focus of the telescope, at the upper end of the tube. Any weight placed there had to be balanced by counterweights ten to twenty times heavier at the mirror end, because the center of rotation of the mounting was so close to the bottom of the tube. It proved impossible to find a way to hang on the telescope all the counterweights that would be required to balance the heavy Bruce spec-trograph. Even worse, it was larger than the space between the end of the tube and the inside of the dome, so the shutter could not be closed with the instrument in place. It had to be removed from the telescope before the dome could be shut.[60]

All this Keeler discovered as soon as he inspected the Crossley. He measured the alignment of the polar axis of the telescope, about which it rotates to counteract the diurnal rotation of the earth, and immediately found that it was not correct.[61] The polar axis was not parallel to the axis of rotation of the earth, so the telescope could not possibly track the stars correctly. Neither Holden nor Hussey had noticed this, although they had the telescope mounted in the dome for over a year before the first director's resignation and departure. The problem arose because the telescope had originally been made for use in England, which is far north of central California. Hence the polar axis, as the mounting had been built, was more nearly vertical than it should have been for Mount Hamilton. To modify it for the Lick Observatory, Holden had ordered a wedge-shaped steel block cast, which was inserted between the telescope pier and the bottom of the mounting. Although Holden had specifically asked Hussey to check the angle carefully, it was wrong.[62] Both of them, and the designer of the wedge, had assumed that the telescope had been built for use in Halifax, Crossley's home in the North of England, but actually it had been made and first used by Andrew A. Common in Ealing, a suburb of London. As a result of this misunderstanding, the angle of the wedge was wrong by approximately two degrees.[60] Apparently no one had measured the actual inclination of the polar axis of the Crossley, mounted in the dome on Mount Hamilton, before Keeler arrived on the scene.

As the director, he could command all the resources of Lick Observatory's little staff. Within a month, Keeler had McDonald, the Mount Hamilton foreman, cut down the pier by two feet, lowering the telescope by a similar amount and thus providing more clearance between it and the dome. At the same time, McDonald finished off the top of the pier with a slight bevel, calculated by Keeler, so that the polar axis was now accurately parallel to the axis of rotation of the earth. Zengeler was put to work on a new drive clock for the Crossley, which would make it follow the diurnal motions of the stars accurately. A telephone line was installed from the Main Building to the Crossley dome, so Keeler could keep in close touch with the progress of the work there. With Palmer assisting him, he "squared on" the mirror, adjusting it so that its optical axis was accurately aligned with the center line of the tube. Keeler had a new low-power finder telescope, for picking up the right area in the sky, installed on the reflector, and a new, lighter-weight tube put on the high-power finder. McDonald and Zengeler straightened and repainted the telescope mounting, and Keeler and Palmer readjusted the polar axis, and touched up the orientation of the mirror with a special collimating eyepiece. Every step cost tremendous effort, but Keeler was willing to spend it to bring the Crossley into operation. He knew that the way to get it done was to do it himself. When Zengeler finished the drive clock and installed it, Keeler and Palmer spent several nights testing, adjusting and improving it before they were satisfied with its operation.

Finally on September 15, Keeler took his first photographs with the Crossley, of a bright star. They were not successes. The images were out of focus. The plateholders were warped. The magnifying power of the eyepiece was too low. The field of the finder telescope needed illumination. The cross hairs were too coarse. The drive oscillated. Keeler analyzed every problem, decided on a solution to it, carried it out himself or had one of his associates do so, and passed on to the next problem. After this round of improvements, on September 27, less than four months after he had taken over as director, he tried his first photographs of the moon with the Crossley. The result, laconically noted in his observing book, was "Plates are fairly good."[61]

Still further adjustments, improvements and modifications were carried out on the basis of this experience. The drive system of the telescope was tightened up. Keeler began experimenting with guiding with a "double-slide" plateholder at the focus of the reflector. The idea is to rate the driving clock so that the telescope follows the diurnal motion of the stars to a very good approximation, but to make the final, finest motions by moving the plateholder itself, in the focal plane of the telescope. The plateholder is firmly clamped so that it cannot move with respect to the guiding eyepiece, but plateholder and eyepiece move together in two coordinates, controlled by fine screws operated by the observer. He turns them so as to keep the guiding eyepiece constantly centered on a "guide star", located just off the edge of the field that is being photographed, and so long as the polar axis is correctly aligned, the result is that all the star images are held fixed on the plate. Current for the light that illuminated the cross hairs was brought down to the Crossley dome by wires from the storage batteries in the Main Building, the only source of electricity on the mountain at that time. Since the field that is being photographed and therefore the plateholder are centered on the optical axis of the telescope, the guide star is slightly "off axis", as the astronomers say. With a relatively fast optical system like the Crossley reflector, the image of the guide star is therefore not perfectly round, but rather is "comatic" or pointed at one end, and Keeler had to teach himself to guide on these elongated images, and to judge and adjust the changing focus of the telescope from their appearance.[60]

Finally, in preparation for serious photographic work, Keeler had the darkroom repainted, and he himself chemically cleaned the Newtonian flat mirror of the telescope, and then resilvered it, making sure to use the latest improvements to Brashear's original method.[63] Probably Keeler would have continued his experiments longer, but in early November Comet Brooks 1898 appeared in the sky, and he decided to photograph it. He got his first exposure, twenty-nine minutes long, in a howling gale. He could not make it longer because the comet, close to the sun, sank so near the western horizon that the telescope could no longer turn to follow it. Since comets move with respect to the background of stars, Keeler had to guide the telescope directly on the nucleus of Comet Brooks, using the long finder telescope instead of the guiding eyepiece.[64] As a result of the comet's motion, the star images were all drawn out into long trails, somewhat irregular because of the wind and the stickiness of the controls. But the comet itself showed a good round image, with a straight, narrow tail projecting from it. He measured the position angle of the tail, and found that it deviated only a few degrees from the prolongation of the "radius vector", the straight line extending from the sun through the nucleus of the comet.[61] This measurement indicated that there is a strong repulsive force from the sun on the particles in the comet's tail, in agreement with many earlier observations of other comets and the theory developed from them by F.W. Bessel. We now understand this repulsive force in straight tails like that of Comet Brooks to result from the "solar wind" of charged particles, streaming out in all directions from the sun.

Keeler, assisted by Palmer, succeeded in getting good plates of the comet on eleven consecutive nights until it went out of sight behind the sun. He measured them all and published the positions of the comet, and the directions of the tail.[65] This was not ground-breaking work, for Barnard and others had photographed many previous comets with small telescopes built around wide-angle lenses, but Keeler's photographs had much larger scale, and showed correspondingly finer detail, since the focal length of the Crossley is so large. This comet paper, the first research Keeler did with the reflector, represents more of a warm-up than an important new pathway in science.

As Comet Brooks disappeared behind the sun, Keeler shifted his attention to the Pleiades, the little cluster of bright stars, "the Seven Sisters", that is so obvious in the autumn skies. It is an area very rich in nebulosity, as was known from earlier visual work, which had revealed the brightest portions of it, and from Barnard's wide-field photographs. Again Keeler's Crossley plates showed very fine detail and especially the filamentary structure that is characteristic of the Pleiades. He succeeded in taking exposures up to an hour and a half long, alternating in guiding with Palmer. On such long exposures the bright stars in the Pleiades produced halation rings on the photographs, and Keeler learned to "back" his plates with a piece of red glass in the plateholder, pressed into optical contact with glycerine.[61] The glass absorbed the blue light that passed through the photographic emulsion coated on the front of the glass plate; without the backing it would have been partly reflected back through the plate a second time and thus produced a diffuse ring around the star image. Keeler's photographs showed that the whole Pleiades region is permeated with nebulosity, and that some features which had previously appeared to be isolated nebulae were in fact especially bright condensations and not separate objects at all. He marveled at how much faint detail the Crossley photographs could show.[66]

Keeler wanted to observe the spectrum of the Pleiades nebula, which he thought "from the character of the involved stars" would be "of the usual gaseous type."[67] In other words, he realized that the emission-line spectra characteristic of planetary and diffuse nebulae like the Orion nebula were connected with the presence of stars with spectra similar to the Orion stars, which we now know are hot stars. With a low-dispersion spectroscope on the 36-inch refractor Keeler tried several times to see the spectrum of the brightest nebula in the Pleiades but never succeeded. It was always too faint.[68] With the advantage of hindsight, we recognize today the reason is that the Pleiades nebulosity has a continuous spectrum, which is spread over all wavelengths, rather than concentrated in a few strong emission lines. Hence although the nebulosity can be photographed easily because of its brightness, in a spectroscope no single wavelength is bright enough to see. The Pleiades stars are just a little cooler than the Orion stars, and therefore do not emit enough ultraviolet light to ionize the nebulosity and make it glow; what we see is the optical starlight scattered by dust particles in the nebulosity. This was first clearly realized by Vesto M. Slipher in 1912.[69]

By mid-November, Keeler could begin photographing the Orion nebula itself. The first plate he obtained, a forty-minute exposure, turned out to be "a superb photograph, showing great extent and at the same time the individual stars in the trapezium." He had guided the telescope for the first twenty-five minutes of this exposure; Palmer for the last fifteen. Then Palmer guided a second, five-minute plate, and even this one "show[ed] a wonderful amount of nebulosity for such a short exposure." Keeler published a reproduction of the forty-minute exposure in the *Publications of the Astronomical Society of the Pacific,* as a first sample of the results he was getting with the Crossley.[70] It was a revelation to most astronomers; the combination of light-gathering power and focal length of the Crossley opened up a whole new world of fine detail in even the faint parts of the nebula. Although Keeler was pleased with the result, he emphasized that the published reproduction did not do full justice to the original; the whole range of brightness could only be seen on a glass transparency.[71]

His main purpose was not pictorial, however. Keeler was trying to observe and understand the differences in the spectrum of the nebula from point to point that Campbell had reported, and that the new director had himself confirmed. To survey the whole nebula, Keeler obtained from John Carbutt, the photographic supplier in Philadelphia, several special light yellow-green glass filters.[72] He had Wright test them with the spectrograph and a laboratory light source, and confirm that, as desired, they completely absorbed the shorter-wavelength radiation and transmitted longer wavelengths. The transition was at about Hβ, which was partly transmitted and partly absorbed; light of slightly longer wavelength, like the two strong green nebular lines (at wavelengths λ4959 and λ5007) was even more strongly transmitted than Hβ. Thus using one of these filters in front of an orthochromatic (green-sensitive) plate, he could take a picture of the nebula very similar to the image seen by the human eye, with the chief contribution coming from the two green "nebulium" lines. By comparing this with one of the ordinary, blue-sensitive plates, taken without a filter and hence showing the nebula chiefly in the light of Hβ and the shorter wavelength lines, he would be able to determine, all over the nebula, which areas were relatively brightest in the "nebulium" lines, and which areas were relatively faintest in those same lines. This he did, but it is much easier to state than it was to do. There were agonizing delays. One batch of orthochromatic plates was received from the G. Cramer Dry Plate Company fogged. The next batch was relatively slow. Keeler finally sensitized some blue plates with erythrocin himself, thus converting them to orthochromatic sensitivity.[58]

His best comparison for the inner part of the nebula was between a four-minute exposure on an ordinary blue plate and a two-hour twenty-minute exposure taken through the filter on the Cramer orthochromatic plate; this was supplemented by a two-hour exposure taken on a different night through the filter on the plate he had sensitized. Coddington assisted him with these photographs, for Palmer had gone down to Berkeley to take graduate courses during the second semester. The main result of the Orion plates they took, which Keeler clearly described, is that the green "nebulium" lines are relatively strong throughout the central, brightest part of the nebula, close to the Trapezium stars. Most of the outer part of the nebula is relatively faint in these same lines. Intermediate regions are intermediate in their properties. Thus Keeler confirmed Campbell's visual observations as far out as they had gone, and extended them to the even fainter, more distant parts of the nebula. One long "streamer," or as we would say today, front, extending to the southeast, is relatively strong in the green lines, while another, shorter, more diffuse feature near to it in projection on the sky is relatively faint in them. Keeler's photograph confirmed that the nebulous region around the star Bond 734, north of the main body of the Orion nebula, is quite faint in the green lines (this is the region in which Runge had confirmed Campbell's visual observations), and found for the first time that the "loops" or "arcs" in the western part of the nebula share this property.

Keeler's report of this work, published in the *Astrophysical Journal,* is an excellent scientific paper. He clearly states his hypotheses, describes just what he did, gives the details of the exposures, tells explicitly the comparisons he made, and declares firmly his conclusions. There is very little theorizing in the paper, but in fact it is a good first reconnaissance of the problem of the excitation of nebular spectra. In a more popular version of the paper, published in the *Publications of the Astronomical Society of the Pacific,* he implicitly attributes the spectral differences in various parts of the nebula to differences in "composition."[73] By this word he meant, as we would, the chemical elements that were present. In speculating about the meaning of his results, he said they were difficult to interpret, but appeared to show that the nebula was nonhomogeneous, with hydrogen concentrated in the outer parts, and "other substances nearer the center."[74] In this he was wrong. If by "composition" he had meant the stage of "excitation," or actually ionization of these elements, he would have been right. As we know today, the green lines are emitted by doubly ionized oxygen, [O III], which is especially strong close to the hottest stars in the Trapezium, that emit copious amounts of ultraviolet ionizing radiation. He could not have been expected to know that, but his observational results were putting him on the track toward it. Keeler's only real mistake was that he believed that the ordinary photographic picture was taken mostly in the blue light of the hydrogen lines Hβ, Hγ, Hδ etc. Actually, by far the strongest line in the photographic spectral region is [O II] $\lambda 3727$, but since all the spectrograms he had seen were taken with refractors, whose glass lenses strongly absorb ultraviolet radiation, he did not realize this. In reality, his "ordinary" photographs, taken with the Crossley reflector, showed a mixture of the hydrogen and singly ionized oxygen radiation. In any case, he was seeing very real physical differences within the nebula.[73]

This paper was widely recognized as a very important step forward. Hale congratulated Keeler on having certainly settled the Campbell-Scheiner controversy with his Orion photographs, and on getting such good results with the telescope that everyone else had seemed to avoid.[75] Only the German astronomers, led by Scheiner, still could not believe anything they had not themselves discovered, particularly on variations in the spectrum of the Orion nebula. Keeler was able to refute all of Scheiner's objections very convincingly, with observational facts and arguments drawn directly from his photographic experience.[76] In contrast to Campbell's published rejoinders on similar occasions, Keeler's reply was polite, friendly and reasonable, and therefore impossible to interpret as a personal attack and to ignore.

In the summer of 1899, Keeler, assisted again by Palmer now that classes were over, took long-exposure photographs of several other diffuse nebulae, M 8 (the Lagoon nebula), M 17 (the Omega nebula), and M 20 (the Trifid nebula).[77] We know them so well today that it is hard for us to realize how sensational his photographs were to the astronomers of his time. They were the first pictures they had ever seen taken with a telescope that had a long enough focal length to show the fine detail, and a fast enough focal ratio to record the faint diffuse nebulosity. They showed much more detail than even the best drawings of the earlier visual observers. Best of all, they were objective, permanent records. The more Keeler studied his photographs, and the more he compared them with the drawings, the less he believed the earlier reports of purported changes in the nebulae.[78]

A little before he took these diffuse nebulae, on April 4, 1899 Keeler tried his first photograph of a spiral nebula. It was a two-hour exposure of M 81 and showed the spiral structure very well, but the object is far to the north, and the polar axis of the telescope was slightly misaligned, resulting in all the stars' images being drawn out into short arcs on the plate. He spent the next seven clear nights adjusting and readjusting the axis until he got it just right.[61] Then in May he took his first plate of M 51, the Whirlpool nebula, another magnificent spiral. He took several photographs of it, the longest a four-hour exposure.[77] On this plate he could see, besides M 51, many smaller nebulae scattered over the field. There was a wealth of detail in M 51, far more than Lord Rosse's drawings showed, or than Isaac Roberts' photographs, that Keeler had seen in Liverpool in 1896. The long exposures under the beautifully clear, steady Mount Hamilton sky, enabled the Crossley reflector to record much fainter features than any astronomer had previously glimpsed. Keeler was very pleased with this photograph, and sent a transparency of it to Hale.[79] When it arrived at Yerkes it created a sensation. According to the ecstatic letter Hale sent back, "everyone in the Observatory considers [this picture] to be far superior to anything of the kind they have ever seen or expected to see. When we get the 2-foot [24-inch reflector] finished we will try our hand at the same object, but it is easy to predict that no such results can be obtained with that instrument."[80]

In July Keeler obtained an excellent series of seven photographs of NGC 6720, the Ring nebula in Lyra.[77] The exposure times ranged from two hours down to thirty seconds, and the nebula was visible on every one of them. It showed best on the ten-minute exposures, but the others helped in making out the structure in the fainter and brighter parts of the nebula, that were under- or overexposed on this picture. All the real features reported by earlier visual observers were shown, as well as many new ones. Keeler could see that actually the nebula was egg-shaped, not elliptical, and that in several regions "faint masses or fringes of nebulosity" projected out from the ring. Furthermore, the ring was made up of many interlacing filaments or "narrow bright rings", and the region within it was crossed by a series of alternately bright and dark bands. Some of these aspects of the structure of the nebula had been vaguely stated by some previous visual observers and contradicted by others, but no one who looked at Keeler's plates could doubt his accurate description. He noted that the central star of the Ring nebula is very blue, as shown by the contrast between its bright photographic image and faint visual magnitude, and he revealed that he was planning a spectrograph (for the Crossley) with which he believed he would be able to study its spectrum.[81]

Hale read Keeler's paper for him at the "Third Conference of Astronomers and Astrophysicists", held at Yerkes Observatory. It was also the first meeting of the very new Astronomical and Astrophysical Society of America (now the American Astronomical Society). The conferences held at the Yerkes dedication in 1897, at which Keeler had given the main address, had been so successful that Hale immediately began laying plans for a second meeting the following summer. Originally he planned to hold it at Williams Bay, but when opposition developed to going back to the same site, and Edward C. Pickering offered Harvard as an alternative, Hale agreed to the change. At this "Second Conference", the idea of forming a professional national astronomical society came out into the open. Simon Newcomb had broached it privately at the dedication, but Hale had managed to stall and keep the issue from being formally discussed there. He and Keeler represented the coming generation, and they both feared that if the older man had his way, the proposed society would be a Washington-based organization that he controlled. Keeler would have preferred no organization at all, but favored the continuation of the research conferences on an informal basis. Hale wanted a truly national society that included physicists as well as astronomers, and he was willing to work hard for it.[82]

By the following summer he had lined up support for his plan of an "American Astronomical and Physical Society." The participants at the Second Conference, which was held in the drawing room of Pickering's house at Harvard College Observatory in late August 1898, voted unanimously either to form an astronomical and astrophysical society or to continue holding annual informal conferences. They voted to put the decision in the hands of a committee consisting of Pickering, Newcomb, Hale, George C. Comstock (who was his close ally), and physicist Edward W. Morley. After one day's deliberation this committee not surprisingly decided on a permanent society and during the next week they signed up the first sixty-one charter members. The same five were named as the first council of the society, and were charged with drafting its constitution and planning its first meeting, to be held at Yerkes Observatory the following summer.[83]

It was at this Third Conference that Hale read Keeler's paper on the Ring nebula, and at which the Astronomical and Astrophysical Society of America (the "compromise" name chosen) was formally organized. Newcomb was elected its first president, Charles A. Young and Hale its vice presidents, Comstock its secretary, classical astronomer Charles L. Doolittle its treasurer, and Pickering, Keeler, Morley, and Ormond Stone of Virginia as members of the council.[84] Throughout this entire period Hale worked twice as hard as anybody else on setting up the society, and made sure that he attended all the meetings and wrote all the reports, no matter who was officially named secretary.[85] It is not surprising that the Astronomical and Astrophysical Society of America came out pretty much as he wanted it to.

In another respect, Hale was not able to enforce one of his "schemes" so successfully. He hoped to organize a small meeting of working astronomical spectroscopists on the day before the Third Conference began, to discuss current radial-velocity programs. The agenda was to be to work out "a general scheme of cooperation in line of sight work." It was essential that Campbell attend, for at Lick Observatory he was the one American astronomer actually carrying out a large and very successful radial-velocity program. Hale was grooming Edwin B. Frost, who had been hired in place of Keeler, to begin a similar program at Yerkes, but so far it had not got started. Partly the small meeting was designed so that Frost and Hale could learn from Campbell's experience, but partly also they wanted to divide up the stars to be observed with him. The Lick astronomer was glad to tell them all he knew, but he would have nothing to do with a joint plan. He was so far ahead of them that cooperating would only slow him down. His program was straightforward – to observe all the stars he could as fast as he could – and since he was doing the work, he naturally thought he ought to make the decisions himself, not act as Hale's observing assistant. This was the message that Keeler politely passed to his old friend, the Yerkes director, on behalf of his premier researcher.[86] The meeting did not take place, but Campbell was always more than generous in helping the Yerkes observers with advice, suggestions, and samples of his own work.[87]

When he mailed his paper to be read at the Third Conference, which began on September 6, 1899, Keeler sent with it a box of transparencies to be exhibited to the assembled astronomers and astrophysicists.[88] The pictures included not only several of the Ring nebula, but also some of the Orion nebula, the Pleiades and its nebulosity, and the spiral M 51. This collection of photographs was a revelation to the participants at the conference. Hale reported that Keeler's three slides, shown on the screen, brought out his points "beautifully", but that the transparencies "created a genuine sensation and showed to many who had been skeptical regarding the advantages of reflectors what the instrument is capable of doing in the right hands. Barnard was simply delighted with the photographs and stood for hours in front of them admiring their details." Hale went on to "congratulate Lick Observatory upon the extraordinary amount of valuable material which it is turning out. Your photographs, together with Campbell's spectroscopic binaries, ought to bring glory enough to the institution for one year at least. Campbell's results upon Polaris, which Frost has just confirmed, are of the very greatest interest and illustrate in a superb way the excellence of his instruments and methods."[89] Keeler was not only doing excellent research himself, but as director was also maintaining the conditions under which the other scientists on the staff could work productively. Brashear had written Campbell, "now that you have my good friend Keeler handy I am sure you fellows will make a *team* that need not be afraid of anybody."[90] Hale's strong vote of confidence in Keeler and in the research pouring out of Lick was too good for the director to keep to himself. He forwarded it to the president of the University of California, along with his own report of the work done at the observatory during the year, and sent a long extract from it to one of the most important regents.[91] Keeler, like Holden, never neglected to inform his superiors in Berkeley of his successes, but the new director had more of them, and more solid ones, than the old one had.

Hale's praise had not been fulsome, either. This was confirmed by the fact that several of the astronomers who saw the transparencies of the

Crossley photographs displayed at the Third Conference, immediately wrote Keeler to beg to be allowed to copy them for their own books or classes. He wanted his work to be widely known, but he was reluctant to distribute copies far and wide before he had had a chance to analyze them himself, and write up his conclusions for publication. He gave permission for his old friend Comstock, and for his fellow Johns Hopkins student Frank P. Whitman to have copies made, but cautioned both of them not to let them go further.[92]

After the display at the conference, Keeler had arranged to have most of these transparencies sent on to the Royal Astronomical Society in London, where they were exhibited. One sparkling picture of the Orion nebula went directly to Crossley, the donor of the telescope. Another went to Holden in his lonely exile in New York City. They got single pictures, but Keeler sent complete sets to the New York Academy of Sciences, and to the American Philosophical Society in Philadelphia. He made sure that the scientific opinion molders saw these spectacular examples of his work.[93]

Everyone recognized these photographs as "very fine ones". Crossley was especially struck by the picture of the Orion nebula, which he called "the finest I have ever seen & I value it much. It proves to me how important it is not only to have a powerful instrument but also a site where it can be used to the greatest possible advantage." Frank Wadsworth was more direct and wrote "I have been reading your recent papers ... on your work with the Crossley reflector with great interest. I always thought that good work could be done with that instrument if the right man got hold of it."[94]

Keeler worked very hard to get this kind of recognition. He observed with the Crossley every clear night in the two-week period around each new moon, when the sky is dark enough to photograph faint nebulae. Although he did not stay up all night long, he usually worked until well past midnight, and then was up the next day to do a full day's work in the office, writing letters, vouchers, reports, and occasionally, if he were lucky enough to find time, a scientific paper. He made every one of the large transparencies of the Crossley pictures that were sent out from Lick Observatory himself, because no one else on the mountain could do it as well as he.[95]

The outside recognition was important in keeping the confidence of the university administration in Lick Observatory high. In the summer of 1899 Keeler lost his greatest supporter, Phelps, who had been a regent since 1880 and chairman of the Lick Observatory Committee since the day the university took it over in 1888. He was seventy-five years old, but still led a healthy, outdoor life, until he was hit by a tandem bicycle speeding down a hill while he was walking near his San Carlos home. The old man suffered a severe head injury, and a few days later he was dead. Phelps was a powerful politician, not at all well educated but extremely shrewd and skillful in his dealings with people. Lick Observatory was his special interest, and he protected it faithfully at the regents' meetings and in the legislature. He had been on the board so long that everyone knew him; he had proved himself so many times that everyone trusted him. To Keeler he had become not only an adviser and protector, but a close personal friend.[96]

Just at this time there was a severe budget crisis in the university, brought on by a huge state deficit. The regents had to cut costs, and they handed the problem over to a faculty committee, made up of the heads of the departments. Keeler, the head of the only one not physically located on the campus, was at a severe disadvantage. Wild suggestions of drastically cutting the Lick budget surfaced at Berkeley. However, Leuschner, by his own account, loyally defended the observatory and reported what was happening to Keeler, something he certainly would not have done if Holden had still been in charge on Mount Hamilton. Keeler hurried to Berkeley, and although at one time he had to agree to give up the salaries for three positions, those of the instrument maker and two of the fellows, in the end the crisis was staved off and all that Lick lost was one fellowship.[97] This far Keeler was willing to go because Ross was having trouble with his eyes, and had decided to drop out of astronomy, resign his fellowship, and devote his full time to studying mathematics. Before the financial crisis Keeler had asked Leuschner to recommend a successor to Ross. The man he nominated, his best student, in Keeler's opinion lacked something, perhaps a sense of proportion, perhaps only a sense of humor. At best he was a "capable ass." Instead of him, Keeler asked the Berkeley professor to name his "second best student." He could not have been too displeased when the fellowship evaporated.[98]

The high scientific reputation of Lick Observatory was the main asset that Keeler used in his attempts to raise money for additional instruments and salaries. He assiduously cultivated the wealthy Phoebe Hearst, and recommended that she be named to the Lick Observatory Committee after Phelps' death. The appointment was not made, and Keeler never got any money out of her for the observatory. He tried very hard when Alice Everitt, a graduate of Cambridge University in England, who had worked at the Royal Greenwich Observatory and at Potsdam, applied for a post at Lick. Keeler wanted to hire her as an assistant astronomer, to work with Campbell in measuring and reducing the radial-velocity spectrograms that were piling up so rapidly. He asked Mrs. Hearst to establish a position for her at the observatory. The wealthy widow replied that although she was very sympathetic to Lick Observatory, Miss Everitt, and Keeler (as she had not been to Holden), she was concentrating all her funds on the George Randolph Hearst Memorial Building for the School of Mines on the Berkeley campus and could not afford to help with the necessary $1,200 a year.[99]

Keeler also tried to raise money for the observatory from D.O. Mills, but the legendary financier breezily replied that he was "unable to respond, as requested" because he was "rather overburden[ed] with donations of various kinds." However, he "[t]rust[ed] that you may have found some one to do the needful" and closed with his "best wishes for the University, and the Lick Observatory."[100]

On October 25, 1899 the University of California inaugurated a new president, Benjamin Ide Wheeler. He brought a breath of fresh air to the campus, which Keeler and the Lick staff welcomed. Keeler and Campbell went to Berkeley for the ceremonies, held in the open alongside a eucalyptus grove. The main speaker was Daniel Coit Gilman of Johns Hopkins, who had been president of the University of California years before Holden's appointment. President Jordan of Stanford University was also present. Keeler must have nodded in approval as he heard his old mentor extol "the most fertile achievement of the nineteenth century," the introduction of measurements of precision. Gilman proclaimed American laboratories, observatories and surveys as "among the best attainments of our countrymen." His whole speech was a call for excellence in research. After the ceremonies, Gilman visited Mount Hamilton with his wife, and saw how well the new director was succeeding in the quest. Gilman was later to describe him as "one of the most brilliant" of the early students at Johns Hopkins.[101]

By this time Keeler knew that he was doing a good job. He considered the organization of the observatory, the staff, and the research they were doing, all "quite satisfactory." The only improvement he would suggest would be more money for operating expenses and income. He was fortunate that Lick was at such a fine site. Not only the many clear nights, but the excellent seeing, made it possible to photograph very faint objects, and would make it possible to obtain spectrograms of them. All mountains, he knew, were not good observing sites; for instance Pike's Peak and Mount Whitney both had much poorer seeing than Mount Hamilton. His experience with the Crossley had convinced him that reflectors were the most effective telescopes for astrophysical research. In the past reflectors had not been adequately mounted, and the Crossley itself had many defects. But he was certain that a large reflector, carefully mounted in an excellent mountain-top site, would be the ideal instrument for astrophysical research, and relatively inexpensive to boot.[102]

These opinions were all firmly based on Keeler's own experience. He continued observing with the Crossley, now photographing mostly nebulae and star clusters.[103] One night in December, however, he took a long exposure on the field of T Tauri and Hind's variable nebula, and when he developed the plate found on it the trail of an asteroid. It was immediately recognizable by its motion. At first Keeler thought it was a new Trojan asteroid, moving in an orbit at the same distance from the sun as Jupiter, but a second exposure, obtained after two intervening cloudy nights, showed this was not so. Many asteroids had previously been discovered photographically with small, wide-field telescopes, especially by Max Wolf in Germany. A few had been found by Coddington at Lick. But this object was much fainter than any previously known asteroid, and Keeler was anxious to follow it up. Two more positions, at intervals of a week or more were necessary to calculate its orbit. By now the typical Mount Hamilton winter weather had set in. It was very cloudy, and if the skies cleared the wind was howling, the Crossley was shaking on its mounting, and the seeing was atrocious. Palmer had gone to Berkeley, and Coddington assisted Keeler at the telescope. Finally on Christmas Eve the skies cleared, the wind died down, and the seeing became fair. They took a thirty-minute exposure at the position predicted for the asteroid by extrapolating its motion, but too much time had elapsed and they did not find it. On Christmas Night they tried again, with a one-hour exposure in a slightly different position. This time Keeler thought the asteroid's trail might be on the plate, but it was so faint he was not sure of it. Finally on December 26, they found it, on another one-hour exposure.

The third position was obtained at the next dark of the moon, after the New Year. By now it was easy to extrapolate the expected position well enough to find the asteroid, but it was moving away from the sun and rapidly becoming fainter. On January 18, 1900, Keeler and Coddington tried to photograph it, but got the right-ascension motion control ropes tangled and had to stop the telescope; by the time they found the field again they could only get a thirty-minute exposure before the moon rose. But the next night they managed to take an hour and a half exposure, and comparing these two plates, found the asteroid again.[104] All these plates were measured by Coddington and Palmer, yielding three accurate positions. From them Palmer calculated a preliminary orbit for the object, which was temporarily named 1899 FD.[105] By the time it next came to opposition, in the spring of 1901, Keeler was dead and no attempt was made to photograph it with the Crossley. Probably no other telescope could have recorded it at that time. It was not photographed again for many years, but on the basis of the preliminary orbit it was given the permanent number (452). By tradition asteroids are named by their discoverers, always with feminine names in those unenlightened days. Since Keeler was dead, the right devolved on his widow. Campbell suggested to her that Cora, her own name, would be most appropriate, but it, Anna (Keeler's mother's name), Isabel[la], (Cora's mother's name) and California had all been preempted. Therefore Campbell coined the name Hamiltonia, and with Cora Keeler's approval it was adopted.[106] However, by the time astronomers were in a position to photograph a seventeenth-magnitude asteroid routinely, the deviation of the actual motion from the preliminary orbit was so large that (452) Hamiltonia was lost. For many years this object was the faintest known asteroid, but it could not be located. It was accidentally photographed, but not recognized, in 1958 at Indiana University and in 1973 at the Crimean Observatory. Then in 1981 fine detective work by L.K. Kristensen, E.C. Bowell and Brian C. Marsden led to the identification of all these images with Hamiltonia, and its orbit is now securely known far into the future.[107]

Keeler's most important work with the Crossley was on the objects we recognize today as spiral galaxies. To him they were spiral nebulae. The first few he photographed in the spring and early summer of 1899, M 81, M 51 and M 101, were objects in which Lord Rosse, with his giant visual reflecting telescope in Ireland, had dimly seen the brightest parts of the spiral pattern. Keeler's long-exposure photographs showed much more of their spiral arms than anyone had ever seen before. He could see these objects were basically double-armed structures, in which the spiral pattern could be traced around, more or less continuously, for several turns. By now he realized that the early visual descriptions were almost useless in predicting what the photographic image of a nebula would look like. He began photographing nebulae more or less at random, to see what they were. On August 3 he took a one-hour exposure of NGC 6946, and as soon as he developed the plate, he recognized that it was another spiral, similar to M 51, M 81 and M 101 in appearance, but much smaller in size. A few nights later he got a 90-minute exposure of NGC 6412, and a 105-minute exposure of NGC 6951. Both were the same. Clearly they were flattened, rotating objects, brightest at their centers, seen in various projections, but all similar to the giant Andromeda nebula, M 31, which was too large for Keeler to photograph with the Crossley, but which he knew well from Barnard's picture. The next night, August 9, Keeler got a fine photograph of NGC 7331, described by Herschel as an elongated nebula, and drawn by Lord Rosse as two parallel rays. Actually, Keeler could see on the plate, it was another spiral, seen nearly edge on. Because spiral galaxies are so flattened, it had appeared elongated to Herschel and Rosse, but Keeler could trace some of the spiral structure and at the same time see the dark band (which we now realize is due to interstellar dust) down its central plane.[104]

From these plates, Keeler concluded that a large fraction of the known nebulae were in fact spirals. Of course, there were exceptions, like planetaries, but he felt almost certain that most of the nebulae described by visual observers as "pretty bright, round, brighter in the middle" nebulae would turn out to be nearly face-on spirals like M 101 or NGC 6946, and most of the "much elongated" ones, edge-on cases like NGC 7331. Although he was not ready to publish this result, he hastened to write Agnes Clerke and inform her of it. As the chief astronomical writer of the English-speaking world, she was an important opinion molder, and Keeler wanted her on his side. He sent her slides of several of his photographs of spirals, and explained how much more they showed than Roberts' pictures.[108]

As Keeler studied his photographs, he could see that on nearly every plate there were many small nebulae scattered over the field. He published a paper listing the positions of seven of them that showed on the plate he had taken centered on M 51.[109] He continued taking one or more long-exposure photographs every clear, moonless night that fall. On a photograph centered on NGC 891, which he recognized as a nearly exactly edge-on spiral, he found thirty-one new nebulae and nebulous objects; on his earlier photograph of NGC 7331, twenty more. Making the very conservative estimate that there were three nebulae per square degree, considerably less than the average he had found on the plates he had taken, he estimated the number of nebulae in the whole sky that were bright enough to photograph with the Crossley as 120,000. Furthermore, a very large fraction of them appeared to be spirals. This he considered highly significant. "We may perhaps regard the spiral form as that which is normally assumed by a compact isolated nebulous mass, and the exception (of which there are many) to somewhat unusual conditions – possibly, in part, to the absence of any definite moment of rotation for the aggregate of the particles constituting the nebula," he wrote.[110]

He was correct in nearly every statement he made. There are very many spiral galaxies in the universe within reach of the Crossley reflector. There are, however, more exceptions than he realized, elliptical galaxies whose "particles'" (stars') total angular momentum apparently adds up to nearly zero, by cancellation. The definition of his photographs was not good enough for him to recognize more than a few of the small ellipticals. Nearly all the large, nearby objects he photographed were spirals.

In a paper he sent to be published in the *Astronomische Nachrichten* Keeler emphasized again the large number of spiral nebulae. He distinguished them from "extended diffuse nebulae" like the Orion and Trifid nebulae (H II regions, large masses of ionized interstellar gas), and the few "isolated compact nebulae that are not spirals," like the Ring and Dumbbell nebulae (planetaries). There were also a few "compact nebulae that might be expected from their resemblance to other forms to have spiral structure," yet in which none can be detected. Keeler mentioned M 32, the round companion of M 31, the Andromeda nebula, as an example, and it is, as we know today, composed of stars like the spiral galaxies, not ionized gas like the H II regions and planetary nebulae. Two other companions of M 31 were listed in his observing book: NGC 205, "Plate good; nebula shows no remarkable structure," and NGC 185, "Plate pretty good, but nebula is not interesting. Trace of spiral structure."[77] (Actually it is elongated, but not a spiral.) In this paper, Keeler stated many important concepts: the large number of spirals, the fact that they are plane objects, seen in all orientations, and especially that they were clearly rotating and that different members of the class very probably had different amounts of rotational angular momentum. He even recognized and stated "that nebulae [spiral galaxies] have a tendency to occur in groups."[111]

In another paper, published in America, Keeler reproduced a plate of one spiral, NGC 7479, and compared it in detail with the drawings and descriptions of the earlier visual observers, Herschel, D'Arrest, Rosse and Tempel. It was very clear that their individual, quite different descriptions were all actually based on the object clearly seen in Keeler's plate, but were incomplete or false because they could not see nearly faint enough to get a complete picture of it. He also quoted Roberts' description of NGC 7479 from his earlier photographs as elliptical, curved at both ends, with no structure visible. From Keeler's picture it was obvious that Roberts' photograph had only recorded the central bright bar of the spiral. The clear implications of the paper were that photographs were much better than drawings for studying nebulae scientifically, and that Keeler's, taken with the Crossley on Mount Hamilton, were much better than Roberts'.[112]

Keeler recognized M 31 as the nearest example of a spiral. He photographed regions in it several times, but because of its large angular size it could not all be recorded on one Crossley plate. He had a small-scale picture of M 31 taken by Barnard, but when he saw a much better photograph, taken by H.C. Wilson at Carleton College with an intermediate-sized camera, he quickly acquired a copy of it. In exchange, he sent Wilson a slide of his own picture of M 51. Wilson asked Keeler what method of backing his plates he used, and the Lick director sent him a lucid description of his technique. Wilson worked with W.W. Payne, who was eager to start publishing some of Keeler's nebular photographs in *Popular Astronomy*. The Lick director fed him just enough results to keep his interest high, and the interest of the amateurs and teachers who read the magazine, but not enough to let his data get out before he had found time to analyze it himself.[113]

To an outstanding scientist who needed the results for his own research, however, Keeler was much more generous. Thomas C. Chamberlin, the great geologist, a few years before had been the first to attack Lord Kelvin's estimate of the age of the sun as a few tens of millions of years. Chamberlin had deduced that it was much older, and had stated that there must be another source of solar energy than gravitation. Thirty years later Robert Atkinson and Fritz Houtermans identified the source as nuclear energy, released in thermonuclear reactions. Now Chamberlin was working on the problem of the origin of the earth, and with it, the origin of the other planets. His collaborator was Forest Ray Moulton, the young celestial mechanics expert. Their basic idea, which became known as the Chamberlin-Moulton hypothesis, was that a star had passed close by the sun and, by tidal interaction, drawn out a streamer of gas from it. This gas, because of the relative motion of the star and sun, would become a rotating nebula, and from this nebula the planets were supposed to have condensed.[114]

Chamberlin, Moulton, Keeler, and nearly every scientist of the time considered spiral nebulae as objects at essentially the same distances as stars. They did not have the concept of them as spiral *galaxies,* star systems as large as the Milky Way itself, far outside its confines. Thus Chamberlin and Moulton thought spiral nebulae might be other, more recent examples of their hypothesis, in which the rotating gas torn out of passing stars had not yet condensed into planetary systems. Both faculty members at the University of Chicago, they had heard of Keeler's remarkable photographs from Hale and others who had seen them at the Third Conference. Chamberlin wrote the Lick director to explain their study and ask if they could use his pictures. He concluded, "Of course, if you yourself contemplate a special study of spiral nebulae and their possible relations to the origin of the solar system, we could not ask you to put matter in our hands for like purposes, but if not, we should esteem it a very great favor to be able to make use of your great harvest of new forms."[115]

Keeler responded handsomely, sending Chamberlin a large set of transparencies of his Crossley photographs of spiral nebulae. He called the geologist's attention to M 31 as the largest example. Keeler said that he intended to discuss the pictures, and the whole subject of nebular evolution, when he published his photographs. He did not think his investigation would overlap Chamberlin and Moulton's but "in any case, I would not, from any personal motive, interfere with any application of the work of our Observatory which will be useful to science." Chamberlin was naturally extremely appreciative. He studied the pictures often in forming his ideas on rotating nebulae, although he confessed that, as a geologist, he felt "a little like the poor heathen Goths of old, who, driven by the Huns and Vandals behind, could not well avoid invading the precincts of the sacred city [astronomy], however reprehensible it was."[116]

Interest in Keeler's photographs was also expressed by T.J.J. See, Hale's and Harper's antagonist, who had now landed a position at the Naval Observatory in Washington. He was impressed by Keeler's papers on the spirals, and intended to study their dynamics theoretically. Keeler emphasized that his photographs showed little resemblance to the drawings by Herschel, and that the analogy of actual spiral nebulae to the forms of rotating fluid masses studied mathematically by Henri Poincaré, which See had mentioned, did not in fact exist. Keeler always was far more interested in observational results than in mathematical theories, however beautiful, that did not apply. Although See said his ideas on nebulae were changing, and expressed great interest in the photographs, Keeler did not offer him any of them.[117]

Antonia C. Maury was a scientist with whom he was far more in sympathy. She was the most creative researcher on spectral classification at Harvard College Observatory, but as she was more of an artist than an organization woman, she left it several times during her career. She asked about getting some slides, and Keeler immediately sent her several he had on hand, followed by a later shipment made especially for her. She was astonished by the superb plate of the spiral nebula M 33, and speculated on its nature and whether star formation was going on in it. When she saw Keeler's spectrum of the Orion nebula, she was particularly interested in the "peculiar ultra-violet line," and wondered what it might be.[118] Three decades later Ira S. Bowen identified it, $\lambda 3727$, as a "forbidden line" of singly ionized oxygen.

By now, Keeler was widely recognized as an outstandingly successful scientist. In December 1898 the Royal Astronomical Society elected him a Foreign Associate, along with Barnard and Burnham.[119] Only a few months later he received the Henry Draper Medal of the National Academy of Sciences for his astrophysical research. The award was made on the basis of the full list of his published papers.[120] Keeler heard that he and Hale had been the two main candidates considered for the award, but his younger friend was generous in his congratulations.[121]

Then in 1900 Keeler was elected to the National Academy of Sciences, the elite group of American scientists. Young, the old pioneer spectroscopist, supported him strongly for election. Newcomb, the most influential astronomer in the Academy, asked Keeler for a list of his papers, and with them as his credentials, the Lick director was elected the first time he was considered.[122]

In the spring of 1900 Keeler turned his attention to photography of globular clusters. He had taken a few exposures of M 13 and M 15 with the Crossley in the summer of 1899. They ranged from ten minutes to two hours, and the long exposures showed myriads of stars.[77] At Harvard College Observatory Solon I. Bailey was beginning his classic photographic studies of variable stars in globular clusters. His observational material came from plates taken with a 13-inch telescope at Harvard's southern station in Arequipa, Peru. Pickering, the Harvard director, was dedicated to converting astronomy to a quantitative science, based on accurate photometric measurements, and he strongly supported Bailey's work.[123] Thus on February 6, 1900, Pickering wrote Keeler and asked him to take some photographs of the cluster M 3, which Bailey knew contained several variable stars. The Lick director agreed to do so, sending at once some transparencies and slides he had on hand as samples of his work. They included the cluster M 13 and several spirals, from the large M 33 down to two enlargements of "very small nebulae." He offered to take a series of short exposures of M 3 with the Crossley, which would permit Bailey to search for rapidly changing variables that he could not detect on the longer exposures necessary to record faint stars with the small Arequipa telescope.[124]

Pickering was enthusiastic, and in May, when the cluster was well placed, Keeler broke off his continuing work on spirals for four nights. He and Palmer took, on successive nights, two long series of ten-minute exposures of M 3, one after another, together with a few longer exposures to show the fainter stars well.[104] These plates were sent off almost immediately to Harvard. Soon Bailey was reporting they were the best plates of M 3 he had ever seen, and with them he was able to make very precise photometric measurements. One star he measured showed the fastest increase in brightness of any variable star then known. The next month Keeler took a series of longer, one-hour exposures, so that Bailey could search for even fainter variables. It was an ideal collaborative research, using the Lick instrument, clear skies, and observational capabilities, and the Harvard photometric experience and skills.[124]

By now Keeler was fully accepted as one of the leading astronomers in America. At the same time as he was corresponding with Pickering, he was writing Langley to give him his ideas on the most profitable types of observations to make at the forthcoming solar eclipse. Keeler noted how the streamers and rays in the corona were distorted near prominences, and thought this phenomenon deserved further study. Within a few days he had written, as an equal, to two of the giant figures of his earliest days in astronomy. Gill, now Her Majesty's Astronomer at the Cape of Good Hope, consulted him on instruments and observing programs. Yet he still had time to answer a query from the manager of a printing supply company, who wanted to know if it would be possible to build a "calculating engine" that could calculate the motions of all the planets back to the year "A.D. 1" and print them out automatically. Keeler's reply, which still sounds modern today, was that "it is a little hard to say what a computing machine cannot be made to do." He thought an approximate solution would be quite possible, but he believed that the full calculation, taking all the perturbations into account, would be beyond the capabilities of the machine, "though it could be made to perform various steps of the process."[125]

By now Keeler had completely won over his own faculty at Lick Observatory. Tucker, initially a skeptic about the new director, summed up his thinking on the last day of 1899:

> Life is peaceful and even here. But the work if one gives it justice, is not easy ... I might absorb from our new associate [Keeler] some level headed ideas, and the steady glow of enthusiasm. He is about as pleasant an all round man as I have had to deal with. A year and a half we have had to rub ideas, and this place throws those engaged pretty well into each others way.[126]

And Barnard, who had been so opposed to Holden and his "no good" reflector, now wrote George Davidson:

> Let me say, for I know you only want the truth – the Crossley reflector has been making some very beautiful photographs. I believe they are really the finest that have been made. This is simply from what I have seen of them.

Then he went on to contrast the excellent Mount Hamilton observing conditions with Williams Bay, which he complained was "a mirey climate for a great telescope and discoveries are few and far between."[127]

By now preparations for the total solar eclipse of May 28, 1900 were well underway. Its path of totality would sweep across the southeastern United States, and nearly every American observatory was planning to send a party to observe it. Keeler had begun getting inquiries about the eclipse from English astronomers as early as 1898, and had decided then that central Georgia would probably provide the best observing sites.[128] The ever-energetic Hale took the lead in organizing a national committee to exchange ideas and coordinate plans for the eclipse. It is interesting to compare the responses he received to his broadside request for suggestions.

Keeler first stated that the committee could not control what individual astronomers did, but could serve as an information clearing house. Ever practical, he said that most observers' programs would be largely governed by the instruments at their disposal. Then he particularly recommended photography of the corona, with long-focus telescopes equipped with rotating sectors, of the type developed and used by his friend Charles Burckhalter at the 1898 India eclipse. The sector was shaped so that the outer, faint part of the corona received a full exposure on the plate all during totality, while the bright light of the inner part was strongly cut down. Thus the steep brightness gradient was reduced in a quantitatively known way, and the overall structure of the corona was recorded on a single plate. Keeler also recommended accurate measurement of the wavelength of the bright green coronal line, to check the result found at the previous eclipse by J. Norman Lockyer and Campbell that it was not the same as the magnesium line seen in absorption in the solar spectrum. Finally, he recommended time-resolved spectra of the edge of the sun, as it was covered and uncovered just at the beginning and end of the eclipse. These were to be taken with an apparatus that moved the photographic plate inside the spectrograph during the exposure, producing a spectrogram that would show how the various lines changed with time, and thus with height in the solar atmosphere. This method had been developed by Campbell and first used by him at the India eclipse two years before. Keeler suggested several specific improvements in the apparatus. His advice was sound and wide ranging, based on his own experience, careful examination of the results obtained at the previous eclipse, and consultation with Burckhalter and Campbell, the experts who had been there.[129]

Campbell, in his letter, emphasized much more strongly that he and the other astronomers could not be controlled by Hale's committee. Only those whose sole observations would be photographing the corona with small telescopes – near-amateurs in Campbell's mind – should be "well distributed" along the path of totality by the committee. He and the other serious professionals would decide for themselves what to do, and he would let the committee know when he had decided. Burckhalter, from little Chabot Observatory in Oakland, sent an extremely practical plan, outlining exactly what he would do himself if he could get to the East. Leuschner sent a very bureaucratic reply, full of numbered paragraphs about committees and subcommittees, and absolutely devoid of scientific ideas. Henry S. Pritchett, who had got the job as superintendent of the Coast and Geodetic Survey in Washington, dwelt mostly on the finances. His letter was bureaucratic, but more reasonable than Leuschner's. He included some scientific ideas, but only in the vaguest and most general terms.[130] Pritchett was the man who went on to have a very successful career as a scientific administrator, in the Survey, as president of the Massachusetts Institute of Technology, and as president of the Carnegie Foundation for the Advancement of Teaching.

Hale's own plan for the eclipse envisaged a vast, well-organized operation. He would try to measure the heat radiation of the corona with a bolometer, while Ernest Nichols of Dartmouth, working with him, would do the same with a radiometer, another energy-sensing device. At the same time Frost would be taking spectrograms of the "reversing layer," the edge of the sun, and Barnard would be taking large-scale direct photographs of the corona. Meanwhile, back at Yerkes, Hale's assistants would be taking spectroheliograms of the sun to map the positions of prominences and other features in the chromosphere on the day of the eclipse. All he needed was the money to finance the expedition.[131]

That was Keeler's problem, too, for neither the Lick nor Yerkes budget included funds for travel to eclipses, and special contributions had to be obtained from wealthy benefactors. Langley was especially interested in the filamentary structure of the inner corona he had briefly glimpsed at the 1878 eclipse, which he had observed from Pike's Peak. Now, as secretary of the Smithsonian Institution, he controlled funds that he could use to equip an expedition to the 1900 eclipse. He heard of Campbell's excellent results with the Schaeberle camera at the 1898 eclipse, and tried to get the expert young Lick observer to join the Smithsonian party. Finally, however, Keeler succeeded in raising the necessary funds from William H. Crocker, brother of the regent who had supported several previous expeditions, and Campbell could go as the leader of a Lick party.[132] They would carry out much of the program suggested by Keeler the previous year.

While Keeler stayed behind at Mount Hamilton, working with Palmer and Coddington and photographing spiral nebulae or globular clusters every night he could, Campbell, accompanied by Charles D. Perrine, set out for Georgia a month before the eclipse. They set up camp at Thomaston, a tiny Southern hamlet on the track of totality. Nearly every astronomer in the eastern United States seemed to be somewhere along the eclipse path, including two young students who were to be leading astrophysicists of the next generation: Henry Norris Russell with the Princeton University group headed by Young, and Joel Stebbins with his professor from the University of Nebraska, G.D. Swezey.[133] Several astronomers, including Stimson J. Brown of the Naval Observatory and Henry C. Lord of Ohio State University, were basing their programs largely on personal advice they had got from Keeler.[134]

Although they were constantly ill from the unaccustomed Southern food and germs, Campbell and Perrine aligned and adjusted all their instruments carefully and practiced their procedures for the eclipse. They were assisted by several volunteers, and Campbell drilled them thoroughly. It was essential to know exactly what to do during every precious second of totality. Perrine was often so sick that even the hard-driving Campbell would not let him get out of bed.

On the eclipse date the skies were clear, and the Lick observers carried out all their tasks with practiced skill. It was so hot that Campbell was afraid to develop their plates in the daytime; even at night the temperature in their primitive darkroom, which they tried to cool with ice, stayed in the mid-seventies. Nevertheless their plates came out well; only one moving-plate spectrograph had jammed. Campbell wanted to leave Thomaston as quickly as possible, but Stone, a classical astronomer who was director of the McCormick Observatory of the University of Virginia, kept him there an extra two days. Stone had photographed the corona at the eclipse but he did not know how to develop his plates. He asked Campbell for help, and as the Lick astronomer reported to Keeler, "it was the appeal of a drowning man and I couldn't refuse." With developing Stone's plates and packing their own instruments for shipment back to California, it was a week before Campbell and Perrine got out of Thomaston. When they finally arrived at Tybee, on the coast of Georgia, according to Campbell "we enjoyed the first square meal we have had since leaving Atlanta on Apr. 30. More than all, the apparent cleanliness of the food here is a pleasant and radical change ... The people of Thomaston did for us the best they know how, but they don't know how." Nevertheless, Campbell said, he and Perrine left "there with everyone respecting, and practically reverencing, the Lick Observatory." In his post-eclipse let-down, Campbell's long report to Keeler mentioned briefly all the successful exposures they had taken at the eclipse, but described the few breakdowns in great detail.[135]

Keeler immediately congratulated Campbell on the group's success at the eclipse. But, knowing that he and Perrine were going on to the second meeting of the Astronomical and Astrophysical Society of America in New York, the director cautioned his younger subordinate:

> While I agree with you that the important results should not be concealed, I would advise dwelling on the successes rather than the failures. If you were to tell a reporter that three plates out of the ten were failures, he would receive a totally different impression from what he would if you gave him the equivalent statement that seven out of ten plates were successes.
>
> I was glad to learn that you found something to eat at last. You must have been in the "Cracker" country. I have been there myself, but never long enough to starve.[136]

As soon as Campbell got to New York, he learned how successful they had in truth been, compared with nearly all the other astronomers who had gone to the eclipse. Langley's Smithsonian party, which had drilled for months in Washington, and run through two complete practice eclipses on the site, plus one full-dress rehearsal the day before the big event, did get results. Their pictures, taken by the Smithsonian Institution photographer with a 135-foot focal-length horizontal telescope, showed about as much coronal detail as the Lick pictures taken with the Schaeberle camera. Charles G. Abbot succeeded in measuring the corona with a bolometer, and found that in heat radiation it was considerably fainter than the dark side of the moon, an unexpected new result. Burckhalter's apparatus worked perfectly and produced some very good coronal photographs.[137]

At the Yerkes camp, located next to the Smithsonian site in Wadesboro, North Carolina, Frost, like Campbell got some very good spectra, and Barnard took a series of excellent coronal photographs, just as he had back in 1889 at Bartlett Springs. But elsewhere, everything was disaster. Without Nichols to help him, Hale failed to get any measurements of the corona with his bolometer. The instrument was damaged on the way down from Williams Bay, but the young director had a new one made on the spot. Working day and night, he succeeded in adjusting it just before the eclipse, but a few moments before totality, the apparatus was knocked out of alignment and it was impossible for him to measure anything. Laboratory spectroscopists Joseph S. Ames, Henry Crew, William G. Humphreys and Lewis E. Jewell all failed to get any data at all with the advanced instruments they had brought to the eclipse. Robert W. Wood, who was with the Naval Observatory party, called it "the Pinehurst fluke ... it was awful," and Ames wrote, "I have come to the conclusion also that at any eclipse a man who was at a previous one is worth ten men who have not been." Hearing all this, Campbell felt much better about the Lick party's accomplishments.[138]

Keeler, between photographing nebulae with the Crossley, was busily emphasizing the positive himself back at Mount Hamilton. He sent a new set of transparencies of his latest pictures to New York, to be exhibited at the astronomy meeting. Most of them were spirals, together with the globular cluster M 13, the Pleiades nebulosity, and M 27, the Dumbbell planetary nebula. After the meeting, Keeler gave instructions that the set of pictures be divided equally between the Columbia and Princeton University astronomy departments.[139] A glowing testimonial came in from Vogel, to whom Keeler had sent a set of transparencies for presentation to the Berlin Academy of Sciences. The Lick director had drawn the old man's attention to the complicated structures of several of the spirals. He pointed out "the arches of nebulosity connected with the secondary nucleus of M 51," or as we would say today, the spiral arm crossing its companion, and "the long thin branches of the spiral [NGC 2903] and the spiral turns at their extremities." Vogel was wonderfully impressed. He wrote back that the pictures were "decidedly the deepest [das Vollkommenste] that have been obtained in this field up to now, and I can only hereby express to you my most sincere congratulations on these magnificent scientific results." Keeler sent this letter to President Wheeler in Berkeley, saying that he had received many letters praising his work with the Crossley, "but Vogel is perhaps the greatest expert in Europe, so that his opinion is of special value."[140]

Keeler spent much of his time in March and April writing a long paper for the *Astrophysical Journal* summarizing his first two years' work with the Crossley reflector. By the end of April he had the first draft completed, and was touching it up for publication.[141] In the paper Keeler described in cool, technical terms all the many improvements he had made in the Crossley to turn it into a productive research tool. He began with the polar-axis alignment, and mentioned lowering the telescope, the wind screen, the new bearings he had installed to replace the mercury flotation system that had been "a delusion," the new drive clock and the new finders that matched the telescope's properties and the observer's capabilities. He mentioned the gears that had been readjusted to iron out most of the irregularities in the drive system. Since the Crossley was used entirely as a photographic telescope, Keeler and his assistant worked with a red lantern faintly illuminating the inside of the dome. The photographic plates they used were not sensitive to red light, and the lower part of the inside of the dome was painted bright red, so the assistant could see what he was doing. In contrast, the upper dome, the telescope, the ladders and the observing platform were all painted dull black, so the observer at the telescope worked in nearly complete darkness and could see the faint guide stars. Keeler described all the minute details of the double-slide plateholder, and the ways in which he could work entirely by feel and get perfect stellar images on the plate. He told of how he could refocus the telescope during long exposures from the appearance of the comatic image of the guide star, the same method Walter Baade used when he resolved thousands of individual stars in M 31, the Andromeda nebula, with the 100-inch Mount Wilson reflector fifty years later. Keeler told of his guiding techniques, how he backed his photographic plates a day or two before use, and how long he could expose them for best results. They are all dry, boring technical details, but they meant the difference between envisaging a program, and actually carrying it out and getting important scientific results.

Keeler wrote that he was using the Crossley to photograph nebulae, from the "great Nebula in Orion" down. In the course of the program, he was discovering "incidentally" many new nebulae, for nearly every plate showed several previously unknown objects. He briefly described his main conclusions to date, that many thousands of unrecorded nebulae exist in the sky, repeating his conservative estimate of 120,000 as the number within reach of the Crossley, that these nebulae exhibit all gradations of apparent size from M 31 down to objects hardly distinguishable from stellar images, and that most of these nebulae are spirals. He went on to say that many of the nebulae described as "double" by Herschel are actually spirals, often of very beautiful, complex structure when seen on well-exposed photographs. This result destroyed the analogy to the figures of bodies of rotating, incompressible fluids obtained mathematically by Poincaré. The actual conditions in these nebulae, Keeler concluded, are much more complicated than those that had been considered in theoretical discussions.

These conclusions, he believed, would have a very direct bearing on many questions of cosmogony. His universe was a small one, however; he speculated that the solar system had evolved from a spiral nebula, and mentioned that Chamberlin and Moulton were already studying this question. Keeler's paper was a very important observational contribution, containing many ideas that proved seminal in the study of galaxies. He sent it off to Hale, and it was published in the June 1900 issue of the *Astrophysical Journal.*[60]

It was the last paper he ever wrote. Keeler had been suffering from what he called a "hard cold" that he could not throw off all that spring and summer. Although he advised his friends, like Barnard [67] and Hale, [142] when they were sick, to rest – "[o]ne does more, in the long run, by taking things just a little easily" – he found it difficult to follow his own advice. Only very rarely does his observing book bear a laconic entry, as on April 27, 1900, when the spiral M 61 was photographed: "[t]his phot[ograph] taken by Mr. Palmer and Mr. Coddington, J.E.K. having a hard cold." Often that spring he would go home from the office in the afternoon, lie down, and tell his wife that he felt very tired and knew he was getting old. In fact he had suffered from some mild heart problem for some years, although he concealed it from his associates. Yet several of them noticed that he frequently had a flushed face, and often breathed in a series of audible sighs. He was a heavy cigar smoker from his student days, and may have had emphysema, perhaps even lung cancer. At the end he could not walk up the hundred-and-fifty-foot rise from the Crossley dome to his house without stopping several times to get his breath back. [143]

Keeler's wife and children went down from Mount Hamilton for a vacation in late June, and just after the Fourth of July he joined them and spent nearly two weeks at sea level trying to recuperate. [144] When he returned, he announced that his doctor had forbidden him to continue working at night "for the present." It was a "great deprivation," for the summer was the best observing season. Publicly, he showed no intimations of mortality, although privately he had drawn up a list of the biographical facts of his life the previous month, before leaving the mountain. [145] Campbell, when he returned on July 27 from the eclipse and the meeting, could see at once that the director was not well. Keeler assured him, however, that his physician had told him there was nothing to fear. On July 30 Keeler left the mountain again. The last letter he wrote from Lick was an exceedingly sympathetic one, full of encouragement, to a telescope maker who had sent him a small lens for testing. It was excellent, Keeler said, analyzing its performance and urging him to go on to bigger objectives. [146]

In San Jose, Keeler saw his doctor again, and was told that he should take another vacation. His illness was diagnosed as pleurisy of the lung, "nothing very serious, but requiring some care," he said. He left for further recuperation at Kono Tayee, where his wife and children were now vacationing with Harry Floyd. Keeler expected to return to the mountain in two or three weeks. [147] However, his condition rapidly deteriorated. A group picture, taken at Kono Tayee in early August, shows him thin and strained, an ever-present cigar between his fingers. On

Fig.20. Group picture at Kono Tayee, a few days before Keeler's death
He is standing against a tree, wearing a cap, an ever-present cigar
between his fingers. Cora M. Keeler is standing in the middle of the back
row, in profile; Harry Floyd is peeking around from behind Keeler.
Henry and Cora F. Keeler are sitting on the bench (1900). (Reproduced
by kind permission of the Mary Lea Shane Archives of Lick
Observatory.)

August 10 he and his wife started back to San Jose to see the doctor again,
but on the train Keeler suffered a slight stroke. Instead of continuing,
they went to a hotel in San Francisco, where his wife summoned specialists to examine him. She located one of his old friends from Johns
Hopkins, William H. Arthur, now a doctor in the Army Medical Corps,
who was in San Francisco waiting a transport for China. He and Dr.
Morris Herzstein, one of the busiest physicians in the city, examined
Keeler. They took him at once to the Waldeck Sanitarium, but they could
do nothing for him. He suffered a second stroke, and died on August 12,

1900.[148] In two short years Keeler had proved, as Campbell wrote, "an ideal Director and investigator," but he was cut off in mid-career; "the loss to Astronomy [was] inestimable."[149]

10

The quality of his voice still rings in my ears

James E. Keeler's unexpected death before his forty-third birthday was a devastating blow to his family and friends. George Ellery Hale, Charles A. Young, and especially John A. Brashear could hardly believe it had happened. Yet Keeler was dead, and his funeral on August 15, 1900 brought the fact home to all his friends and colleagues in the Bay area. The services were held in Grace Episcopal Church on Nob Hill in San Francisco, just as Captain Richard S. Floyd's had been, nine years before. Nearly everyone from Mount Hamilton was present. The pallbearers included President Benjamin Ide Wheeler of the University of California, several regents, Keeler's fellow astronomers W.W. Campbell and Charles D. Perrine from Lick Observatory, Armin O. Leuschner, and two other professors from Berkeley. Keeler's remains were deposited in the Floyd family vault in Laurel Hill Cemetery.[1]

Campbell suggested to Cora Keeler that her husband's body be cremated, and the ashes placed in the base of the 36-inch refractor on Mount Hamilton, where James Lick had been entombed. She consulted Keeler's mother and his sister, and all three agreed that it would be a "fitting and worthy disposition of the remains." However, when the regents were approached, they did not accept the idea. They gave as their opinion that "[t]hough from a sentimental point of view the plan would be a good one, yet the great telescope should stand sentinel over James Lick's body alone." Campbell next suggested that Keeler's body be buried on a quiet rise a little to the east of the Main Building, not far from the present site of Lick Observatory's largest telescope, the 120-inch Shane reflector. Cora Keeler, like many recent widows, found it hard to reach a decision. She wrote Campbell that she did not want to ask again and be rejected; the regents did not ask her.[2]

Ultimately, Keeler's body was cremated and entombed not at Lick, but at the new Allegheny Observatory. Just a few weeks before Keeler died, Brashear had decided the time was ripe to call in the pledges and begin building at the new site in Riverview Park, out of the smoke and grime of the city. The basic plan was Keeler's. Frank Wadsworth, the new director, added specifications for a large battery of specialized astrophysical instruments. The cornerstone was laid on October 1, 1900. However, Wadsworth proved just as difficult to get along with as he had been at Chicago and Yerkes, and was soon quarreling with Brashear and the rest of the Observatory Committee.[3] In 1904 he resigned and became the manager of a plate-glass company in Pittsburgh. At Allegheny Wadsworth was succeeded by Frank Schlesinger, who brought the new observatory to completion in 1914. It contains the James E. Keeler Memorial Telescope, a 31-inch reflector that was completed in 1906, as well as the Thaw Memorial Telescope, a 31-inch refractor especially designed for photographic astrometry. The Thaw refractor was named for William Thaw, Samuel P. Langley's patron who financed Keeler's year of graduate study in Germany, and for his son William Thaw, Jr. In addition, the old 13-inch refractor that Keeler had used was moved to the new site, and dedicated for public use, as Brashear had always dreamt.[4] Keeler's ashes were brought to Allegheny soon after the reflector was completed, and were deposited in its pier.[5] Brashear was the leading spirit in entombing his friend's remains there, no doubt partly in imitation of Johns Hopkins University, where Henry A. Rowland's ashes had been walled up in his laboratory a few years previously. It was an age more conscious of death than ours.

Within a few weeks of her husband's funeral Cora Keeler left Mount Hamilton. The children, seven-year-old Henry and six-year-old Cora, were too young to handle their grief in approved adult style. With their little friends, they acted it out in games of "death" and "funeral" that shocked the inhibited bachelors on the mountain. With her children Cora went first to Kono Tayee, where they stayed with her cousin Harry Floyd for a time, then to her family home in Louisiana. Ultimately she settled in Berkeley.[6] Her husband had not been rich, and had only a little insurance, but it and the small legacy she had inherited from Harry's mother were all she had to live on. It was hard to make ends meet. She received Keeler's salary until the end of the year, but after that there was nothing from the University of California.[7]

Phoebe Hearst kept in touch with Cora, sending her gifts for the children every Christmas, and grapes from her estate near Atherton at frequent intervals. When Henry finished high school, Phoebe Hearst arranged for him to spend a month at her summer retreat near Mount

Shasta, and then to be given a job on her vast San Simeon ranch. Like his father, he was tall and thin, and the healthy outdoor work built up his body at the same time it improved his self-confidence.[8] Then Brashear and Campbell arranged for Andrew Carnegie to finance Henry's college education. He was a good student, and wanted to become an engineer, so it was natural for Carnegie to send him to the new Carnegie Technical School he had founded in Pittsburgh.[9] There Henry was a very good civil engineering student, and impressed his teachers as "an exceptionally fine fellow" and "a most agreeable and pleasant young man." At the end of his freshman year he went home to Berkeley for a vacation, but the next summer he worked as a surveyor on a railroad line in Texas, and the following summer as an engineer in Pittsburgh. By the time he graduated in 1914, Henry was considered "one of the most promising men" in the class.[10] He had inherited not only his father's brains, but his urge for travel as well. He got a job with the Standard Oil Company that took him to China. There he worked for four years. He learned Chinese, and got along well with the local people as well as with his fellow American co-workers. His future seemed assured. But like his father, Henry Keeler was struck down before his time. At the age of twenty-five, while stationed in Soochow, he suffered an attack of appendicitis, and although he was operated on, he died a few days later. Again his mother was left with her grief.[11]

Her daughter Cora finished high school and went to work. She had the wanderlust too, and moved to Honolulu for a time. Young Cora married Charles H. Moore, an Army officer, and lived on into her eighties. Cora Keeler, her mother, had died in Berkeley in 1944.[12] Like Keeler's mother and grandmother, the women in his family were long-lived.

Though Keeler was dead, his memory lived on in the hearts of his friends. Many admiring obituary articles appeared. The most balanced accounts of his life and scientific work are those by Campbell and by Hale; the most emotional, that by Brashear.[13] The best picture of him as a human being was provided by Charles S. Hastings, his first physics teacher at Johns Hopkins and his longtime friend. Two of his colleagues at Lick Observatory, Charles D. Perrine and Richard H. Tucker, added specific details and insights based on long association with him.[14] Although memorial biographies are notoriously unreliable guides to a recently deceased scientist's personality, these articles are so uniformly positive and glowing that it is clear that the writers greatly admired him as a scientist and loved him as a human being. Keeler's last paper, on his epochal work with the Crossley reflector, was reprinted as a memorial to him, first in the *Publications of the Astronomical Society of the Pacific*, then in a separate volume of the *Publications of the Lick Observatory* luxuriously illustrated with his photographs of nebulae and clusters.[15]

The one American astronomical prize Keeler did not get while he was alive was the Catherine Bruce Medal of the Astronomical Society of the Pacific. Edward S. Holden had persuaded the old spinster to put up the money to found it just before the final upheaval that led to his ouster. The medal was intended to honor excellent service to astronomy.[16] The first recipient, in 1898, was Simon Newcomb. By the terms of the award, candidates were nominated by the directors of six specified major observatories, including Lick and Yerkes, and a choice from this slate was made by the board of trustees of the society. For the next year Keeler nominated Arthur Auwers, a distinguished German classical astronomer, William Huggins, and "Otto von Struve, of St. Petersburg [Russia]", the noted double-star observer. Auwers was chosen by the board. The following year Keeler nominated David Gill, longtime astronomer at the Cape of Good Hope, Edward C. Pickering, and Hermann Vogel, the outstanding German spectroscopist. Hale had nominated Keeler for the 1898 medal, and wanted to nominate him again for 1900, but the Lick director urged him to support Pickering instead. Hale did as his friend asked, but Gill was chosen by the board. By the time the selection process rolled around for the next year, Keeler was dead. There was no provision for awarding the medal posthumously, but the board, meeting less than a month after his death, decided not to give it to anyone in 1901.[17] Undoubtedly, in their minds, no one else deserved the Bruce medal as much as Keeler.

At Lick, Keeler's memory was revered and he was almost deified. In published scientific papers from the observatory for many years he was only referred to as "Director Keeler." Kenneth Campbell, who grew up on Mount Hamilton, remembered that his father and mother never said anything against Holden, but that in any social gathering there were likely to be some stories told at the first director's expense. Young Ken soon concluded that he had not been a favorite person on the mountain. Keeler, on the other hand, was considered absolutely "tops," both professionally and personally. The Campbells and everyone else on the mountain thought of him as an ideal person, scientist and director.[18] Joel Stebbins, who came to Mount Hamilton as a graduate student just a year after Keeler's death, reported to George C. Comstock in Madison:

> Perhaps the only thing that every one here ever agreed upon is with regard to Keeler. He was a favorite with every one on the mountain and is still talked of a great deal. They all think he was the perfect man for the place.[19]

At the fiftieth anniversary of the beginning of research at Lick Observatory, Joseph H. Moore, who had arrived at Mount Hamilton from Johns Hopkins himself in 1903, wrote just as enthusiastically of Keeler, and twenty-five years later, C. Donald Shane did so again. Shane's first memory of astronomy was hearing, as a small boy, that "nobody could be as good as a director" as the recently deceased Keeler. [20]

And William H. Wright, who worked under the first four Lick directors before he himself became the fifth, testified to his feelings for Keeler in the words

> His death took place more than twenty years ago, but the imprint of his personality remains so strong upon my memory that his presence seems a matter of yesterday, and the quality of his voice still rings in my ears. The climax of Keeler's career was probably reached in the Directorship of the Lick Observatory, and the faculties which made for his remarkable success during the short period of his incumbency were both scientific and administrative ... To me the outstanding characteristic of Keeler's mind was the simplicity and orderliness of his thought ... It was characteristic of Keeler's method of work that he was always keenly alive to what he was doing[;] he never became the blind slave of routine ... Keeler's attitude toward the young scientific man who aided him in his investigations was such as to inspire in him the keenest enthusiasm for the work and the warmest loyalty to his chief ... [H]is "direction," so far as I was able to observe, took the form of tactful suggestions whose intrinsic merit forced their adoption. Thus without the visible exercise of authority was his control of the activities of the Observatory accomplished. [21]

In spite of Keeler's death, science went on without a break at Lick Observatory. Campbell, now the senior astronomer on the staff, had been temporarily in charge when the director left the mountain; with Keeler dead he remained in the post until a new director could be selected. The astronomers went on observing, getting data, and writing up their results for publication. Campbell assigned Perrine to the Crossley reflector, and on September 15, 1900 he and Harold K. Palmer got the first plate taken with it after Keeler's death. [22]

The great experimental physicist Robert W. Wood came to Mount Hamilton just then. He was at the beginning of his career, and had conceived a new method to observe the solar corona without an eclipse. He was on the faculty of the University of Wisconsin, but was visiting in California that summer, and had arranged with Keeler to come to Lick to try his new method. Since the radiation of the corona is polarized, Wood's plan was to observe in polarized light at the wavelength of a solar absorption line. This would give the maximum suppression of scattered

Fig.21.James E. Keeler, second director of Lick Observatory. (Reproduced by kind permission of the Mary Lea Shane Archives of Lick Observatory.)

sunlight, and perhaps reveal the corona. Wood brought his polarization detector (a Savart plate with a Nicol quartz prism) with him, but needed to borrow the other equipment to form an image of the sun. Benjamin W. Snow, Keeler's boyhood chum from Illinois, introduced Wood with a letter to the director. "Even if you do not find the corona, you will learn to know a thoroughly jolly and companionable man, and one who has won

for himself a very enviable place among the investigators of our faculty,"
he wrote.

Keeler was glad to invite Wood to the mountain, but before the young
physicist arrived the director was dead. Campbell, who had met Wood at
the Astronomical and Astrophysical Society of America meeting in New
York, then repeated the invitation. Wood did come and try the polarization method, but he did not succeed in seeing the corona any more than
Hale had. Wood perceptively noted that the coronal light might be
sunlight "reflected" (we would say scattered) by particles with such large
velocities that the Doppler effect nearly completely wipes out the absorption lines of the solar spectrum of the corona; as we know today, Wood's
"particles" are free electrons at a temperature of millions of degrees. The
visit reinforced Wood's astronomical interests, and he continued to in-
vent original new types of observations.[23]

Less than a week after Keeler's death President Wheeler wrote to
leading astronomers around the country, asking for their advice on a
successor.[24] He was determined to nominate one candidate to the regents; he did not want another public, free-for-all election. Wheeler
regarded the Lick directorship as the best astronomical position in the
world, and he wanted the best man available to succeed Keeler. There
was no question in anyone's mind who that would be. Everyone recognized Campbell as the logical choice. He was the senior astronomer on
the Lick staff, widely recognized as an outstanding research scientist.
Letters of support flocked in from everywhere. Newcomb, Pickering,
Hale and everyone else consulted recommended Campbell strongly. All
the members of the Lick staff supported him. His own position was that
he would not seek the post, but if he were appointed, he would gladly
accept. "I am in love with this place. I know the instruments, the climate,
the staff, and their great possibilities. That is all I shall say in my own
behalf." On the other hand, if someone else were brought in as director,
Campbell would loyally serve under him and be glad to have more time
for his own scientific work.[25]

Campbell conveyed this information, along with a brief statement of
the qualifications the ideal candidate should have – which sounded remarkably like his own qualifications – to Hale, Newcomb and others. He
was worried that Leuschner might be working secretly against him, and
some die-hard sentimentalists brought up John M. Schaeberle's name
again, but there was never any doubt that Campbell would be chosen. On
December 11, 1900 he received a telegram announcing his appointment
and immediately accepted.[26]

Campbell's main research was the measurement of stellar radial velocities. It provided hard numerical data for understanding the motions of the stars, their arrangement in space, and even clues as to their origins. Particularly to classical astronomers like Newcomb these accurate measurements were the most important new information the spectrograph could provide. Keeler had urged the older man to nominate Campbell for the Henry Draper Medal for his astrophysical discoveries, but instead Newcomb put him forward for the Nobel Prize. It had recently been established, and the Washington astronomer, a foreign member of the Swedish Academy, had been asked to nominate an American scientist for the new award. There was no prize in astronomy, but Newcomb believed that Campbell's spectroscopic work could be considered as physics. He encouraged the Lick astronomer to get out a paper on his radial-velocity work. His earlier astrophysical papers on nebulae, Wolf-Rayet stars, novae and spectroscopic binaries could serve as back-up material.[27] Campbell did not receive the Nobel Prize, but his nomination by Newcomb indicated the great importance of his research.

Campbell remained as director of Lick Observatory for thirty years. Although in his youth he had been a scientific pioneer, he became increasingly rigid in his ideas and concentrated the efforts of the observatory more and more on the radial-velocity problem. He successfully observed four more solar eclipses, and at the last, in 1922 at Wallal, Australia, he and Robert Trumpler obtained plates with which they confirmed to high accuracy the light deflection predicted by Einstein's general theory of relativity. Returning from this eclipse, Campbell was met at the dock in San Francisco by a delegation of regents, who persuaded him to accept the presidency of the University of California. He served in that position from 1922 until 1930, retaining the nominal directorship of Lick Observatory, although Robert G. Aitken, as associate director, handled the day-to-day supervision. Campbell retired as president and director in 1930, and then was elected president of the National Academy of Sciences. Very conservative, he clashed repeatedly with President Franklin D. Roosevelt. After his term at the Academy ended, Campbell returned to California, but he was old, tired, blind in one eye and losing the sight in the other, and suffering increasingly from aphasia. He committed suicide on June 14, 1938.[28]

During his long reign as director, Campbell became more and more dictatorial. The tributes to Keeler by Wright and Moore, and the near reverence in which he was held by the whole Lick staff, must be recognized as based partly on the contrast between him and Campbell. They themselves realized this, as did Campbell. Wright was particularly critical of him. On one occasion, many years later, he told a young assistant that he had once said, in Campbell's presence, that Keeler was the best director Lick Observatory ever had – and that Campbell had been distinctly cool to him for some time after that.[29]

Holden, on the other hand, became increasingly benign with age. As librarian at West Point, he taught a generation of cadets, and many young officers as well, how to use indexes, how to research topics, and how to organize their term papers. Brashear, visiting a friend at the Military Academy, met Holden for the first time in years and had a long talk with him. He reported to Henry Crew, "He is much mellowed and does not seem like the same Holden to me."[30] Campbell, after he had sat in the seats of the mighty a few years himself, came to realize that sometimes brash young scientists make unfounded criticisms of wise middle-aged directors. Perhaps he had been too harsh in supporting Hussey when he refused to follow Holden's orders to put the Crossley reflector into operation. They became reconciled, and although there was some stress and strain over just what the memorial tablet that Campbell wanted to erect for Holden in the halls of Lick Observatory should say, they remained on friendly terms until the first director's death in 1914. Campbell ordered the flags on Mount Hamilton to half mast the moment he heard the news.[31]

One of Keeler's main interests was always in the graduate-student assistants who worked with him. After Keeler's death, Henry Harrer, his assistant at Allegheny, whom he helped to publish the paper on the spectrum of α Orionis, dropped out of astronomy. Harrer liked observational work best, but did not have much background in advanced physics and mathematics. He had tried to study on his own after Keeler had left for California, and worked with Wadsworth for a time, but before long he decided to get a better-paying job as an engineer.[32]

At Lick, Ernest F. Coddington was the first of the fellows to leave. He managed to get together enough money to go to Germany to study at the University of Berlin. Part of it came from friends who sponsored his education, and part of it he earned himself observing and computing orbits for the asteroids originally discovered by James C. Watson, who had left most of his estate for this purpose. German graduate training was still important to many young scientists, although the situation was gradually changing. Coddington departed two months before Keeler died. He did his thesis at Berlin in celestial mechanics, and received his Ph.D. in 1902. Then Coddington returned to Ohio State University, where he had been an undergraduate. He taught mathematics and mechanics there for many years, but did no more research in astronomy.[33]

Coddington's anticipated departure quickened Keeler's search for a new graduate-student fellow to help with the observing work at Mount Hamilton. The radial-velocity program was growing rapidly, and Campbell and Wright badly needed an assistant. Keeler was far less receptive than Holden had been to special students who only wanted to stay the summer. Quarters on the mountain were chock full, and Keeler could not waste precious rooms or apartments on visitors unless he was sure they were highly qualified and would make real contributions. Thus Professor Arthur Raumm, of the University of Washington in Seattle, who wanted to come in Keeler's first summer as director, and Professor A.M. Mattoon, of Park College in Missouri, in his last, were politely told there was no room available for them. Keeler used their requests to put pressure on the university administration for more housing at Mount Hamilton.[34]

However, another prospective special student, A. Frank Maxwell, got much different treatment. He was a graduate of the University of Minnesota, who had taken a full course in astronomy and had a lot of experience with telescopes. Professor Francis P. Leavenworth recommended him strongly, and Keeler was eager to have him at Mount Hamilton for the summer of 1900. Maxwell was a school teacher, who was hoping to get a job in California, and the director no doubt hoped, if he worked out, to find a way to keep him at the observatory. However, at the last minute Maxwell's father became seriously ill, and the young teacher stayed with him and never came to the mountain.[35]

One long-term prospect was Hiram Bingham, Jr., the son and grandson of American missionaries in Hawaii. After graduating from Yale in 1898, young Bingham had returned to Honolulu to head a mission himself, but soon decided that that life was not for him. He worked briefly as a chemist for a sugar company, but then in the summer of 1899 decided to come to California to study astronomy. Keeler approved his interest, but warned him it was not a remunerative profession:

> Openings are few, and only those who have a real love and aptitude for the study are likely to succeed. Success in fact depends largely upon a certain originality of mind and ability to do – qualities which are inborn and not acquired.

Bingham's courses at Yale had prepared him to be a minister, not a scientist, and Keeler advised him he would have to spend at least a year on the Berkeley campus first, taking the undergraduate subjects he had missed. Bingham was delighted at the opportunity and sailed for the mainland. He welcomed the challenge, he wrote Keeler:

> My father long ago assured me that Astronomy was not a remunerative profession, but I see no reason in that for giving up my idea. I believe it to be the noblest of all the sciences, and the most unselfish. Whether I

> shall ever achieve success in it remains very much in the misty future –
> but it will not be because I have not put forth earnest and sincere effort in
> that direction. Hardships are no obstacle. Work is no bugbear. If I can
> only be given something to *do*, something to *accomplish*, and a fair
> chance to succeed, I shall be perfectly happy.

Bingham came to Mount Hamilton and visited Keeler, but he also called
on Leuschner at the campus, and learned the realities of the mathematics
and orbit computation he would have to wade through before he would be
allowed to work at Mount Hamilton. Probably wisely, Bingham decided
to study history instead. Astronomy's loss was history's and government's
gain, for after receiving his M.A. at Berkeley in 1900, Bingham went on
to Harvard and earned his Ph.D. in the new field of South American
history. He became a long-term faculty member at Yale, then entered
politics and was elected lieutenant governor, governor, and senator from
Connecticut, and ended as chairman of the Loyalty Review Board under
President Harry S. Truman.[36]

   With Bingham out of the picture, the best prospect Keeler had to
replace Coddington was Herbert M. Reese, who had been recommended
by Joseph S. Ames, the laboratory spectroscopist at Johns Hopkins.
Reese had already earned his Ph.D., but had been unable to get a job in
physics, so he was willing to consider the Lick assistantship. He had
absolutely no experience in astronomy, except that he had helped Ames
in his abortive attempt to observe the spectrum of the corona at the May
28, 1900 eclipse. Nevertheless, Keeler had no options and offered him the
job. After months of indecision, while he was no doubt trying to find a
better position somewhere else, Reese finally turned down the offer,
saying he could not live on $600 a year.[37] Keeler did not succeed in filling
the job before he died, but in September Reese reconsidered, and Campbell desperate for help, hired him. However, Reese showed very little
initiative, and drifted out of astronomy in a few years.[38]

   The person whom Keeler should have hired for the fellowship, but did
not, was Heber D. Curtis. He was another Michigander, born in Muske-
gon and ten years younger than Campbell and William J. Hussey. Curtis
had studied at the University of Michigan while they were on its faculty,
but he had never taken a course from either of them, or even entered the
observatory. He had been a classical student, and had learned Latin,
Greek, Hebrew, Sanskrit and Assyrian at Ann Arbor. After earning his
A.B. in 1892 and A.M. in 1893 he taught in the Detroit High School for
six months, and then became the professor of Latin and Greek at little
Napa College in California. It had an eight-inch refractor and through it
he became interested in astronomy. When Napa College merged with the

University of the Pacific in 1896, Curtis moved to its campus, close to Mount Hamilton between San Jose and Santa Clara. There was another telescope at the University of the Pacific and when the chance came, he switched from teaching Latin and Greek to mathematics and astronomy.[39]

Curtis first became acquainted with Holden when the University of the Pacific awarded the Lick director an honorary degree in 1896, and the young professor of classics was given the job of drawing up the Latin diploma. Holden encouraged his interest in astronomy, and Curtis spent six weeks in the summer of 1897 as a special student at Mount Hamilton.[40] In spite of the upheaval that was going on, this stay confirmed his now strong interest in astronomy. By the following winter Curtis had decided to quit his job at the University of the Pacific and become a graduate student at Lick. It was a difficult road to follow, because he was married and had a family to consider, and his undergraduate preparation in science was meager indeed. Nevertheless, Curtis applied for admission to study for the Ph.D. degree. Leuschner, however, blocked his candidacy. He needed more mathematics courses, and in any case would have to wait until Keeler arrived before a decision could be reached, the Berkeley professor said. Curtis could not wait; he had to commit himself to teaching another year or resign his position at the University of the Pacific. He gave up the idea for the time being, and kept his job, but spent his whole summer vacation at Mount Hamilton as a special student again. He worked especially on astronomical photography. The next summer, in 1899, he went back to Ann Arbor and studied celestial mechanics, and the following winter, he put in his two-week Christmas vacation at the Lick Observatory library, finishing up an orbit computation.[41]

In February, Curtis formally applied for a Lick Observatory fellowship for the following year. Keeler told him frankly that if a vacancy occurred he would be considered, but that the greatest need of the observatory was "for an assistant skilled in astrophysical and spectroscopic work, to take part in the observations of motions of stars in the line of sight, and my first recommendation will have to be made with reference to this need." Curtis wisely shopped around, and was offered a fellowship at the University of Virginia under Ormond Stone. His work would be purely in celestial mechanics, and he could earn his Ph.D. in two years. But Curtis preferred to do observational work at Lick, and asked Keeler for a definite answer. "Is it any more likely now that a fellow will be appointed, and if so, do I stand a chance? I am quite anxious to secure it," he wrote. Keeler's answer was even more negative this time. The greatest need was for a spectroscopic observer, he said, and that would govern his recommendation for the next vacancy. After that the prospects were so uncertain, Keeler wrote, that Curtis "would do well to accept Stone's offer, if you consider it a desirable one." Doubtless Curtis' abilities and experience qualified him for something better than the Virginia fellowship, but it at least might lead to a career in research after he got his degree. It was clear to Curtis that he had no chance for a Lick Observatory fellowship, and he accepted the Virginia one as a bird in the hand.[42]

It was just at this time that Campbell and Perrine were leaving Mount Hamilton for their eclipse camp in Georgia. Curtis, who was heading east anyway, asked permission to join them as a volunteer assistant. Keeler was glad to accept him in this role, and Curtis was soon at Thomaston, helping Campbell and Perrine set up the equipment.[43]

At the eclipse site, Curtis proved himself an expert with instruments. In his summers at Mount Hamilton he had evidently trained himself well, for he could get any telescope or spectrograph adjusted and into working condition. With Perrine ill much of the time, Campbell depended heavily on Curtis' help. After the successful observations were completed, the Lick astronomers wrote that they did not see how they could have dispensed with Curtis' services, nor how anyone could have met their "exacting demands" better than he had. The tight-fisted Campbell was so pleased that he offered to pay Curtis' traveling expenses retroactively – from Atlanta, not from San Jose.[43] More importantly, he promised to do his best to get Curtis a job as an assistant at Lick as soon as he had earned his degree.

In 1902 Curtis completed his work at Virginia, and Campbell immediately hired him. At first he worked as an assistant on the radial-velocity program, and showed himself to be a skilled, reliable observer. Before long he was using nearly every instrument on the mountain, and getting important results with them.[39] In the case of Curtis, Keeler had paid too much attention to his deficiencies in formal course work (and probably, to his classical education) and not enough to his drive, determination, and mechanical abilities.

By the time Reese finally had made up his mind to turn down the Lick fellowship in early July 1900, Curtis was already firmly committed to the University of Virginia. Keeler desperately wrote Hale, to ask if he could recommend anyone for the post, and in particular if Walter S. Adams, then a graduate student at Yerkes Observatory, would take it. It turned out that Adams was going abroad for a year's study in France and Germany; otherwise he could have come to Lick. He returned to continue with Hale after his year in Europe, and eventually became the long-time director of Mount Wilson Observatory.[44]

A few days after learning that Adams would not accept the fellowship, Keeler received an application from Elizabeth Wylie, of Gardiner, Montana. She was a recent graduate of Wellesley College, where she had studied mathematics and astronomy. Although she was probably only a little less qualified than Palmer, R. Tracy Crawford, and Frank Ross had been when they started two years before, Keeler wrote her:

> So far, all students here have been men. Our quarters are very limited, and are not arranged with reference to the accommodations of ladies. For this reason I fear that there will be little chance for you at Mt. Hamilton at present.

He recommended she apply to the Students' Observatory in Berkeley, which had "good equipment for purposes of instruction" and "excellent courses in practical astronomy." The same day Keeler turned down Elizabeth Wylie's application, he was writing Ames at Johns Hopkins to ask if he had another candidate, now that Reese was out of the picture. "I shall be much obliged if you will reply as soon as convenient, as we have already lost much time over this appointment." It is not fair to blame Keeler for failing to break out of the conventions of his time, but Elizabeth Wylie's father ran a dude ranch and pack-train outfit at Yellowstone National Park, and she was probably at least as capable of survival at Mount Hamilton as most of the male fellows.[45]

One other woman, Adelaide M. Hobe, had been turned down for a fellowship on similar grounds. She was a Berkeley product, who had completed the full undergraduate astronomy program in 1899. According to Leuschner she was especially well qualified on the theoretical (celestial mechanics) side. After graduating, she worked for him as an assistant and proved herself a "rapid, neat and accurate computer" who also showed "ability as a librarian and secretary." In recommending her for a position at Yerkes Observatory, Leuschner wrote

> She will undoubtedly do better than many a man though she has no experience in measuring spectrograms for lack of necessary instruments. Professor Keeler would have gladly appointed her at Mount Hamilton, if it had been possible for her to arrange for board and room on the mountain. Some of the Fellows at Mount Hamilton are messing [eating] with the workmen, which, of course, would never do for a young lady.

Hale did not hire Adelaide Hobe at Yerkes either, but later Campbell did bring her to Mount Hamilton as a full-time assistant. She worked there for many years, largely measuring and reducing spectrograms for the radial-velocity program. Eventually she left for a better job in the Food Research Institute at Stanford.[46] Keeler never did fill the vacant fellowship; after his death when Reese came he proved capable of messing with the workmen, if not of being a rapid, neat and accurate computer.

The first of the graduate students Keeler had brought to Lick received their Ph.D.'s in 1901. Ross had switched to mathematics after his eyes gave him so much trouble in his first year on Mount Hamilton. His thesis was on differential equations, but he retained his interest in astronomy and after leaving California moved through a long series of research positions. For ten years he was in the research division of the Eastman Kodak Company, and then for fifteen a professor at Yerkes Observatory. His most important contributions were in lens designs and astronomical photography.[47] Crawford, who graduated at the same time as Ross, received the first earned Ph.D. in astronomy ever awarded by the University of California. His thesis, on the atmospheric refraction at Mount Hamilton, had been suggested by Keeler and supervised by Tucker. Crawford became a long-term faculty member at Berkeley, where he calculated many orbits of comets, asteroids and satellites. He succeeded Leuschner as chairman of the Berkeley astronomy department in 1936, and retired himself in 1946.[48]

Keeler's insistence on getting graduate-student assistants who had some training in physics as well as in mathematics and astronomy was finally satisfied the year after his death. The two new Lick fellows who replaced Coddington and Crawford were Stebbins and Ralph H. Curtiss. Curtiss was a Berkeley product, "one of the brightest men we have ever turned out," who had prepared in spectroscopy as well as in mathematical astronomy.[46] Stebbins, a graduate of the University of Nebraska, where he had taken a general course in science, specializing in mathematics and astronomy, also had one year as a graduate student at the University of Wisconsin. There he observed and worked on orbit theory with Comstock, and also took advanced physics courses in electricity and optics. As Lick fellows Stebbins and Curtiss were both excellent students, and then went on to long and successful research careers afterward.[49]

Just before he was named director in December 1901, Campbell had a marvelous success in obtaining a new large telescope for Lick Observatory. His radial-velocity program was providing important new information on the motions of the stars. But because of the observatory's location, the stars he could observe were restricted to the northern hemisphere of the sky, and to regions not too far south of the celestial equator. For statistical purposes, it was important to have similar data for stars spread all over the southern hemisphere, but there was no southern observatory with a large telescope to observe them. In the best do-it-yourself tradition, Campbell decided that Lick Observatory should build its own southern station in Chile or Australia, where the climate and observing conditions fairly well mirror those in California.

Campbell had seen Keeler's successes with the Crossley reflector close at hand, and had been as much impressed as anyone. He realized that a reflecting telescope would be much cheaper, and quicker to build, than a refracting telescope of the same aperture. Campbell planned carefully – he became a notorious skinflint in his later years as director – and decided that to build a 36-inch reflector, with the same light-gathering power as the 36-inch refractor on Mount Hamilton, equip it with a spectrograph, send it to Chile and erect it there, and operate it for two years would cost $24,000. Campbell added a ten per cent contingency fund, making a total of $26,400 ("[w]ith this I could promise a surplus"), and sent off a request for the funds to D.O. Mills in New York.

Campbell succeeded where Holden and Keeler had failed before him. Something about his letter, or the testimonials from Newcomb and Vogel that he had enclosed, caught Mills' attention. He wrote back

> I am greatly interested in your letter of the 17th ult. and the object of it, and feel inclined to furnish the money for the expedition to observe the southern skies as set forth by you. If this work can be accomplished in two years for $24,000.00, I shall take pleasure in furnishing the funds, the money to be drawn as required in the work.
>
> I should be pleased to have you give me a little more in detail your plans for the expedition, and also whether it can be carried out within the above sum.

Campbell gave the financier no time for second thoughts, but immediately accepted his "generous offer", forwarded the letter to President Wheeler, released the news to the press, and then sent Mills a little more detail on his plans. Mills' gift was the largest to the observatory since Lick's. Campbell was very pleased. He believed that nothing was too good for Lick Observatory.[50]

The 36-inch reflector, except for the optics, was built in California. Campbell provided the general design, a San Francisco engineering firm converted it to detailed construction blueprints, and the Fulton Engine works in Los Angeles built it. It was a Cassegrain reflecting telescope, with a primary mirror 36 inches in diameter. The secondary, hyperbolic mirror, mounted near its focus, directed the light back through a hole in the primary to the focal point of the telescope, where the spectrograph was mounted. Brashear furnished the optics for the telescope and the spectrograph, and the whole outfit was assembled and tested at Mount Hamilton, disassembled, boxed, and shipped off to Chile in little more than two years from the time Mills said yes.[51]

Campbell had planned to go to Santiago himself and superintend the installation of the telescope. However he slipped and fell, breaking his leg, while testing the telescope mounting, and was unable to do so. Wright went instead, assisted by Palmer, the Lick fellow who had worked most closely with Keeler on the Crossley reflector. He had done his thesis on the spectra of faint planetary nebulae, taken with the Crossley reflector. Palmer was an excellent observer, but a very independent soul who did not like deadlines. Campbell allowed him far more freedom than the other students, probably because he reminded him of Keeler. Palmer did not want to go to Chile, tried to get various other jobs, and put off writing up his thesis several times. Finally, saying he had been too lazy to turn up any other job, he agreed to go, but only if he were paid a salary of $900 for the first year and $1,100 for the second, "which means that I will go for $900.00 but not for $899.00 or less." He wanted to postpone his final examination for the Ph.D. until after his return. Campbell and Leuschner practically forced him to take it on the day before he sailed with Wright for Valparaiso. Campbell could not come to Berkeley for the oral examination because of his broken leg, but sent a list of questions to Leuschner, and designated Wright to participate as his representative. Palmer passed with flying colors and left for Chile a Doctor of Philosophy.[52]

Wright, Palmer and the main parts of the telescope reached Valparaiso on April 18, 1903. They found a strike of the launch boatmen in progress, but somehow managed to get the equipment ashore themselves. Wright located a site on Cerro San Cristobal in the outskirts of Santiago, and the ground was broken for the observatory on May 27. The first observation of a star was made on September 11, within a month twenty stellar spectrograms had been obtained, and by November the systematic program was well underway. The whole "Mills Expedition", as it was called, was very successful. The telescope stayed in operation under Lick auspices for twenty-five years, and over 10,000 spectrograms were obtained. All the Lick systematic studies of radial velocities of various groups of stars were complete over the whole sky because of it. It blazed the trail for Cerro Tololo Interamerican Observatory, the United States national observatory with its 158-inch reflector, built in the Chilean Andes fifty years later.[53]

Campbell brought the observatory in under budget, for $24,000 for the telescope, dome, spectrograph and all, and had it in operation within less than three years from the date the money became available. It had the same aperture as the 36-inch refractor, which had cost by far the major part of James Lick's $570,000, and had taken twelve years to complete.

For big telescopes the advantage was all with the reflector, as several astronomers and telescope makers had earlier predicted, as Keeler had demonstrated with the Crossley, and as Campbell had recognized and followed up on. The day of the refractor was over, and although a few more intermediate-sized ones were built, no American professional astronomer ever thought seriously of building a very large telescope as anything but a reflector, after Keeler's work with the Crossley. Mount Wilson, McDonald, Palomar, Lick Observatory (now), Kitt Peak and Cerro Tololo are all built around large reflecting telescopes.

Keeler and Hale had discussed reflecting telescopes many times, and the young Yerkes director had, if anything, even more advanced ideas than his friend on their suitability for astrophysical research. What was more, Hale was able to command the resources to translate his thinking into action. He decided, long before Yerkes Observatory was completed, that a "five-foot" reflecting telescope would be just the thing to complement its 40-inch refractor. Clearly neither President William R. Harper of the University of Chicago nor Charles T. Yerkes, the street-railway magnate, would provide funds to start working on another telescope before the largest refractor in the world was even finished. In this emergency Hale turned once again to his wealthy father, William E. Hale. He agreed to provide the money to buy the large glass blank in France from which the mirror was to be ground. In early 1896 William E. Hale ordered the glass, and by 1897 it had been delivered at Yerkes Observatory. He also personally provided the salary with which his son hired optician George W. Ritchey to grind and figure the disk to a parabolic reflector. Thus, at the dedication ceremonies in October 1897, the visiting astronomers saw Ritchey working on the mirror in the optical shop in the basement of Yerkes.[54]

The 60-inch reflector was a long-term project, but to get started, Hale decided to erect a smaller reflector as soon as he could. From regular University of Chicago funds he paid Ritchey $200 for a 24-inch parabolic mirror, which the optician had completed in Chicago in 1896 before he went to work for Hale. It became the basis for a reflecting telescope which was erected in one of the small domes at Yerkes Observatory. The mounting for this reflector was started by Wadsworth, but after he left Yerkes Ritchey completed it, and the telescope was ready for operation in November 1901, over a year after Keeler's death. The mechanical system was far superior to the Crossley, and with the 24-inch Ritchey was able to obtain in 1901 and 1902 even better photographs of some nebulae than Keeler had taken at Lick.[55]

Well before this, however, Hale reported to Harper that the 60-inch mirror was approaching completion. "All" that remained was to provide the mounting for the telescope, and to construct a building and dome to house it. Since he had bought the glass blank and paid Ritchey to figure and polish it into an exact paraboloid, the mirror belonged to William E. Hale. He valued it at $10,000, but he was willing to give it to the University of Chicago, on condition it provide the building, dome, mounting and all necessary auxiliary apparatus for the complete telescope. Furthermore, if a donor came forward, the new reflector could be named for him; William E. Hale was willing to remain anonymous.

Actually the 60-inch mirror was not close to being completed. William E. Hale was mortally ill, and died a few months later; apparently Hale wrote this letter to get the facts of the mirror's ownership, which Harper knew very well, on record before his father's death.[56] Hale was very deeply attached to his father and dependent upon him, not least emotionally. His death, though anticipated, was a devastating blow. As Brashear wrote Keeler

> The death of Mr. Hale has also taken a good man from among us. Poor George feels it very keenly, and well he may. Few there are who have had such privileges and such chances as those given our good friend by his father, but George deserved all the trust that his father placed in him.

Hale could not accept that he would never see his father again. He grasped desperately for some return of the belief in immortality in which he had been raised by his parents. Six months later Hale's mother followed her husband in death. Although he certainly had not wanted it, for the first time in his life Hale was on his own, without the protective, supportive parents in whose house he and his wife still lived whenever they were in Chicago.[57]

Very soon after his father's death, Hale revealed for the first time his idea of erecting the 60-inch reflector not at Williams Bay, but in Southern California. He had come to realize clearly how much better the weather was for observing there than in Southern Wisconsin. Hale envisaged operating the telescope as a remote station in California, sending out astronomers (including himself) from Yerkes Observatory, each of whom would stay for a few months at a time. He estimated the cost of building and housing the telescope, with its auxiliary instruments, as $85,000. In addition he was seeking $50,000 as an endowment fund for the new station. He would contribute the 60-inch mirror, now his own personal property, which he valued at $20,000. Hale approached N.B. Ream, a wealthy Chicagoan, in January 1899, but did not succeed in getting the $135,000 from him. A few months later Harper and Hale tried Yerkes, but he did not want to finance a new branch observatory either.

Keeler invited Hale, if he raised the money for the new telescope, to put it on Mount Hamilton, where he could use the already developed road, water supply and other facilities. The Yerkes director replied that nothing would give him greater pleasure than to do so, and that he could imagine nothing more satisfactory than making an annual trip to Lick, where he and Keeler could work together. But the observing weather in the winter was better in Southern California, so that was where he wanted to put the reflector, although as yet he had no good prospects for the money. Hale was correct in his assessment of the weather, but no doubt he also prized his independence far more than the financial and logistical advantages of locating at a developed site. If he took the new reflector to Mount Hamilton, he and it would always be in the shadow of Keeler and Lick, he must have thought.[58]

After his mother died, Hale openly threatened to leave the University of Chicago. Harper had frequently treated him like a bad child, chastising him severely over matters like the repairs Hale insisted should be made to a staff member's house, and the omission of the name of the University of Chicago from Yerkes Observatory news releases. As long as his parents were alive, it would have been impossible for Hale to go. Now he could get even with his "harsh stepfather," Harper, by taking his 60-inch mirror and going away – or at least threatening to do so. Hale reminded the president that a year had elapsed since he had said the university would accept the mirror, but no money had yet been raised to meet his conditions. Hale said he would like Chicago to have the mirror, but rather than see it unused he would take it elsewhere. "The remarkable photographs of nebulae obtained by Professor Keeler with the much smaller reflecting telescope of the Lick Observatory, and exhibited here during the astronomical conference, gave some idea of what may be expected from our instrument in one department of research," he wrote Harper.[59]

In 1901 Hale tried very seriously to get the Massachusetts Institute of Technology to raise an endowment to build a telescope and an observatory around his 60-inch mirror, but this scheme fell through.[60] His thoughts turned again to California, and by December 1903 Hale had moved his family to Pasadena. He was enchanted by the warm sun and the clear skies, the orange trees and flowers. His children were healthy again, playing outside in their shirtsleeves. He could not help but contrast it with the $-20\,°F$ temperatures at Williams Bay the previous week. Hale was trying to get money from Carnegie to build a solar observatory on Mount Wilson, which he could see from his bedroom window on Palmetto

Street. He was almost certain that it was *"the* place to continue my solar work."

Two months later his daughter was sick again, but Hale had met Alicia Mosgrove, the attractive young woman whose company he came to find increasingly more stimulating than his wife's. He had decided that "'home' must be here, and that I must pull up stakes at the Y[erkes] O[bservatory]." He was in a quandary as to how to do it, but do it he must. Scheme after scheme came to his mind for raising financial support and for bringing Ferdinand Ellerman, E.E. Barnard, Ritchey, and Adams from Yerkes to work with him.[61] Before long he did get financial support from Carnegie, sever his connection with the University of Chicago (Harper died soon afterward, of cancer) and build the Mount Wilson Solar Observatory.

During this period Palmer, Keeler's former assistant, worked briefly for Hale. In Chile Palmer had done a large share of the observing, but came to feel that he could not stand the continual night work at the telescope. He turned down the staff position Campbell had offered him at Lick because it would mean more nights without sleep. At Mount Wilson, Palmer became a daytime solar observer, but soon tired of it and quit astronomy. One of the most creative of the early Lick students, he ended up a hydraulic engineer, but always kept a soft spot in his heart for Mount Hamilton.[62]

By 1908 the 60-inch telescope was completed on Mount Wilson. It was the first large reflector built embodying Ritchey's idea of "flotation" of the mirror by balancing weighted levers, which provided near-freedom from flexure together with high mechanical stability.[63] It was the start of a succession of large reflecting telescopes that pushed the frontiers of optical astronomy ever outward. Next came the 100-inch Hooker telescope, also built at Mount Wilson, under Hale's directorship, then the 200-inch, for which he provided the initial impetus, completed at Palomar Mountain in 1948.

Hale lived on until 1938. His solar research earned him many well-deserved awards. He became increasingly a statesman of science, a promoter and organizer of vast schemes. Among other accomplishments, he converted little Throop Polytechnic Institute into the California Institute of Technology, he organized the National Research Council and through it science's contributions to America's World War I effort, and he revived the National Academy of Sciences and built its marble palace on Constitution Avenue in Washington. He was a major figure in transforming the International Union for Cooperation in Solar Research into the International Astronomical Union. In Europe Hale played an important part in the organization of the International Research Council, now the International Council of Scientific Unions. Yet the contradictions between his public image and private doubts and anxieties haunted him increasingly. He suffered from severe depression. At the moments of his greatest apparent triumphs, he became a recluse and an absentee director. He tried many psychiatrists and cures, and lived out the end of his life in the garden of his private solar observatory in San Marino.[64]

What would Keeler have done if he had lived another thirty years, as Campbell and Hale did? It is impossible to be sure, for history is a single track of what *did* happen. But almost certainly Keeler would not have become the president of a university, nor the organizer of national and international science. His whole life was tied up in doing scientific research. Very probably, if he had lived, Keeler would have made important contributions to our understanding of nebulae and galaxies, such as those made by Curtis and others at Lick, by Vesto M. Slipher and by Edwin P. Hubble in the next quarter century.

When Keeler started his program of direct photography of nebulae, he saw it as a temporary diversion from the spectroscopic work he wanted to do with the Crossley reflector. It would allow him to get the bugs ironed out of the telescope, until the Bruce spectrograph that Campbell had designed, and that had been constructed by the Lick Observatory instrument makers Percy Urmy and Emil Zengeler, was fully operational. As we have seen, Keeler began by finding that when the spectrograph was mounted on the telescope, the dome shutter could not be closed. He solved this design flaw by having the telescope lowered two feet. But then he discovered the fundamental problem. Campbell had designed the spectrograph for radial velocity work, with three prisms to disperse the light, and collimator and camera lenses to make it parallel, and then refocus it on the plate. He had ordered three-element lenses from Brashear, to give "a *reasonably large field.*" No one had done any numerical calculations to see if such lenses would in fact form sufficiently sharp images to produce a high-definition spectrogram, as any optical designer would today. Campbell could not make such computations, but he should have had an optical expert do them for him. At the very least he should have got the lenses and tested them before building the spectrograph, rather than vice versa as he did.[65]

His assumption that lenses could be obtained that would satisfy the necessary optical requirements had been fulfilled with the Mills spectrograph of the 36-inch refractor that he had designed earlier. It, like all the refractors that were the work-horse telescopes of the time, was a relatively slow optical system. Its focal ratio was F/19, so the collimator and camera lenses for the Mills spectrograph had these same proportions. At this focal ratio, all aberrations are minimized, and the lenses produced high-quality spectrograms. However, the Crossley reflector was much faster, with a focal ratio F/5.7. That is what made it so effective for direct photography of nebulae. Even the best triplet lenses of that day were incapable of providing high definition at that focal ratio, over the field necessary to get good radial-velocity spectrograms.

Very soon after Keeler arrived at Lick in 1898 he ordered a new reflecting-type slit for the "nearly completed" Bruce spectrograph, and inspected the instrument for the first time. To his practiced eye, the problem of the fast focal-ratio lenses was immediately apparent. He wrote Brashear that he had not yet tried the spectrograph, but that he was "morally certain" that with those triplet lenses its field of good definition would be too small to be usable. Keeler suggested that a more compli-cated four-element lens, composed of two cemented doublets, might be substituted for the triplet in the camera. He asked Brashear to check this idea with Hastings, his optical design expert, before proceeding further. The report from Hastings was disquieting. Although some improvement would be provided by the compound lens Keeler had suggested, it would probably take an even more complicated system to produce really satis-factory spectrograms. Keeler was willing to order it, if Hastings could come up with a satisfactory solution. However, his final design called for a five-element lens, using four different kinds of glass, a system that would be extremely difficult to make and that would have very large light losses. Keeler, who was then just starting to get important new results by direct photography with the Crossley, decided to let spectroscopy wait for the time being, and did not order the lens.[66]

However, Keeler continued to turn the problem over in his mind. By the spring of 1899 he apparently had decided to abandon the Bruce spectrograph and start anew with a design for a much faster instrument. He wrote Vogel:

> Our 3-ft Crossley reflector promises to be very efficient for certain kinds of spectroscopic work, though not for motions in the line of sight. A spectrograph for it is being made here, but it will be some time before this instrument is ready for use.[67]

In fact it was not "being made," but Keeler clearly had the basic idea in mind at that time. He was too busy then to put a design down on paper. But in mid-July Keeler was galvanized into action. He had taken his series of photographs of the Ring nebula in Lyra, and saw how bright its central star was in blue, photographically active light. Keeler wanted to see the star's spectrum, and tried by holding a small, direct-vision spectroscope at the focal plane of the Crossley. The spectrum was too faint to be observed that way. Once again Keeler could see that a low-dispersion photographic spectrograph, as efficient as possible, was what he needed. This time he was determined to go ahead with it. The Bruce spectrograph was a white elephant, he decided, and he would junk it once and for all. He removed the mounting ring and rods that had been installed on the telescope to hold the spectrograph in place some day. With them he could also take off "an immense quantity of lead," the counterweights on the telescope. At the same time Keeler wrote Brashear for a quotation on a 45° quartz prism for the new spectrograph he wanted to build. He asked if, until he bought the quartz prism, Brashear could lend him one, "as I have some important experiments to make the next time the moon gets out of the way."[68]

Keeler's idea was to make a slitless spectrograph, which would give very low dispersion, for the Crossley. In its simplest form there would not even be a collimator or a camera lens, just the prism in the converging light beam from the primary mirror. Keeler was not certain that it would really work in this form, but if the aberrations were too large, he could add two small single quartz lenses, one in front of the prism as a collimator, the second behind it as a camera.[69] In either case the whole small spectrograph would be mounted in the plateholder that was otherwise used for direct photography. That way the observer could use the whole guiding apparatus, just as if he were taking a picture of a nebula.[70] Because of the short focal length, low dispersion, quartz optics, and absence of a slit, the whole system would be extremely efficient. For the same reasons the resulting spectrograms would contain only a minimum of spectral information. They would not be suitable for precision measurement of radial velocities; indeed without a slit the wavelength scale would be very poorly determined. But these spectrograms would be all that was needed to reveal the basic physical nature of faint objects, like the central stars of nebulae, and of the nebulae themselves. That was what Keeler intended to do with the spectrograph as soon as it was completed.[71]

One of the objects Keeler wanted to investigate spectroscopically was M 31, the Andromeda nebula. In Germany Julius Scheiner had obtained a spectrogram of it, which he reported showed a continuous spectrum with absorption features at two of the wavelengths corresponding to solar absorption lines.[72] If he were correct, it would mean this largest spiral nebula was not a gas cloud forming into a star or stars, as Keeler and nearly all other astronomers of the time believed. It was an exceedingly interesting result to Keeler. As he wrote,

[I]t is difficult to interpret Scheiner's spectroscopic observations unless we are to regard the nebula as in fact an assemblage of small [faint] stars – a view which has great difficulties of its own. It also seems to me possible that there is a thin "reversing layer" around the central condensed part of the nebula, and that no lines would be found in the spectrum of the outer portions. It might be possible to investigate this question experimentally, though the observations would be difficult. The Crossley reflector is an efficient instrument for the purpose, and I [may] take up this and other spectroscopic work with it later, but at present I am using it for other purposes.[73]

The other purposes were the direct photographs of planetary nebulae, globular clusters and spirals Keeler was taking night after night that summer and fall with the Crossley. The results were new and important, and Keeler thought there would be time to convert to spectroscopy later. Brashear did not have a small piece of quartz on hand. He was just going on vacation, and promised to do something about it later.[74] Nothing happened on the spectrograph for months. Keeler never did make his experiments.

In November he had a few days to think about the fast spectrograph for the Crossley again. He ordered a quartz prism, specifying that it should be one inch high and have an apex angle of 45°. He told Brashear that the exact dimensions were not important, and that he would design the instrument to fit the prism the optician sent. A few days later, however, after further design studies, Keeler wrote his friend again, saying the prism should be a little larger, and that the apex angle should be 50° instead of 45° if possible, to provide just a little more dispersion. In December he sent a firm order for the prism. Large pieces of quartz were hard to find, but Brashear finally located one. Keeler finished the design of the spectrograph. He became impatient, and tried to hurry Brashear up, but his shop was swamped with orders from astronomers everywhere for the forthcoming solar eclipse. Keeler himself was urging him to send the special prisms and lenses Campbell and Perrine needed for the Lick eclipse instruments. The quartz prism had to wait.[75]

Finally on March 31, 1900 Brashear finished all the optics in the Lick order, including the quartz prism, and shipped them to California. When Keeler opened the box, he found there had been a breakdown in communication. Brashear had made the prism with an apex angle of 45°. Keeler had completed the design, however, assuming the angle would be 50°. No doubt Zengeler, the Lick instrument maker, had already begun cutting and assembling the metal pieces of the spectrograph. Rather than redesign and rebuild the whole instrument, Keeler sent the prism back to Allegheny, asking Brashear to rework it to an angle of 50°. Again, rush orders for the eclipse had first priority, and Brashear did not modify the quartz prism and send it back to California until nearly the end of June.[76] By then the spectrograph was nearly completed.[69] Zengeler finished it on July 30, 1900, the very day Keeler left Mount Hamilton for the last time.[70] He never got to use the spectrograph on a nebula.

A few weeks after Keeler's death, Campbell and Palmer tried the slitless spectrograph for the first time. They found that, as Keeler had feared might be the case, the aberrations were too large with the instrument in this lensless form and they could not get a satisfactory focus. Nothing further was done for a year, and then Campbell had Palmer design changes in the instrument which allowed quartz lenses to be inserted in front of and behind the prism, as Keeler had planned to do if necessary. In this form the little spectrograph was quite efficient, especially for emission-line objects. Palmer used it through the summer of 1901, and took many spectrograms of planetary nebulae, novae, and Wolf-Rayet stars. After Palmer left Mount Hamilton for Berkeley, Stebbins took some more spectrograms for him to help him finish his Ph.D. thesis. Although the dispersion was only about 500 Å/mm, the spectrograph was extremely efficient for qualitative studies of the general spectral features of faint objects. With its all-quartz optics, it was particularly fast in the ultraviolet spectral region. Palmer identified the ultraviolet lines he called λ337, λ345 in several planetary nebulae, lines we now recognize as [Ne V] λλ3345, 3425. He noticed the great strength of λ373 ([O II] λ3727) in the Ring nebula and he saw that its central star had a continuous spectrum, while the central stars of some other planetary nebulae showed the emission bands characteristic of Wolf-Rayet spectra.[70] Stebbins took several spectrograms of Nova Persei 1901 with the instrument, and found nearly all the nebular emission lines, including the extreme ultraviolet ones, strong in this object.[77] It was exciting, pioneering research, prospecting in completely new areas of astrophysics, but by now Campbell was far more interested in precision radial-velocity research. He put Stebbins to work on the spectrum of the long-period variable star Mira Ceti with the 36-inch refractor, and assigned no one to work with the slitless spectrograph for many years. Only much later did Keivin Burns use it to study the structure of the Ring nebula in the light of different nebular emission lines, a problem exactly in Keeler's line.[78]

In 1907 Edward A. Fath, another Lick graduate student, made an observational study of the spectra of spiral nebulae. He began by trying to check Scheiner's statement that M 31, the Andromeda nebula, had absorption lines in its spectrum. Fath's first observations were made with the quartz slitless spectrograph. They were promising, but showed that a slit spectrograph would be necessary to learn much about these objects. He constructed a small instrument of this type, also for use at the Crossley. Its dispersion was only about 400 Å/mm, and its resolution was not very good, but with it he clearly saw the absorption lines of solar-type stars in the spectrum of M 31 and several other spirals. Their spectra were very faint – Fath estimated it would have taken 450 hours exposure time to take a satisfactory spectrum of the brightest part of M 31 with the radial-velocity spectrograph on the 36-inch refractor – but with exposures of several nights with the Crossley and his little spectrograph he could see the only possible interpretation was that these spiral "nebulae" were composed of many, many stars, similar in a general way to the sun. As a test of his picture, Fath took spectrograms with the same equipment of several globular clusters, which he and everyone else knew were certainly composed of stars, because they could see many individual ones in them. The spectra of the globular clusters showed the same solar-type absorption lines as the spectra of the spirals, confirming Fath's idea beautifully. He recognized that since he could not see individual stars in the spirals, they must be much farther away than the globular clusters, but he did not quite have the courage, or the support of his director, to state boldly that this result must be true. Instead, he wrote that his "hypothesis" would "stand or fall" on the parallaxes (distances) of the spirals.[79]

The true scientific successor of Keeler was neither his last students Harrer and Palmer, both of whom dropped out of astronomy, nor his first assistants Campbell and Leuschner, who went such different ways. His heir was really Curtis, to whom Keeler would not give a fellowship, because he thought the former classics professor lacked sufficient formal scientific training. After his return from the Lick southern station in Chile in 1910, Curtis took over the Crossley reflector and continued the program of observations of nebulae that Keeler had begun. Curtis practically rebuilt the telescope, improving its mechanical stability still further. He took many photographs of spiral nebulae, planetary nebulae, and diffuse nebulae. He studied them carefully and came to understand their nature very clearly indeed.[39] In a series of three outstanding papers in the *Publications of the Lick Observatory* he described his conclusions on the nature of planetary nebulae and spirals.

Curtis recognized the dark bands in edge-on spirals as the result of "occulting matter" due to "the same general cause" that produces occulting effects in our own galaxy like "coal sacks" and Barnard's "dark nebulae." It was the best description of the effects of interstellar dust, without using those words, that was ever published. In addition, Curtis could see that same occulting matter present in all spirals, not just the edge-on ones. They all belonged to one family. He believed that essentially all the nebulae that were not planetaries or diffuse were spirals. In this he was wrong, though, for he thought that the objects we now recognize as elliptical galaxies were unresolved spirals.

Curtis had many more years to observe with the Crossley than Keeler. On the basis of his more complete data, he raised the estimated number of nebulae within reach of the reflector to one million. Furthermore, he found that none of them were near the plane of the Milky Way; all were clustered near its poles. This, he realized, was not their true distribution; it resulted from the absorption (or "occulting") of light by dark matter in the Milky Way.[80] In a compelling paper he firmly stated the conclusion that the spiral nebulae were "island universes" ... "similar to our galaxy in extent and number of component stars."[81]

In 1921 the Swedish astronomer Knut Lundmark came to Lick Observatory as a visitor. He had studied carefully the direct photographs of spirals, particularly those of M 31 and M 33 that Ritchey had taken with the 60-inch reflector on Mount Wilson. Lundmark could see that especially in the outer parts of their spiral arms there were numerous nearly stellar condensations. Ritchey had called them nebulous stars, but Lundmark formed the opinion that they were groups, clusters and associations of stars, too faint and too close to one another to resolve individually, but that together gave the impression of nebulous objects. He used the Crossley to take long-exposure spectrograms of the spiral arms, just as Keeler had been preparing to do in 1900. On the resulting plates Lundmark could see that the spectrum of the arms was very similar to the solar spectrum. It was continuous, with absorption lines at K, H and G, just as in the spectra of hundreds of ordinary stars. Lundmark concluded that the spiral arms probably did consist of stars, clusters, and some "nebular matter," for which we today might read gas and dust.[82] He was absolutely correct.

Much of what Palmer, Burns, Fath, Curtis and Lundmark accomplished, Keeler would have very probably done himself if he had lived. They carried out exactly the programs he had planned, and for which he had designed the spectrograph that was finally completed just as he died. As director Keeler would have pushed the study of the nebulae much harder than Campbell did. He would have brought the resources of the observatory to bear on this problem. Keeler himself was a far better trained, more experienced spectroscopist than any of them. No doubt he would have reached the conclusion that the spirals were galaxies of stars even before Fath's tentative and hesitant statement of this conclusion, long before Curtis and Lundmark's firm results.

358   James E. Keeler, pioneer American astrophysicist

Quite probably Keeler would also have discovered the very large radial velocities of the spiral nebulae as Slipher did at Lowell Observatory in 1913. Fath had not seen this, or if he had suspected it, had not published it, because his spectrograph had no provision for inserting a laboratory comparison spectrum. Keeler undoubtedly would have started just as Fath had, without a comparison source, but would have had the instrument maker install one just as soon as he began getting usable spectrograms. Slipher knew very little about spectroscopy when he began his study; it is hard to believe that Keeler would not have discovered with the Crossley what the Lowell astronomer found with a 24-inch refractor.[83] The large velocities Slipher measured for the spirals were part of the evidence Curtis used to establish that they could not be within our galaxy, but must be separate, independent systems.

The person who really completely solved the riddle of the spirals was Hubble. With the 100-inch reflector on Mount Wilson, he was able to photograph individual highly luminous stars in several of them. Among these stars he identified regularly pulsating Cepheid variables, and from them determined the distances to the spirals, which were thus revealed to be truly galaxies at distances of millions of light years from the earth.[84] This Keeler would never have been able to do with the Crossley reflector, but he would have prepared the way for Hubble. In fact, Keeler's demonstration of the power of the reflector for direct photography of nebulae, and his discovery of the large number of spirals throughout the universe, out to the furthest observable distances, were the starting points for Hubble's work. Nicholas U. Mayall, who otherwise idolized Hubble, thought that his one short-coming was that he omitted any reference to Keeler's "epochal work" from his book, The Realm of the Nebulae.[85]

History is always written by the survivors. By his contemporaries Keeler was considered the outstanding astrophysicist of his generation in America. With a little telescope in a smoky city, and with the biggest telescopes of the time, on one of the best sites in the world, he had made important discoveries concerning the nature of the planets, stars, nebulae and galaxies. He had led the way in the design and use of spectroscopes and spectrographs for astrophysical research. He had outlined the problems that he and his generation would attack – many of which are still very live research subjects today. He had made great progress himself, and had influenced the direction of research followed by many astronomers all over this country. But Keeler was cut off at the peak of his career. He had not accomplished half of what he could have done had he been granted a normal life span. Human memory is fallible. Gradually Keeler's name was dropped from the discoveries he made. Today when we read in a textbook about the forbidden lines in the spectra of the nebulae, that do not arise in any terrestrial laboratory source, or about the orbital velocities of the tiny particles in Saturn's rings, or about the large numbers of galaxies in the universe, we do not associate these observations with him. Yet astronomy is the richer because of Keeler. No scientist is indispensable, and if he had never lived, other astronomers would by now have made all the discoveries he made. But in his life, he did lead the advance, and by his death, our knowledge of the universe was retarded. A skilled, intelligent research worker, he was at the same time a successful administrator, an outstanding graduate teacher and a warm, well-loved human being, a type rare indeed in the annals of science.

References

The main primary sources for this book are contemporary letters, memoranda, statements and the like, contemporary newspaper or magazine articles, and published scientific papers. In addition published historical and biographical articles, and published autobiographies, biographies and histories have also been used to supplement them.

In the chapter notes that follow, the few individuals for whom there are a great many citations are referred to by their initials only, as given in the table below. Published papers are identified by the author's name or initials, journal name (in abbreviated form, as given in the second table below, for the most frequently cited ones), volume number, page number, year of publication. Letters are cited by the names of the writer and addressee, the date (in square brackets if not explicitly given on the letter, and preceded by a ˜ symbol if only approximately known), and the archives in which it was consulted. The names of most of the archives are given by initials, as in the third table below. For archives explicitly divided into various collections, the names of those consulted are given in the table. Many of the letters may be found in two archives, the original in one, the copy kept by the sender in another. For instance copies of most, but not all, of the letters between Keeler and Hale are both in the Yerkes Observatory Archives and in either the Hillman Library or the Lick Observatory Archives. Only one source is given in the references, usually the one where I first happened to consult the letter.

Most of the newspaper articles are from papers for which files or microfilms are readily available, for instance in public and university libraries, state historical libraries or societies. In these cases no source is listed. But for clippings from obscure newspapers, the location of the clipping that I consulted is stated.

Biographical information from the *Dictionary of American Biography*,

360

Dictionary of Scientific Biography, National Cyclopedia of American Biography, Who Was Who in American History – Science and Technology, various editions of *American Men of Science,* and of *Who's Who* and *Poggendorf Biographisch-Literarisches Handwörterbuch zur Geschichte der Exacten Wisserschaften* have been used without citation.

The books that were useful for background information and general reference material are listed in the bibliography, after the individual chapter references.

Journals

AA   *Astronomy and Astro-Physics*
AJ   *Astronomical Journal*
AJS   *American Journal of Science*
AN   *Astronomische Nachrichten*
Ap. J   *Astrophysical Journal*
BMNAS   *Biographical Memoirs, National Academy of Sciences*
CR   *Comptes Rendus*
JHA   *Journal for the History of Astronomy*
LOB   *Lick Observatory Bulletin*
LO Cont.   *Lick Observatory Contribution*
MNAS   *Memoirs, National Academy of Sciences*
MNRAS   *Monthly Notices of the Royal Astronomical Society*
MRAS   *Memoirs of the Royal Astronomical Society*
PA   *Popular Astronomy*
PASP   *Publications of the Astronomical Society of the Pacific*
PLO   *Publications of the Lick Observatory*
SM   *Sidereal Messenger*

Individuals

EEB   E.E. Barnard
JAB   John A. Brashear
SWB   S.W. Burnham
WWC   W.W. Campbell
GD   George Davidson
RSF   Richard S. Floyd
ESH   Edward S. Holden
GEH   George Ellery Hale
WJH   William J. Hussey
WRH   William Rainey Harper
JEK   James E. Keeler
WFK   William F. Keeler
AOL   Armin O. Leuschner
SPL   Samuel P. Langley
SN   Simon Newcomb
JMS   John M. Schaeberle
MAT   Mary A. Tucker
RHT   Richard H. Tucker
SDT   Sidney D. Townley

Archival sources

AIP   American Institute of Physics, New York
Henry Crew Papers (now in the Northwestern University Archives, Evanston, Illinois)
Thomas C. Mendenhall Papers
AO   Allegheny Observatory, University of Pittsburgh (This material is now incorporated in HL)
BHL   Michigan Historical Collections, Bentley Historical Library,

University of Michigan, Ann
Arbor
William J. Hussey Family Papers

BL   Bancroft Library, University of
California, Berkeley
George Davidson Papers
Richard S. Floyd Papers
Phoebe A. Hearst Papers
Eugene W. Hilgard Papers
Armin O. Leuschner Papers

CO   Chabot Observatory Science
Center, Oakland, California

DCL   Dartmouth College Library,
Hanover, New Hampshire
Charles A. Young Papers

FHA   Ferdinand Hamburger, Jr.
Archives, Johns Hopkins
University, Baltimore,
Maryland

HCO   Records of the Harvard College
Observatory, Harvard
University Archives,
Cambridge, Massachusetts

HHL   Manuscripts Department, Henry
E. Huntington Library, San
Marino, California
George Ellery Hale Collection

HL   Archives of Industrial Society,
Hillman Library, University
of Pittsburgh
Allegheny Observatory Records

HPM   George Ellery Hale Papers,
Microfilm Edition, California
Institute of Technology,
Pasadena

LC   Manuscript Division, Library of
Congress, Washington
Simon Newcomb Papers

LO   Lick Observatory Plate Vault,
Mount Hamilton, California

MEL   Milton S. Eisenhower Library,
Johns Hopkins University,
Baltimore, Maryland
Special Collections Division

RAS   Royal Astronomical Society
Archives, Burlington House,
London

RL   Special Collections, Regenstein

Library, University of Chicago
Presidents' Papers, 1889 - 1925
William Rainey Harper Papers

SHSW State Historical Society of
Wisconsin, Madison
George H. Paul Papers
William F. Vilas Papers

SIA   Smithsonian Institution
Archives, Washington
Records of the Office of the
Secretary, 1887–1907

SLO   Mary Lea Shane Archives of Lick
Observatory, University
Library, University of
California, Santa Cruz

SUA   Stanford University Archives,
Palo Alto, California
David Starr Jordan Papers

TFP   Townley Family Papers
Sidney D. Townley Papers

UCA   University of California
Archives, Bancroft Library,
Berkeley

UW   University of Wisconsin
Archives, Memorial Library,
University of Wisconsin,
Madison
College of Letters and Science,
Department of Astronomy

VU   Special Collections, Vanderbilt
University Library, Nashville,
Tennessee
Edward Emerson Barnard
Papers

YOA   Yerkes Observatory, Archives,
Williams Bay, Wisconsin
Office of the Registrar,
Carnegie-Mellon University,
Pittsburgh
Firestone Library,
Princeton University, Princeton,
New Jersey
Henry Norris Russell Papers
Western Pennsylvania Historical
Society, Pittsburgh

Chapter 1

1 *Pittsburgh Dispatch*, Mar 11, 1898, HL.

2 T.R. Ball, *Johns Hopkins Alumni Magazine, 14*, 410, 1926.

3 JEK to J. Boughton, Feb 23, 1893, JEK, Data for Biographical Notice, 1895, HL; J. Boughton, *A Geneology of the Families of John Rockwell of Stamford, Conn., 1641, and Ralph Keeler of Hartford, Conn., 1639.* New York, William F. Jones, 1903.

4 "Two Northfield Boys," clipping from unnamed church magazine, Aug 1886, WFK journal, SLO.

5 Letter, WFK to *Florida Times-Union*, June 14, 1885, SLO.

6 *Aboard the USS Monitor: 1862. The Letters of Acting Paymaster William Frederick Keeler, U.S. Navy to His Wife Anna*, Ed. Robert W. Daly. Annapolis, U.S. Naval Institute, 1964.

7 *New Haven Register*, Aug 25, 1883, SLO.

8 *Aboard the USS Florida: 1863-65. The Letters of Acting Paymaster William Frederick Keeler, U.S. Navy to His Wife Anna*, Ed. Robert W. Daly. Annapolis, U.S. Naval Institute, 1968.

9 C.S. Hastings, *BMNAS, 5*, 231, 1903.

10 JEK, Johns Hopkins University application, Dec 20, 1877, FHA.

11 JEK to J.M. Imbrie, Feb 1, 1897, HL.

12 *Florida Dispatch*, Aug 11, 1884, SLO.

13 WFK journals, 1878-1886, SLO.

14 *Florida Times-Union*, May 21, 1885, SLO.

15 *Florida Times-Union*, June 16, 1885, SLO.

16 JEK Record Book, Sep 10, 1875 - Jan 24, 1876, (copy) SLO.

17 JEK to G.F. Davidson, Dec 11, 1889, BL.

18 C.D. Perrine, *PA, 8*, 409, 1900.

19 JEK, *PASP, 12*, 167, 1900.

20 *New Haven Journal and Courier*, Aug 23, 1877.

21 *New York Times*, Jan 3, 1904; *New York Tribune*, Jan 3, 1904.

22 *Chronicle of the Historical Society of the Tarrytowns*, No. 5, Jan 1960, p. 3; W.W. Payne, *PA 12*, 262, 1904.

23 C.M. Keeler to WWC, [~Sep 1, 1900], SLO.

24 C.H. Rockwell to D.C. Gilman, Nov 24, 1877, FHA.

25 G.J. Brush to C.S. Hastings, Dec 17, 1877, FHA.

26 C.H. Rockwell to D.C. Gilman, Dec 7, 1877, FHA.

27 A.P. Rockwell to E.C. Pickering, Oct 13, 1877; C.H. Rockwell to E.C. Pickering, Oct 20, Dec 10, 1877, HCO.

28 B.Z. Jones & L.G. Boyd, *The Harvard College Observatory–The First Four Directorships 1839-1919*. Cambridge, Harvard University Press, 1971, pp. 427-429.

29 A. Flexner, *Daniel Coit Gilman: Creator of the American Type of University*. New York, Harcourt, Brace & Co., 1946; F. Cordasco, *Daniel Coit Gilman and the Protean PhD*. Leiden, E.J. Brill, 1960; H. Hawkins, *Pioneer: A History of the Johns Hopkins University 1874-1889*. Dartmouth, Cornell University Press, 1960.

30 H. Crew to R.S. Woodbury, Dec 4, 1935, AIP.

31 F. Schlesinger, *Ap.J*, 76, 149, 1932; H.S. Uhler, *BMNAS*, 20, 273, 1939.

32 D.C. Gilman to SN, Feb 7, 1876, LC.

33 JEK passport, issued Apr 9, 1883, a little over five years later.

34 W.C. Day, letter to *Baltimore American*, Apr 18, 1895, FHA.

35 JEK, Johns Hopkins University Report of Examinations, May 1881, FHA; *La Salle County Press*, Jan 24, 1880, SLO.

36 JEK to D.C. Gilman, May 19, 1893, FHA.

37 ESH, *U.S. Naval Observatory Washington Observations*, 1876, Part 2, Appendix 3, p. 145, 1880.

38 ESH, *U.S. Naval Observatory Washington Observations*, 1878, Appendix 1, p. 33, 1882.

39 *Weekly Rocky Mountain News*, July 10, 1878; *Denver Daily Tribune*, July 30, 1878.

40 JEK, Eclipse Expedition (diary), July 19-24, 1878, SLO.

41 *Jacksonville Sun and Press*, Sep 19, 1878, SLO.

42 *Central City Daily Register Call*, July 29, 1878.

43 ESH, *MNAS*, 2, 5, 1884.

44 *La Salle County Press*, Aug 10, 1878, SLO.

45 Partial transcript of JEK courses, Johns Hopkins University.

46 F.P. Whitman to JEK, Oct 10, 1899, F.P. Whitman to WWC, Dec 18, 1900, SLO.

47 Allegheny Observatory Daily Journal, May 16, 1881, AO.

Chapter 2

1 SPL to ESH, Jan. 30, 1885, SLO.

2 C. Adler, *Bull. Phil. Soc. Washington*, 15, 1, 1906.

3 A. Starrett, *Through One Hundred and Fifty Years, The University of Pittsburgh*. Pittsburgh, University of Pittsburgh Press, 1973, p. 289 et seq.

4 *Denver Daily Tribune*, July 30, 1878.

5 JAB, *The Autobiography of a Man Who Loved the Stars*. Boston & New York, Houghton Mifflin Co., 1925.

6 W.R. Beardsley, *Samuel Pierpont Langley – His Early Academic Years at the*

Western University of Pennsylvania, PhD Thesis, University of Pittsburgh, 1978; W.R. Beardsley, *Western Pennsylvania Historical Magazine, 64*, 345, 1981.

7 SPL, *Researches on Solar Heat and Its Absorption by the Earth's Atmosphere: A Report on the Mount Whitney Expedition.* Washington, Government Printing Office, 1884.

8 Report, SPL to Observatory Committee, May 31, 1881, AO.

9 C.S. Hastings, *BMNAS, 5,* 231, 1903.

10 Letter, SPL to E.C. Pickering, Dec 1880, quoted in B.Z. Jones and L.G. Boyd, *The Harvard College Observatory. The First Four Directorships.* Cambridge, Harvard University Press, 1971, p. 427.

11 *La Salle County Press,* June 25, 1881, SLO.

12 *Baltimore Sun,* July 21, 1881, SLO.

13 Oscar Lewis, *George Davidson. Pioneer West Coast Scientist.* Berkeley, University of California Press, 1954.

14 WWC, *PASP, 26,* 28, 1914.

15 SPL to GD, July 19, 1881, BL.

16 W.C. Wyckoff, *Harper's, 67,* 81, 1883.

17 *Baltimore Sun,* undated, [~Oct 15, 1881], SLO.

18 *Mount Whitney, California* Quadrangle Map, U.S. Coast and Geodetic Survey.

19 *La Salle County Press,* Sep 24, 1881, SLO.

20 JEK to W.M.Hayes, May 21, 1895, HL.

21 SPL, Report of the Director of the Observatory, Board of Trustees Minutes, June 4, 1883, AO.

22 JEK to JAB, July 31, 1890, F.W. Very to C. Abbe, Aug 4, 1892, HL.

23 SPL, *AJS, 27,* 169, 1884.

24 SPL to ESH, Nov 14, 1881, SLO; SPL to ESH, Dec 12, 1881, UW.

25 JAB, *PA, 8,* 476, 1900.

26 JEK to D.C. Gilman, Apr 22, 1882, FHA.

27 W. Thaw to SPL, June 14, 1882, quoted in *In Memoriam, William Thaw.* Pittsburgh, J. Eichbaum and Co., 1891; see also reference 5, p. 78.

28 JEK to D.C. Gilman, June 15, 1882, FHA.

29 WFK journals, 1878-1886, SLO.

30 ESH and SWB, *AJS, 23,* 48, 1882.

31 *Pittsburgh Telegraph,* Dec 6, 1882, SLO.

32 JEK, *SM, 1,* 292, 1883; SPL, *MNRAS, 43,* 71, 1883.

33 C.S. Hastings, *SM, 1,* 273, 1883.

34 *Baltimore Sun,* May 11, 1883, SLO.

35 *Baltimore Sun,* June 30, 1883, SLO.

36 C.D. Perrine, *PA, 8,* 409, 1900.

37 C. Reid, *Hilbert.* Berlin, Springer Verlag, 1978, p. 10.

38 *Baltimore Sun,* Sep 14, 1883, SLO.

39 JEK to D.C. Gilman, Nov 4, 1883, FHA.

40 *Baltimore Sun,* Nov 19, 1883, SLO.

41 *Baltimore Sun,* Oct 29, 1883, SLO.

42 WWC, *Ap.J, 12,* 239, 1900.

43 W.F. Meggers, *BMNAS, 37,* 33, 1964.

44 H. Crew, *Ap.J, 94,* 5, 1941.

45 D.M. Livingston, *The Master of Light: A Biography of Albert A. Michelson.* Chicago, University of Chicago Press, 1967.
46 JEK to J. Scheiner, Dec 15, 1892, HL.
47 JEK, *AJS, 28,* 190, 1884.
48 SPL to W. Grunow, June 12, 1884, HL.
49 SPL to W. Thaw, June 24, 1884, HL.
50 *Baltimore Sun,* Dec 29, 1884, SLO.
51 JEK to GEH, July 8, 1895, HL.
52 SPL, *MNAS, 3,* 13, 1884.
53 Pittsburgh newspaper, [~Oct 7, 1885], SLO; *Florida Times-Union,* Oct 20, 1885, SLO.
54 JEK, *SM, 4,* 311, 1885; *5,* 222, 1886.
55 JEK to D.C. Gilman, Sep 23, 1884, FHA.
56 SPL to ESH, Nov 1, 1885, SLO.
57 JEK to ESH, Mar 12, 1886, SLO.

Chapter 3

1 Rosemary Lick, *The Generous Miser: The Story of James Lick of California.* Menlo Park, Ward Ritchie Press, 1967.
2 W.H. Wright, *PASP, 50,* 143, 1938.
3 H.E. Mathews, "Reminiscences of the James Lick Trust" (handwritten manuscript), Feb 1920, SLO.
4 G. Madeira to ESH, July 14, 1887, SLO.
5 Oscar Lewis, *George Davidson. Pioneer West Coast Scientist.* Berkeley, University of California Press, 1954.
6 GD to SN, Nov 11, 1873, LC.
7 A. Clark to J. Lick, Nov 3, 1873, SLO.
8 W.W. Payne, *SM, 6,* 250, 1887.
9 A. Clark & Sons to J. Lick, Dec 16, 1873, SLO.
10 GD, "Memorandum on J. Lick," 1892, BL; GD, *University of California Magazine, 5,* 131, 1899.
11 G.W. Hill, *BMNAS, 6,* 241, 1908.
12 SN, *The Reminiscences of an Astronomer.* Boston & New York, Houghton, Mifflin & Co., 1903.
13 D.O. Mills to ESH, Oct 19, 1874, SLO; SN to H.E. Mathews, Feb 10, 1891, LC; ESH, *SM, 7,* 49, 1888.
14 M.H. Newcomb to J. Le Conte, July 14, 1897, UCA.
15 RSF, *PLO, 1,* 1, 1887.
16 F.J. Neubauer, *PA, 58,* 318, 1950.
17 S.E. Tillman, *Ann. Report Association Graduates U.S. Military Academy, 52,* 1915.
18 WWC, *BMNAS, 8,* 347, 1919.
19 SN to D.O. Mills, Oct 8, 1874, SLO; ESH, *PLO, 1,* 4, 1887; RSF to E.B. Mastick, Sep 8, 1888, LC.
20 F.J. Neubauer, *PA, 58,* 201, 1950.
21 *San Francisco Examiner,* [~March 1891], SLO.
22 ESH, *PASP, 2,* 309, 1890.

23 SWB to SN, July 18, 1879, Nov 5, 1879, LC; SWB, *PLO, 1,* 13, 1887.
24 SPL to ESH, May 21, 1885, UW.
25 ESH to C.A. Young, Feb 4, 1881, DCL; C.H. Rockwell to ESH, Mar 2, 1881, UW.
26 Fauth & Co. to ESH, July 27, 1881, ESH to R. Friedlander & Sohn, July 13, 1882, UW, are two samples of many similar letters.
27 ESH Bill to the Lick Trustees, May 1, 1888, SLO.
28 ESH to R.S. Floyd, Oct 15, 1884, Jan 20, 1885, SLO.
29 J.S. Hager to SN, June 18, 1885, SN to J.S. Hager, June 19, 1885, LC.
30 J.S. Hager to SN, June 30, 1885, LC; SN to ESH, Sep 22, 1885, SLO.
31 ESH to RSF, July 20, Sep 17, 1885, ESH to SWB, Sep 18, 1885, SLO.
32 *The Berkleyan,* Jan 12, 1886.
33 S. Bull to ESH, Nov 28, Dec 16, 1885, S. Bull, Report, July 14, 1885, SLO.
34 S. Bull to ESH, Apr 6, 1885, SLO.
35 RSF to SN, Nov 4, 1885, LC; ESH to RSF, July 17, 1887, SN to RSF, Feb 4, 1888, SLO.
36 SPL to ESH, Nov 14, 1881, SLO.
37 SPL to ESH, Feb 5, 1885, UW.
38 Allegheny Observatory Daily Journal, Mar 10, 1885, AO.
39 JAB to ESH, Apr 22, May 19, 1885, UW.
40 JEK to ESH, Apr 16, 1885, UW; JEK to ESH, May 25, 1885, SLO.
41 SPL to ESH, Oct 2, 1885, SLO.
42 ESH to JEK, Nov 4, 1885, SLO.
43 JEK to ESH, Nov 11, 1885, SLO.
44 SPL to ESH, Feb 8, 1886, SLO.
45 JEK to ESH, Feb 19, 1886, SLO.
46 Allegheny Observatory Daily Journal, Mar 2, 1886, JEK to SPL, Mar 6, 1886, AO.
47 WFK Journal, SLO.
48 JEK to ESH, Mar 12, 1886, SLO; JEK to SPL, Mar 12, 1886, AO.
49 JEK to ESH, Mar 26, 1886, SLO.
50 SPL to ESH, Nov 1, 1885, SLO.
51 ESH to RSF, July 3, 1886, SLO.
52 JEK to ESH, Apr 3, Apr 4, 1886, ESH to JEK, Apr 3, 1886, SLO.
53 Allegheny Observatory Daily Journal, Apr 10, 11, 1886, AO.
54 SPL to ESH, Apr 4, 1886, SLO; ESH to SPL, Apr 16, 1886, AO.
55 C.R. Cross to SPL, Feb 12, 1886, C.S. Hastings to SPL, Mar 4, 1886, HL.
56 JEK to ESH, May 1, 1886, T.E. Fraser, "Log", Apr 25 - July 20, 1886, SLO; T.E. Fraser to RSF, May 2, 1886, BL.
57 JEK, *SM, 6,* 233, 1887.
58 RSF to G.C. Comstock, Oct 11, 1887, UW.
59 JEK to ESH, May 21, May 26, June 15, 1886, SLO.
60 JEK to T.R. Ball, June 21, 1886, FHA; RSF to H.E. Mathews, Mar 21, 1888, SLO.
61 JEK to ESH, June 25, 1886, SLO.
62 SDT, *PASP, 46,* 171, 1934; J. Stebbins, *BMNAS, 20,* 161, 1939.
63 ESH to G.C. Comstock, June 17, 1886, SLO.
64 G.C. Comstock to RSF, Aug 14, Sep 19, 1887, SLO.

65 A. Clark to RSF, Sep 16, 1886, A. Clark to James Lick Trust, Oct 18, 1886 (telegram), SLO.

66 SN to RSF, Oct 29, Nov 20, 1886, A. Hall to SN, Nov 19, 1886, SLO.

67 *SM, 6,* 87, 1887.

68 T.E. Fraser, "Log", Dec 27, 1886, SLO.

69 JEK to ESH, Jan 12, Jan 23, 1887, SLO.

70 T.E. Fraser, "Log", Jan 7-10, 1887, SLO; *San Francisco Daily Examiner,* Jan 10, 1887.

71 JEK to W. Thaw, Feb 1, 1887, Western Pennsylvania Historical Society.

72 JEK to JAB, July 26, Aug 24, Oct 23, 1887, etc., SLO.

73 G.C. Comstock to RSF, Jan 6, 1888, SLO.

74 JEK, *SM, 10,* 433, 1891; JEK, *AA, 11,* 140, 1892.

75 JEK to ESH, Jan 28, 1887, SLO.

76 JEK to ESH, Oct 2, 1887, SLO.

77 JEK to ESH, Dec 4, 1887, SLO.

78 RSF to H.E. Mathews, Dec 24, 1887, SLO.

79 RSF to G.C. Comstock, Dec 24, 1887, UW.

80 RSF to H.E. Mathews, Dec 29, 1887, SLO.

81 JEK to ESH, Jan 3, 1888, SLO.

82 RSF, "Notes L.O." (a memorandum), Dec 31, 1887 - Jan 2, 1888, SLO.

83 JEK to ESH, Jan 6, 1888, SLO.

84 RSF, undated memorandum, SLO.

85 [JEK], *San Francisco Examiner,* Jan 10, 1888.

86 JEK, *SM, 7,* 79, 1888.

87 JEK, *AJ, 8,* 175, 1889.

88 D.E. Osterbrock, *Science, 209,* 444, 1980; D.E. Osterbrock and D.P. Cruikshank, *Icarus, 53,* 165, 1983.

89 JEK to ESH, Jan 14, 1888, SLO.

90 JEK to ESH, Feb 2, 1888, SLO.

91 R.G. Aitken, *PASP, 35,* 252, 1923.

92 *San Francisco Chronicle,* Feb 18, 1888.

93 *San Francisco Chronicle,* Feb 19, 20, 1888.

94 RSF to E.B. Mastick, Feb 21, 1888, SLO.

95 J.R. Daugherty to RSF, Feb 28, 1888, BL.

96 *St. Louis Globe-Democrat,* Feb 19, 1888.

97 V. Gadesden to RSF, [Feb 18], [~Feb 24, 1888], ESH to RSF, Feb 22, 1888, J.T. Boyd to RSF, Feb 28, 1888, BL; RSF to ESH, Feb 24, 1888, SLO.

98 E.B. Mastick to RSF, Feb 20, 1888, SLO.

99 *San Jose Daily Times,* Feb 22, 1888; *Daily Alta California,* Feb 25, 1888.

100 E.B. Mastick to RSF, Mar 8, 1888, SLO.

101 JEK to ESH, Mar 1, Mar 18, 1888, SLO.

102 A. Carnegie to L. S[t]anford, Jan 9, 1888, SLO.

103 RSF to H.E. Mathews, May 3, 1888, SLO; P. De Vecchi to RSF, May 23, 1888, BL.

104 C.S. Floyd to H.E. Mathews, May 18, 1888, SLO; JEK to D.C. Gilman, May 29, 1888, MEL.

105 E.B. Mastick to RSF, Mar 13, 1888, SLO.

106 JEK, *PASP, 11,* 164, 1899; *San Francisco Examiner,* June 12, 1899.

107 JEK to RSF, Apr 22, 1888, BL.
108 ESH to RSF, Apr 22, 1888, BL.
109 JEK to ESH, Apr 15, Apr 22, 1888, SLO.
110 JEK, *AJ, 8,* 73, 1888.
111 ESH to H.E. Mathews, May 2, 1888, etc., SLO.
112 ESH to A.S. Hallidie, May 11, 1888, UCA.
113 "Inventory of Property at Mount Hamilton . . .", April 1888, SLO.

Chapter 4

1 WWC, *BMNAS, 8,* 347, 1919
2 S.E. Tillman, *Annual Report of Association of Graduates of the U.S. Military Academy,* 52, 1915.
3 *The Berkeleyan, 20,* 78, 87, 1886.
4 *San Francisco Evening Post,* Sep 25, 1888; *San Francisco Chronicle,* [~Mar 18, 1895], SLO.
5 EEB, *PA, 29,* 309, 1921; E.B. Frost, *Ap.J, 54,* 1, 1921.
6 SWB, *PLO, 1,* 13, 1887.
7 ESH & SWB, *AJS, 23,* 48, 1882.
8 SWB to RSF, Feb 23, 1882, BL.
9 SWB to RSF, May 31, 1888, BL.
10 ESH to C.A. Young, Jan 28, 1882, DCL.
11 ESH to RSF, Nov 24, 1886, SLO.
12 ESH to SWB, Apr 25, 1888 (telegram), SLO.
13 ESH to G.C. Comstock, June 17, 1886, SLO.
14 G.C. Comstock to ESH, Sep 12, 1886, Mar 23, 1887, SLO.
15 SPL to ESH, Oct 2, 1885, SLO; ESH to B.A. Gould, Oct 5, 1885, UW; ESH to G.C. Comstock, Dec 16, 1886, Mar 15, 1887, SLO.
16 G.C. Comstock to ESH, June 1, 1887, SLO.
17 ESH to RSF, June 17, 1887, SLO.
18 ESH to RSF, Oct 15, 1884, SLO.
19 G.C. Comstock to ESH, Aug 20, 1887, SLO.
20 WJH, *PASP, 36,* 309, 1924; RHT, *PASP, 36,* 312, 1924.
21 ESH to JMS, Mar 19, 1887, Jan 23, 1888, SLO.
22 R.H. Hardie, *Leaflets ASP, 9,* Nos. 415, 416, 1964.
23 H.N. McTyeire to EEB, Mar 5, 1883, EEB, undated autobiographical fragments, VU; SWB, *PA, 1,* 193, 341, 441, 1894; EEB to SN, Apr 27, 1891, Oct 12, 1892, LC.
24 E.B. Frost, *BMNAS, 21,* 14th mem., 1, 1924.
25 EEB to ESH, Oct 26, 1885, SLO.
26 EEB to ESH, Oct 31, Nov 12, 1885, Apr 17, 1886, SLO.
27 ESH to EEB, Apr 26, 1886, VU.
28 EEB to ESH, May 7, 1886, SLO.
29 ESH to EEB, Mar 19, July 14, 1887, VU.
30 EEB to ESH, July 26, 1887 (telegram), SLO.
31 EEB to ESH, July 28, 1887, SLO.
32 EEB to ESH, Aug 15, Aug 20, 1887 (1st letter), SLO.
33 EEB to ESH, Aug 20, 1887 (2nd letter), SLO.

34 EEB to ESH, Oct 26, 1887, SLO.

35 JEK to EEB, Nov 27, Dec 24, 1887, VU.

36 RSF to EEB, Mar 19, 1888, EEB to RSF, Mar 24, 1888, SLO.

37 EEB to ESH, Mar 28, 1888, EEB to RSF, Apr 18, 1888, JEK to ESH, Apr 9, 1888, SLO.

38 EEB to H.E. Mathews, May 20, 1888, SLO.

39 F.J. Neubauer, *PA, 58*, 318, 1950.

40 ESH to C.B. Hill, Jan 10, 1888, SLO.

41 *Annual Report of Sec. Board Regents*, 1888, p. 26.

42 ESH to JEK, July 15, 1887, JEK to RSF, Oct 29, 1887, SLO.

43 *Annual Report of Sec. Board Regents*, 1889, p. 32.

44 ESH to H. Davis, Sep 1, 1888, UCA.

45 ESH to RSF, Oct 25, 1887, SLO.

46 RSF to SN, Feb 13, 1885, LC.

47 ESH to SWB, July 27, 1887, SLO; SWB to J.H.C. Bonté, Sep 12, 1889, UCA.

48 ESH to RSF, Nov 28, 1887, JEK to ESH, Apr 1, 1888, SLO; ESH to SWB, JMS, JEK, EEB & C.B. Hill, June 1, 1888, VU.

49 JEK to JMS, Jan 21, 1893. HL.

50 SDT Diary, 1892, TFP; J. Stebbins to H.[ome] F.[olks], Oct 6, 1901, SLO.

51 ESH to "My Dear Sir," Mar 15, 1891, UCA.

52 JEK to H.E. Mathews, July 11, July 25, 1888, SLO.

53 ESH to EEB, June 10, 1890, June 6, 1891, SLO.

54 ESH to JEK, Jan 7, 1889, SLO.

55 JEK, *PASP, 2*, 3, 1890.

56 JEK, *PASP, 1*, 65, 1889; JEK, Repsold Altazimuth Observing Book III, LO.

57 *San Jose Mercury*, [~Dec 1889], SLO.

58 JEK, *The Lick Observatory*, handwritten manuscript, [1888], SLO.

59 JEK, *PASP, 2*, 25, 1890.

60 ESH, *Suggestions for Observing the Total Eclipse of the Sun on January 1, 1889*, Sacramento, State Office, 1888, BL.

61 JEK to ESH, Oct 28, 1888, SLO.

62 D. Alter, *PASP, 65*, 269, 1953; P. Herget, *BMNAS, 49*, 129, 1978; ESH to H. Davis, Oct 1, 1888, UCA.

63 ESH to H. Davis, Sep 18, 1888, UCA.

64 ESH to H. Davis, Dec 10, 1888, UCA.

65 ESH to C.W. Slack, Mar 23, 1895, SLO.

66 AOL to R.G. Aitken, Mar 11, 1931, SLO.

67 EEB to ESH, Dec 23, 1888, SLO.

68 JEK, *LO Cont., 1*, 31, 1889.

69 JEK to ESH, Dec 26, 1888, SLO.

70 EEB to ESH, Dec 31, 1888, SLO.

71 C.S. Hastings, *MNAS, 2*, 102, 1883.

72 ESH, *PASP, 6*, 245, 1894.

73 RSF to SN, Jan 28, 1889, LC.

74 EEB, *LO Cont., 1*, 56, 1889.

75 JEK to RSF, Jan 13, Jan 17, 1889, BL.

76 ESH to JEK, Feb 9, 1889, SLO.

77 JEK, *LO Cont., 1*, 104, 1889; JEK, Repsold Altazimuth Observing Books I, II, LO.
78 JEK, 36-inch Equatorial Observing Books 1a - 9, LO.
79 JEK, Drawings of Jupiter and Mars, SLO.
80 EEB, *MNRAS, 51*, 543, 1891.
81 JEK, *PASP, 2*, 15, 286, 1890; JEK, *MNRAS, 51*, 31, 1890; JEK, *J. Brit. Astron. Assoc., 1*, 435, 1891; JEK to A.S. Williams, Sep 13, 1890, SLO; JEK to A.S. Williams, Dec 16, 1889, Dec 15, 1891, HL.
82 EEB, *MNRAS, 52*, 7, 1891; G.W. Hough, *AA, 13*, 89, 1894.
83 EEB to SN, Sep 11, 1894, Mar 11, 1896, LC.
84 ESH, JMS, JEK, *PASP, 2*, 299, 1890.
85 C.S. Hastings to JEK, Mar 1, Mar 23, 1893, HL.
86 JEK, *AJ, 10*, 89, 1890.
87 JEK to A. Marth, July 3, Aug 2, Sep 26, 1890, JEK to B.A. Gould, Sep 26 , 1890, SLO.
88 JEK, *AN, 122*, 401, 1889.
89 JEK, *PASP, 1*, 36, 1889.
90 JEK, *Scientific American*, Nov 10, 1888.
91 JEK, *PASP, 1*, 81, 1889.
92 ESH, *PASP, 1*, 9, 1889.
93 *PASP, 2*, 37, 1890.
94 ESH to H. Davis, Jan 14, 1890, ESH to T.G. Phelps, Jan 24, 1890, UCA.
95 *San Francisco Examiner*, [~Feb 19, 1890], SLO.
96 ESH to H. Davis, Feb 20, 1890, UCA.
97 EEB to GD, Mar 3, 1890, BL.
98 ESH to T.G. Phelps, Feb 20, Feb 25, 1890, SLO.
99 ESH, *PASP, 2*, 50, 1890.
100 ESH to J.H.C. Bonté, Oct 1, Nov 9, Dec 14, 1889, UCA.
101 JEK, *AA, 12*, 350, 1893.
102 JEK, *AA, 12*, 361, 1893.
103 JEK, *PASP, 1*, 80, 1889.
104 JEK, *SM, 10*, 433, 1891.
105 JEK, *PASP, 2*, 129, 1890.
106 JEK to D.C. Gilman, Nov 4, 1883, FHA; JEK to F.E. Ross, Apr 3, 1899, SLO.
107 W.H. Wright, *BMNAS, 25*, 35, 1949.
108 M.E. Cooley to SPL, May 5, 1886, AO.
109 ESH to WWC, Oct 26, 1889, SLO; WWC to ESH, Nov 4, 1889, ESH to H. Davis, Nov 9, 1889, UCA.
110 *Annual Registers of the University of California* 1890-1 through 1897-8; ESH to S.J. Cunningham, Feb 3, 1893, SLO.
111 JEK to WWC, Sep 30, 1890, SLO.
112 J.N. Lockyer, *Proc. Roy. Soc. Lon., 48*, 167, 1891.
113 W. Huggins & M. Huggins, *Proc. Roy. Soc. Lon., 46*, 40, 1889.
114 JEK to H.A. Rowland, Feb 12, 1890, SLO.
115 JEK to H.A. Rowland, Mar 28, 1890, SLO.
116 W. Huggins to ESH, Mar 15, 1890, SLO.
117 JEK to W. Huggins, Apr 3, 1890, SLO.
118 JEK to W. Huggins, May 12, 1890, SLO.

119 JEK to W. Huggins, June 14, 1890, W. Huggins to ESH, June 30, 1890, SLO.
120 JEK to H. C. Vogel, Aug 18, Sep 29, 1890, SLO.
121 JEK to AOL, Sep 27, 1890, BL.
122 JEK, *PASP, 2,* 265, 1890; JEK to H.A. Rowland, Sep [~8], 1890, HL.
123 JEK to W. Huggins, Aug 31, 1890, HL; JEK to W. Huggins, Oct 7, 1890, SLO.
124 JEK to F. Soulé, July 21, 1890, SLO.
125 JEK to JAB, Sep 17, 1890, SLO.
126 ESH to JEK, Sep 15, 1890, SLO.
127 JEK to ESH, Sep 16, 1890, SLO.
128 JEK to W. Huggins, Oct 20, 1890, SLO.
129 JEK to W.H. Pickering, Oct 20, 1890, SLO; JEK to A. Clerke, Nov 1, 1890, JEK
     to W.W. Payne, Nov 1, 1890, HL.
130 JEK, *Observatory, 14,* 52, 1891.
131 E.W. Maunder, *Observatory, 13,* 253, 1890.
132 W. Huggins & M. Huggins, *Proc. Roy. Soc. Lon. A, 48,* 202, 1891.
133 JEK to W. Huggins, Feb 21, Feb 23, 1891, SLO.
134 JEK to W. Huggins, Mar 31, 1891, SLO.
135 JEK, *Proc. Roy. Soc. Lon. A, 49,* 399, 1891; JEK to W. Huggins, Apr 13, 1891,
     SLO.
136 *Observatory, 14,* 210, 1891.
137 JEK to Editor of the *English Mechanic,* [~Feb 22, 1892], HL.
138 JEK to J.N. Lockyer, Jan 8, 1891, JEK to W. Huggins, Jan 13, 1891, SLO.
139 JEK to C. Braun, Jan 14, 1891, JEK to N.C. Duner, Jan 27, 1891, HL.
140 C.A. Young to ESH, Oct 9, 1890, SLO.
141 ESH to T.G. Phelps, Aug 26, 1890, UCA.
142 ESH to C.A. Young, Oct 1, 1890, DCL; ESH to W. Huggins, Feb 8, Apr 4, 1890,
     SLO.
143 ESH to D.C. Gilman, Oct 16, 1889, FHA.
144 I. Remsen to ESH, Oct 23, 1889, SLO.
145 ESH to I. Remsen, May 16, 1890, ESH to D.C. Gilman, Jan 19, 1891, SLO.
146 D.C. Gilman to ESH, Feb 14, 1891, SLO.

Chapter 5

1 JEK to ESH, Apr 1, 1888, SLO.
2 JEK to WWC, Sep 30, 1890, SLO.
3 SDT to F. Wright, Aug 14, Sep 25, 1892, TFP.
4 *Annual Report(s) of the Secretary to the Board of Regents of the University of
   California for Year(s) Ending June 30, 1888, 1889, 1890.*
5 H. Hawkins, *Pioneer: A History of the Johns Hopkins University 1874-1889.*
   Ithaca, Cornell University Press, 1960, p. 129.
6 JEK to Mr. Sinclair, May 15, 1889, HL.
7 JEK to H. Crew, July 13, 1891, HL.
8 JEK to E.M. Preston, Mar 1, 1889, JEK to Mr. Pratt, Mar 1, 1889, SLO.
9 JEK to RSF, Jan 17, 1889, Feb 2, 1889, BL.
10 RSF to SN, Jan 28, 1889, LC.
11 JEK to SPL, June 11, 1889, HL.

12 JEK to T.G. Phelps, Jan 8, 1890, HL; T.G. Phelps to C. Bartlett, Jan 10, 1890, UCA.

13 JEK, *AA, 11,* 840, 1892.

14 ESH to W. Huggins, Nov 12, 1888, ESH to E.B. Knobel, Apr 10, Sep 23, 1890, RAS; ESH to JEK, Sep 15, 1890, ESH to W. Huggins, Feb 8, 1890, SLO.

15 JEK to G.C. Comstock, Nov 2, 1887, UW.

16 H.H. Bancroft, *Chronicles of the Builders of the Commonwealth, Volume I.* San Francisco, The History Co., 1891.

17 JEK to I.M. Scott, Mar 6, 1890, HL.

18 JEK to D.C. Gilman, Oct 4, Nov 2, 1889, FHA.

19 W. Thaw to SPL, Nov 29, 1887, AO.

20 C. Adler, *Ann. Report Smithsonian Institution,* p. 515, 1906; JAB, *The Autobiography of a Man Who Loved the Stars* Boston, Houghton Mifflin Co., 1925.

21 JAB to ESH, Apr 5, 1889, July 11, 1889, SLO.

22 JEK to JAB, July 31, 1890, HL.

23 JEK to W. Thaw, Jr., July 31, 1890, HL.

24 JEK to C.A. Young, Oct 9, 1890, DCL.

25 *San Francisco Examiner,* [~Nov 4, 1890], SLO.

26 *San Francisco Evening Bulletin,* Jan 3, 1891; *San Jose Record,* Jan 5, 1891.

27 *SM, 6,* 295, 1887; JEK, *SM, 7,* 9, 1888; JEK to W.W. Payne, June 23, 1890, HL.

28 E.A. Fath, *PA 36,* 267, 1928; D.L. Leonard, *The History of Carleton College.* Chicago, Fleming H. Revell Co., 1904; L.A. Headley & M.E. Jarchow, *Carleton. The First Century.* Northfield, Carleton College, 1966.

29 E.B. Frost, *BMNAS, 7,* 91, 1910.

30 C.A. Young to ESH, Mar 4, 1891, SLO.

31 JEK to W. Thaw, Jr., Feb 5, 1891, AO.

32 *San Francisco Examiner,* Feb 1, 1891.

33 A.L. Starrett, *Through One Hundred and Fifty Years: The University of Pittsburgh.* Pittsburgh, University of Pittsburgh Press, 1932.

34 San Francisco newspapers, Dec 30, 1890, SLO.

35 JEK, 36-inch Equatorial Observing Books, 8-9, LO.

36 JEK to A. Clerke, Jan 29, 1891, HL.

37 JEK to W.W. Matthews, Jan 27, 1891, SLO.

38 *Pittsburgh Dispatch,* Mar 11, 1898.

39 JEK to W. Thaw, Jr., Feb 25, 1891, AO.

40 JEK to SPL, Feb 24, 1891, HL.

41 JEK to D.C. Gilman, Feb 21, 1891, FHA.

42 San Francisco newspaper, Feb 27, 1891, SLO.

43 *San Francisco Examiner,* [~Mar 10, 1891], SLO.

44 JEK to W. Thaw, Jr., Mar 12, 1891, AO.

45 JEK to SPL, Mar 12, 1891, HL.

46 JAB to W. Thaw, Jr., Apr 7, 1891, AO.

47 JAB to W. Thaw, Jr., Apr 15, 1891, AO.

48 W.J. Holland to W. Thaw, Jr., Apr 14, 1891, AO.

49 SPL to W. Thaw, Jr., Apr 18, 1891, AO.

50 JAB to W. Thaw, Jr., Apr 20, 1891, AO.

51 W.J. Holland to W. Thaw, Jr., Apr 20, 1891, AO.

52 JEK to D.C. Gilman, Apr 21, 1891, MEL.

53 JEK to SPL, Apr 21, 1891, HL.

54 JEK to W. Thaw, Jr. Apr 22, 1891, AO.

55 JEK to W.J. Holland, Apr 28, 1891, HL.

56 JEK to W. Huggins, Apr 21, 1891, JEK to M.W. Harrington, May 1, 1891, JEK
    to A. Clerke, May 18, 1891, HL.

57 ESH to WWC, Dec 30, 1890, SLO.

58 WWC to ESH, Jan 24, 1891, SLO.

59 WWC to ESH, Feb 16, 1891 (telegram), ESH to WWC, Feb 21, 1891, SLO.

60 ESH to WWC, Mar 26, 1891, SLO.

61 WWC to ESH, Apr 28, 1891, SLO.

62 JEK to H.C.G. Brandt, Aug 1, 1890, HL.

63 JEK to M. Kellogg, Apr 21, 1891, SLO.

64 ESH to M. Kellogg, Apr 21, 1891, ESH to T.G. Phelps, Apr 21, 1891, SLO.

65 ESH to WWC, Apr 22, 1891, SLO.

66 JEK to WWC, Apr 22, 1891, SLO.

67 JEK to WWC, May 7, 1891, SLO.

68 ESH to WWC, May 14, May 15, 1891, SLO.

69 *San Francisco Chronicle,* May 12, 1891; *San Jose Mercury,* May 13, 1891; *San
    Jose Record,* May 13, 1891.

70 ESH to E.W. Maunder, Apr 19, 1891, SLO.

71 ESH to T.G. Phelps, Apr 13, 1892, SLO.

72 W.W. Matthews, Wedding Announcement, June 16, 1891, SLO.

73 C.K. Moore to H. Crawford, Feb 28, 1961, SLO; C.K. Moore to author, June 9,
    1977.

74 JEK to ESH, July 9, 1891, SLO.

Chapter 6

1 S.S. Lorant, *Pittsburgh – The Story of an American City.* Garden City, Doubleday & Co., 1964. R. Lubove, *Pittsburgh.* New York, New Viewpoints, 1976.

2 W.R. Beardsley, *Western Pennsylvania Historical Magazine, 64,* 213, 1981.

3 *Catalogue of the Western University of Pennsylvania for the Year Ending June 30,
   1891.*

4 *Annual Register of the Western University of Pennsylvania for the Year Ending
   June 1890.*

5 A.L. Starrett, *Through One Hundred and Fifty Years, The University of Pittsburgh.* Pittsburgh, University of Pittsburgh Press, 1932.

6 JEK to D.C. Gilman, July 9, 1891, MEL; JEK to EEB, Aug 19, 1891, VU.

7 JEK to W. Thaw, Jr., Apr 22, 1891, JEK to JAB, May 12, 1891, HL.

8 JAB to W. Thaw, Jr., Apr 20, 1891, F.L.O. Wadsworth, Report to Board of
   Trustees June 30, 1901, AO.

9 JEK to ESH, July 9, 1891, SLO; JEK to W. Thaw, Jr., July 17, 1891, HL;
   Pittsburgh and Allegheny Blue Book, 1892.

10 SPL to W. Thaw, Nov 28, 1887, AO.

11 L.O. Howard to F.W. Very, June 29, July 10, 1889, HL.

12 G.W. Ritchey to F.W. Very, Dec 15, 1887, G.W. Ritchey to SPL, Jan 4, 1888,
    AO.

13 F.W. Very to W. Thaw, Jr., May 1, May 2, 1891, AO.

14 JEK to F.W. Very, May 8, 1891, May 26, 1891, HL.

15 JEK, printed circular, July 15, 1891, AO.

16 JEK to M.W. Harrington, Dec 12, 1891, June 8, 1892, JEK to F.W. Very, Jan 29, 1892, HL.

17 JEK to H.A. Rowland, Sep 8, 1892, JEK to G.F. Barker, Sep 8, Oct 2, 1892, HL.

18 F.W. Very to C. Abbe, Aug 4, 1892, JEK to F.W. Very, Aug 10, 1896, F.W. Very to W.L. Moore, Aug 31, 1897, HL; JAB to JEK, Mar 6, 1899, SLO.

19 JEK to W.R. Waugh, Aug 12, 1891, HL.

20 JEK to G.B. Goode, July 22, 1891, W.C. Winlock to G.B. Goode, Aug 10, 1891, SIA.

21 JEK to SPL, Sep 21, 1891, HL.

22 JEK to W. Thaw, Jr., Apr 26, 1892, HL; W. Thaw, Jr., to SPL, Apr 27, 1892, SPL to W. Thaw Jr., May 4, 1892, SIA.

23 JEK to SPL, May 9, Aug 5, 1892, W.C. Winlock to JEK, Aug 9, Aug 15, 1892, SIA.

24 JEK to W. Huggins, July 15, 1891, HL.

25 JEK to L. Boss, Oct 19, 1891, HL.

26 JEK to ESH, Sep 7, 1891, HL.

27 JEK to C. Piazzi Smyth, June 27, 1892, HL; JEK, *AA, 12,* 40, 1893; JEK to GEH, June 4, 1900, HPM.

28 JEK to V. Schumann, Dec 9, 1891, HL.

29 JEK to G.A. Douglass, Nov 23, Nov 30, 1891, HL.

30 JEK to EEB, May 5, June 28, 1892, VU.

31 JEK to L.E. Jewell, Mar 9, Mar 30, 1892, L.E. Jewell to JEK, Mar 8, Mar 23, 1892, HL.

32 JEK to ESH, Jan 29, 1892, SLO.

33 JEK to ESH, Sep 24, 1891, SLO.

34 W.E. Lincoln to JEK, Feb 12, 1892, JEK to W.S. Pier, Apr 23, 1892, W.S. Pier to JEK, Apr 27, Aug 25, 1892, E.M. O'Neill to W.S. Pier, July 7, 1892, JEK to Warner & Swasey, July 24, Oct 24, 1893, HL.

35 JEK to ESH, June 16, 1892, SLO.

36 JEK to EEB, Jan 18, 1892, VU; JEK to GEH, Dec 21, 1891, Jan 10, 1892, JEK to J.M. Rees, Jan 7, 1892, HL.

37 ESH to JEK, Aug 26, 1891, GEH to M.W. Harrington, Dec 31, 1891, M.W. Harrington to JEK, Jan 2, 1892, JEK to GEH, Jan 7, 1892, GEH to JEK, Jan 12, 189[2], HL.

38 JEK to M.W. Harrington, Feb 18, 1892, JEK to GEH, Feb 18, 1892, HL.

39 ESH to JEK, Apr 11, May 2, 1892, HL.

40 C.A. Young to JEK, May 23, 1892, HL.

41 JEK to J. Scheiner, May 24, 1892, JEK to C.A. Young, May 25, 1892, HL.

42 E.B. Frost to C.A. Young, June 25, 1892, DCL; E.B. Frost to JEK, July 18, 1892, HL.

43 JEK to E.B. Frost, Aug 2, Dec 14, 1892, E.B. Frost to JEK, Dec 6, Dec 23, 1892, HL.

44 E.B. Frost to JEK, Dec 16, 1893, Jan 21, 1894, HL.

45 WWC to JEK, Apr 19, 1892, HL; WWC to C.A. Young, Apr 19, 1892, DCL.

46 WWC to JEK, Sep 14, 1892, HL.

47 JEK to C.C. Buel, Sep 28, Oct 8, 1892, HL; JEK, *Century, 50,* 455, 1895.

48 JEK, *MRAS, 51,* 45, 1893.

49 W.G. Hoyt, *Lowell and Mars.* Tucson, University of Arizona Press, 1976.

50 JEK to WWC, Apr 5, 1892, SLO; JEK to J.B. Doyle, Aug 1, 1892, JEK to W.S. Pier, Aug 26, 1892, HL; S.D. Townley to F. Wright, Oct 9, 1892, TFP.

51 ESH, *AA, 11,* 663, 1892; EEB, *AA, 11,* 680, 1892.

52 EEB to JEK, Sep 21, 1892, Dec 29, 1893, HL.

53 N.E. Green to JEK, June 18, 1893, W.H. Pickering to JEK, Dec 28, 1893, B.A. Gould to JEK, Dec 26, 1893, E. Knobel to JEK, Jan 27, 1894, P. Lowell to JEK, May 12, 1894, HL; SPL to JEK, Jan 10, 1894, SIA.

54 JEK to C.A. Young, Oct 31, Nov 7, 1892, DCL; JEK to ESH, Aug 4, 1894, SLO.

55 JEK to G.H. Clapp, Nov 4, 1891, Feb 25, 1892, HL.

56 JEK, *AA, 11,* 567, 768, 1892.

57 JEK to E.B. Knobel, May 1, 1891, RAS; JEK to ESH, July 20, Aug 6, 1891, SLO.

58 JEK to ESH, Nov 23, Dec 3, 1891, SLO.

59 JEK to ESH, Mar 10, 1892, SLO.

60 ESH to JEK, Jan 26, 1893, SLO; JEK to ESH, Aug 17, 1893, JEK to GEH, Aug 17, 1893, HL; JEK, *AA, 12,* 733, 1893.

61 JEK to ESH, Feb 6, 1893, HL.

62 ESH to JEK, Feb 15, May 10, Aug 10, 1894, JEK to ESH, May 18, 1894, SLO.

63 JEK, *PLO, 3,* 161, 1894.

64 ESH to B. Hasselberg, Aug 14, 1894, SLO; B. Hasselberg, *PASP, 7,* 17, 1895.

65 A.M. Clerke to JEK, Oct 26, 1894, H. Crew to JEK, Oct 6, 1894, D. Gill to JEK, Nov 14, 1894, W. Huggins to JEK, Oct 21, 1894, E.B. Frost to JEK, Oct 10, 1894, HL.

66 JEK, *PASP, 2,* 160, 1890.

67 JEK, *AA, 11,* 824, 1892.

68 JEK to ESH, Jan 17, 1893, JEK to GEH, Feb 16, 1893, HL; JEK, *AA, 12,* 361, 1893.

69 JEK to WWC, Oct 24, 1892, SLO.

70 JEK to L. Becker, Mar 6, 1893, HL.

71 JEK to J. Scheiner, Jan 23, 1894, JEK to S.J. Brown, June 30, 1894, HL.

72 JEK to G. Cramer, Dec 21, 1892, JEK to R.F. Newall, Apr 17, 1893, JEK to A. Seed, Sep 18, 1893, HL.

73 JEK to GEH, May 8, 1893, JEK to WWC, Oct 20, 1893, HL.

74 JEK to J.S. Ames, Oct 20, 1891, May 19, 1894, JEK to H. Crew, Mar 6, 1894, JEK to H. Kayser, Mar 14, 1894, HL; JEK to H. Crew, May 22, 1895, Mar 25, 1896, AIP.

75 JEK to E.B. Frost, Jan 25, 1894, E.B. Frost to JEK, Feb 9, 1894, F. Ellerman to JEK, Jan 26, 1894, HL.

76 H. Crew to JEK, Mar 26, Apr 10, 1894, E.C. Pickering to JEK, Nov 24, 1894, HL.

77 JEK to G.A. Douglass, Jan 26, 1895, JEK to H.C. Vogel, Apr 13, 1895, HL.

78 JEK to R. Bache, Aug 24, 1896, HL.

79 JEK, *AA, 13,* 59, 1894.

80 JEK, *AA, 13,* 688, 1894.

81 JEK, *AA, 13,* 660, 1894.

82 JEK to G.F. Barker, Mar 5, 1898, HL.

83 JEK to GEH, Mar 4, 1896, YOA; JEK to W. Huggins, Mar 2, 1897, HL; B.L. Welther, *Bull. American Astronomical Society*, *13*, 816, 1981.

84 JEK, *Ap.J*, *6*, 423, 1897.

85 JEK to GEH, Feb 17, 1898, YOA; JEK to GEH, June 22, 1898, SLO.

86 JEK, *PA*, *1*, 9, 1893; *1*, 102, 1893; *1*, 169, 200, 1894; *2*, 20, 1894.

87 JEK to JAB, Jan 15, 1894, JEK to WWC, Jan 20, 1894, HL.

88 JEK to D. Gill, Dec 13, 1894, Nov 13, 1895, HL.

89 W.H. Pickering to JEK, Nov 21, 1893, JEK to W.H. Pickering, Nov 24, 1893, JEK to GD, May 4, 1894, HL.

90 JEK to W.J. Holland, July 25, 1891, HL.

91 JEK to EEB, July 26, 1891, VU.

92 *Catalogues of the Western University of Pennsylvania*, 1892-1899.

93 J.N. Hart to JEK, May 12, May 28, 1892, JEK to J.N. Hart, May 18, June 6, 1892, AOL to JEK, June 3, 1893, JEK to AOL, June 5, 1893, HL.

94 JEK to W.M. Burge, Nov 9, 1893, JEK to E. Stratton, Mar 27, 1894, JEK to H.C. Emmert, May 29, 1896, HL.

95 JEK to V.L. Grassle, Feb 21, 1898, HL.

96 JEK to WWC, May 16, 1892, HL.

97 JEK to GEH, Oct 28, 1895, Apr 7, 1896, YOA; JEK to GEH, Nov 7, Dec 6, 1895, Jan 7, 1896, HL.

98 JEK to W.H. Dodds, Nov 20, 1892, HL.

99 A.M. Deens to JEK, Apr 29, 1893, JEK to W.H. Dodds, May 5, 1893, HL.

100 EEB to W.W. Payne, Aug 23, 1892, HL.

101 JEK to P.S. van Mierop, Jan 27, 1892, P.S. van Mierop to JEK, Jan 31, 1892, HL.

102 JEK to GEH, July 11, 1896, YOA.

103 JEK to P.S. van Mierop, Feb 1, 1892, HL.

104 JEK to I.M. Heysinger, Jan 17, 1895, HL; R. d'E. Atkinson & F.G. Houtermans, *Zs. f. Physik*, *54*, 656, 1929.

105 JEK to I. Heising, Mar 21, 1895, HL.

106 E.W. Keeler to JEK, Apr 15, 1895, JEK to E.W. Keeler, Apr 22, 1895, HL.

107 JEK to B. Thaw, May 31, 1893, HL.

108 H. Plotkin, *Proc. Am. Phil. Soc.*, *122*, 385, 1978.

109 JEK to B. Thaw, Jan 3, 1894, B. Thaw to JEK, Jan 5, Jan 19, 1894, F.L. Sheppard to JEK, Jan 11, 1894, HL; W.J. Holland to B. Thaw, Jan 4, 1894, AO.

110 JEK to B. Thaw, July 1, 1894, Observatory Cash Balances [~Sep 1, 1895], [~July 7, 1896], [~July 31, 1897], JEK to H. Harrer, Oct 28, 1895, HL.

111 JEK to W.M. Stevenson, Dec 8, 1894, JEK to J. Barker & Co., Jan 20, 1895, HL.

112 JEK to WWC, Aug 17, 1893, HL.

113 JEK to W. Huggins, June 12, 1894, W. Huggins to JEK, July 29, 1894, HL.

114 JEK, *AA*, *13*, 476, 1894.

115 WWC to JEK, Mar 18, 1892, HL.

116 R.G. Aitken, *PASP*, *50*, 204, 1938; W.H. Wright, *BMNAS*, *25*, 35, 1949.

117 WWC to JEK, Sep 29, 1892, HL.

118 L. Boss to JEK, June 3, Dec 8, 1891, HL.

119 WWC to C.A. Young, Sep 19, 1892, DCL.

120 WWC to JEK, Mar 14, 1893, HL.

121 WWC to JEK, Sep 7, 1893, HL.

122 WWC to JEK, Feb 15, 1894, HL.

123 JEK to WWC, Mar 5, 1894, HL.

124 WWC to JEK, Mar 17, 1894, HL.

125 JEK to WWC, Mar 27, 1894, HL.

126 WWC to JEK, Apr 4, 1894, HL.

127 JEK to W.W. Payne, Apr 13, 1894, JEK to WWC, Apr 18, 1894, HL.

128 JEK to W.W. Payne, May 1, 1894, JEK to WWC, May 1, 1894, JEK to W.W. Payne, May 3, 1894, HL.

129 JEK to WWC, May 4, 1894, JEK to E.B. Frost, May 19, 1894, HL.

130 WWC, *AA, 13,* 384, 494, 1894.

131 WWC, *AA, 13,* 448, 1894.

132 JEK to W.W. Payne, Apr 27, 1894, JEK to J.S. Ames, May 3, 1894, HL.

133 ESH to JAB, Apr 17, 1893, SLO.

134 JAB to WWC, Oct 22, 1892, ESH to JEK, Jan 23, 1894, SLO; WWC to JEK, June 15, 1893, JEK to ESH, Jan 31, 1894, HL.

135 WWC to JEK, May 10, 1894, HL.

136 WWC to JEK, Mar 1, 1895, June 1, 1896, HL.

137 WWC, *Ap.J, 8,* 123, 1898.

138 W. Huggins, *AA, 13,* 568, 1894; WWC, *AA, 13,* 695, 1894.

139 WWC to GEH, Oct 9, 1894, SLO.

140 JEK, *AA, 13,* 772, 1894.

141 WWC and JEK, *AA, 13,* 857, 1894.

142 JEK to WWC, Nov 20, 1894, HL.

143 JEK to WWC, Dec 11, 1894, HL.

144 JEK to WWC, May 27, 1895, HL.

145 JEK, *Ap.J, 1,* 248, 1895.

146 JEK to ESH, July 27, 1891, Jan 29, 1892, SLO.

147 ESH to JEK, Jan 28, 1893, SLO.

148 JEK to JMS, Jan 21, 1893, JEK to S.J. Brown, Sep 17, 1894, HL.

149 AOL to JEK, Aug 14, 1893, GEH to JEK, Sep 14, 1893, HL.

150 JEK, "Biography", June 20, 1900, SLO.

151 JEK to WWC, Apr 5, 1892, JEK to E.B. Frost, Dec 9, 1896, HL.

152 JEK to GEH, Nov 6, Nov 11, 1896, YOA.

153 H.E. Mathews to JEK, July 2, 1895, HL.

154 C.S. Hastings, *BMNAS, 5,* 231, 1903; JAB to C. Burckhalter, Aug 24, 1900, CO; JAB to GD, Oct 1, 1900, BL.

155 JEK to J.R. Macfarlane, Dec 4, 1893, JEK to W.M. Hayes, May 21, 1895, HL.

156 G.E. Lumsden to JEK, Dec 26, 1893, JEK to G.E. Lumsden, Nov 19, 1894, JEK to GEH, Nov 19, 1894, HL; JEK, *Ap.J, 1,* 101, 1895.

157 W. Huggins to JEK, Feb 17, Mar 20, 1895, JEK to W. Huggins, Mar 8, 1895, HL. JEK, *Ap.J, 1,* 350, 1895.

158 JEK to G.F. Barker, Nov 1, 1893, HL.

159 JEK, *Ap.J, 1,* 416, 1895.

160 H. Deslandres, *CR, 120,* 417, 1895.

161 JEK, *Ap.J, 1,* 352, 1895.

162 W.H. Wright to W.F. Kellogg, Nov 9, 1922, SLO.

163 JEK to GEH, Apr 10, 1895, YOA.

164 JEK to GEH, Apr 13, 1895, HL.

165 GEH to JEK, Apr 12, 1895, HL.

166 JEK to GEH, Apr 18, 1895, YOA; JEK to J.M. Cattell, Apr 22, 1895, GEH to JEK, Apr 23, 1895, HL.

167 WWC to JEK, June 3, 1895, HL; WWC, *Ap.J, 2*, 127, 1895.

168 *New York Times*, Apr 14, Apr 17, Apr 18, 1895.

169 *Philadelphia Public Ledger*, Apr 4, 16, 1895.

170 EEB to JEK, Apr 29, 1895, JEK to EEB, May 11, 1895, JEK to A.M. Clerke, Apr 29, 1895, HL.

171 JEK, *PASP, 7*, 154, 1895; JEK, *PA, 2*, 443, 1895.

172 JEK to E.W. Maunder, May 11, 1895, HL; JEK, *MNRAS, 55*, 474, 1895.

173 E.W. Maunder to JEK, June 12, 1895, HL.

174 JEK to GEH, May 10, 1895, HL; JEK, *Ap.J, 2*, 63, 1895.

175 H. Deslandres, *CR, 120*, 1155, 1895.

176 JEK, *Ap.J, 2*, 163, 1895.

177 WWC to JEK, May 8, 1895, W. Huggins to JEK, Apr 18, 1895, C.H. Rockwell to JEK, May 4, 1895, A.M. Clerke to JEK, May 26, 1895, HL.

178 G.E. Lumsden to JEK, Aug 21, 1895, HL.

179 *Proc. Am. Acad. Arts & Sci., 34*, 626, 1898.

180 GEH to WRH, Apr 24, 1894, RL.

181 C.A. Young to SN, Apr 15, 1897, Dec 22, 1898, LC; H.N. Russell, Class Notes, "Astronomy", Fall 1895, Firestone Library, Princeton University.

182 J.A. Parkhurst to JEK, Dec 31, 1897, Jan 8, Apr 21, 1898, JEK to J.A. Parkhurst, Jan 8, 1898, HL.

183 F.J. Engelhardt to JEK, Apr 14, 1895, JEK to F.J. Engelhardt, Apr 19, 1895, HL.

184 L. Brenner to JEK, Aug 8, 1895, HL.

185 JEK to C.A. Young, Apr 26, 1895, HL.

186 C.A. Young to JEK, May 2, 1895, W.H. Collins to JEK, Apr 30, 1895, JEK to C.A. Young, May 4, 1895, JEK to W.H. Collins, May 4, 1895, HL.

187 JEK to W.H. Collins, May 7, June 8, 1895, JEK to GEH, June 8, June 27, 1895, HL.

188 JEK to GEH, Aug 6, Sep 27, 1895, JEK to C.S. Hastings, Nov 4, 1895, July 2, 1896, JEK to H. Jacoby, Mar 3, 1896, HL.

189 W.A. Herdman to JEK, Feb [~15] 1896, JEK to W.A. Herdman, Mar 25, 1896, HL.

190 JEK to GEH, Mar 19, 1896, YOA; JEK to H. Crew, July 24, 1896, GEH to JEK, Mar 26, 1896, H.C. Lord to JEK, May 11, June 21, 1896, H. Crew to JEK, July 31, 1896, HL.

191 JEK to GEH, Mar 27, Apr 27, 1896, YOA.

192 D.M. Drysdale to JEK, July 1, 1896, M. Schamberg & Co. to JEK, July 30, 1896, JEK to H.G. Guiness, Dec 1, 1896, HL; JEK to GEH, Aug 16, 1896, YOA.

193 *Report of BAAS*, 699, 1896.

194 I. Roberts, *Report of BAAS*, 707, 1896.

195 JEK to H.F. Griffiths, Nov 26, 1896, HL.

196 JEK, *Report of BAAS*, 729, 1896.

197 W. Huggins to JEK, Sep 25, 1896, D. Gill to JEK, "Sunday" [1896], HL.

198 C.M. Keeler to GEH, Sep 29, Oct 4, 1896, YOA; JEK to Hasley and Emmert, Nov 6, 1896, HL.

199 JEK to L. Brenner, Nov 24, 1896, JEK to H. Kreutz, Dec 1, 1896, JEK to J.S. Ames, Jan 15, 1897, JEK to W. Weinrich, June 2, 1897, HL.

Chapter 7

1 GEH passports, June 20, 1891, Oct 28, 1893, YOA.

2 GEH to H.M. Goodwin, July 9, 1890, HHL.

3 GEH to H.A. Rowland, July 3, 1890, MEL.

4 GEH to H.M. Goodwin, July 30, 1890, HHL.

5 GEH, Autobiographical Notes, undated, HPM.

6 JEK, Observing Book, 36-inch Refractor, July 10, 1890, LO.

7 H. Wright, *Explorer of the Universe: A Biography of George Ellery Hale.* New York, E.P. Dutton & Co., 1966. Many of the facts of Hale's life and career not otherwise referenced have been taken from this excellent book.

8 JAB to ESH, June 18, Sep 19, Oct 18, 1889, Aug 13, 1891, SLO.

9 JAB to ESH, Sep 25, Oct 26, 1886, Dec 8, 1886, Jan 18, 1888, SLO.

10 GEH & F. Ellerman, *Pub. Y.O., 3,* Part 1, 1904.

11 GEH, Thesis for B.S. in Physics, M.I.T, 1890 (reprinted in H. Wright, J.N. Warnow & C. Weiner, *The Legacy of George Ellery Hale.* Cambridge, M.I.T. Press, 1972, pp. 117-139.

12 W.E. Hale to GEH, Apr 25, 1888, YOA.

13 GEH to H.M. Goodwin, Aug 10, Sep 19, 1890, HHL.

14 JEK to GEH, Jan 29, 1891, YOA.

15 GEH, *SM, 10,* 23, 1891.

16 JEK to GEH, Feb 25, 1891, YOA.

17 ESH to GEH, Apr 29, May 6 (telegram), 1891, GEH to ESH, May 6 (telegram), May 15, 1891, SLO.

18 JEK to GEH, July 9, 1891, YOA.

19 T.W. Goodspeed, *William Rainey Harper: First President of the University of Chicago.* Chicago, University of Chicago Press, 1928.

20 R.J. Storr, *Harper's University: The Beginnings.* Chicago, University of Chicago Press, 1966.

21 WRH to G.A. Douglass, Feb 11, 1891, WRH to GEH, Feb 19, Apr 28, 1891, YOA; GEH to WRH, Apr 30, 1891, RL.

22 WRH to GEH, May 13, May 19, 1891, YOA.

23 GEH to WRH, May 30, 1891, WRH to GEH, June 5, 1891, YOA.

24 GEH to WRH, Apr 29, 1892, RL.

25 A. Hall to WRH, June 13, 1892, RL.

26 GEH to WRH, June 30, 1892, W.E. Hale to WRH, July 1, 1892, WRH to GEH, July 27, 1892, RL.

27 A.A. Michelson to WRH, July 3, 1892, Aug 17, [1892], RL.

28 T.J.J. See to WRH, May 9, 1892, W.B. Smith to WRH, May 29, 1892, RL.

29 T.J.J. See to WRH, June 7, 1892, RL.

30 WRH to T.J.J. See, undated agreement [~June 27, 1892], T.J.J. See to WRH, July 14, 1892, RL.

31 T.J.J. See to WRH, Aug 13, Aug 15, Aug 18, Sep 13, 1892, T.J.J. See to E.B. Hulbert, Aug 30, 1892, W.B. Smith to WRH, Sep 4, Sep, 14, 1892, G.D. Purinton to WRH, July 26, Sep 8, 1892, RL.

32 GEH, "Plan of instruction", July 3, 1892, RL.

33 H.S. Miller, *ASP Leaflet* No. 479, 1969.

34 WRH to F.T. Gates, Tues. 9 p.m. [~May] 1892, RL.

35 GEH to WRH, Sep 23, 1892, RL.

36 WRH to F.T. Gates, Oct 10, 1892, RL.
37 *Chicago Daily Inter Ocean,* Oct 12, Oct 13, 1892; *Chicago Daily Tribune,* Oct 12, Oct 13, 1892; JEK to H.E. Mathews, Dec 20, 1892, HL.
38 GEH to JEK, Oct 12, 1891, JEK to GEH, Oct 15, 1891 (2), HL.
39 GEH to JEK, Nov 19, 1891, HL.
40 GEH to JEK, Dec 22, 1891, Jan 13, 1892, HL; JEK, *AA, 11,* 140, 1892.
41 JEK to GEH, June 28, July 1, Aug 8, 1892, YOA; GEH to JEK, June 29, 1892, HL; SWB to C.A. Young, July 12, 1892, DCL.
42 GEH to JEK, Oct 23, 1892, HL.
43 GEH to JEK, [Aug] 3, 1892, HL; JEK to GEH, Oct 28, Nov 16, Nov 17, 1892, YOA.
44 H.H. Bancroft, *The Book of the Fair.* Chicago, Bancroft Company, 1893; R. Johnson, *A History of the World's Columbian Exposition Held in Chicago in 1893.* New York, D. Appleton & Co., 1897.
45 *Chicago World's Congress Auxiliary. Programs & Addresses.* A three-volume bound set of the reports, announcements, and programs of all the Congresses, collected by C.C. Bonney, presented by his estate to the John Crerar Library Center, and now preserved in the Midwest Inter- Library Center. These volumes are hereafter referred to as CWCAPA.
46 *Preliminary Announcement of Committees,* [~Oct 30, 1890], CWCAPA; GEH to ESH, Mar 23, 1891, SLO.
47 *First Report of the Auxiliary,* Feb 23, 1891, CWCAPA.
48 H. Crew, *Ap.J, 30,* 68, 1909; [H. Crew] *PASP, 21,* 39, 1909.
49 *Preliminary Address of the General Committee of the World's Congress Auxiliary on Mathematics and Astronomy,* March 1892, CWCAPA.
50 GEH to ESH, Mar 1, 1893, GEH to WWC, Mar 1, 1893, SLO; GEH, *AA, 12,* 640, 1893.
51 *PASP, 4,* 156, 1892; *AA, 11,* 616, 1892; *San Francisco Examiner,* June 16, 1892.
52 GEH, *AA, 260,* 364, 1893.
53 GEH to "My esteemed Assistants" [M. Hale and W.B. Hale], July 5, 1891, YOA.
54 GEH to JEK, Mar 23, May 27, 1893, HL.
55 JEK to GEH, Mar 27, 1893, HL; JEK to GEH, May 8, 1893, YOA.
56 GEH to WRH, May 30, June 9, 1893, RL; A.A. Cable to WRH, June 3, 1893, WRH to GEH, June 5, June 10, 1893, YOA; W. Thaw, Jr. to JEK, Nov 14, 1891, JEK to F. Thomson, Jan 9, 1893, JEK to S.M. Prevost, May 20, 1898, HL.
57 GEH to JEK, June 6, 1893, JEK to J.I. Buchanan, June 12, 1893, HL; JAB to GEH, June 8, June 14, 1893, YOA.
58 JAB to GEH, June 24, 1893, YOA.
59 GEH, *AA, 13,* 662, 1894; GEH to JEK, June 9, 1893, JEK to WWC, July 13, 1893, JEK to GD, Aug 10, 1893, HL.
60 GEH to JEK, July 24, 1893, HL.
61 JEK to GEH, July 29, 1893, JEK to ESH, Aug 17, 1893, HL.
62 JEK to GEH, Aug 17, 1893, HL.
63 *Chicago Tribune, Chicago Inter Ocean,* and *Chicago Daily News,* August 21-26, 1893. Later descriptions of the Fair and Congress are also from these sources.
64 JEK to AOL, Aug 15, 1893, BL; JEK to W.W. Payne, Aug 15, 1893, HL.
65 AOL to JEK, Aug 14, 1893, HL.

66 *Bulletin of the Congress of the Department of Science and Philosophy*, Aug 21, 1893, CWCAPA.
67 *Report of the President to the Directors of the World's Columbian Exposition*, Chicago, 1892-3, Chapter 3.
68 G.A. Johnson, *Science*, *22*, 116, 1893.
69 C.C. Bonney, *Opening Address*, May 15, 1893, CWCAPA.
70 GEH, *AA*, *12*, 743, 1893. Later descriptions of the Congress are also from this source.
71 *Programme of Congress on Psychical Science*, Aug 12, 1893, CWCAPA.
72 G.W. Hough, *AA*, *13*, 89, 1894.
73 F.H. Bigelow, *AA*, *12*, 706, 1893.
74 *Science*, *32*, 145, 1893.
75 JAB to GD, Mar 2, 1892, BL.
76 JEK to S.W.B. Moorhead, June 8, 1893, Warner & Swasey to JEK, June 22, 1893, HL.
77 B.C. Truman, *History of the World's Fair*. Chicago, E.C. Morse & Co., 1893.
78 JEK to ESH, Oct 21, 1893, SLO.
79 *Final Report of the California World's Fair Commission*, Sacramento, State Office, 1894.
80 *Programme of Conference on Mathematics and Astronomy*, Aug 12, 1893, CWCAPA.
81 JEK, *AA*, *12*, 733, 1893.
82 W.P. Fleming, *AA*, *12*, 683, 1893.
83 JEK to C.A. Young, Oct 31, Nov 7, 1892, DCL.
84 JEK to WWC, Oct 20, 1893, HL.
85 JEK to GEH, Sep 4, 1893, HL.
86 JEK to GEH, Oct 19, 1892, YOA.
87 WWC to JEK, Sep 7, 1893, JEK to J.P. Hall, Sep 22, 1893, HL.
88 GEH to JEK, Sep 13, 1892, HL.
89 JEK to GEH, Sep 19, 1892, YOA.
90 GEH to JEK, Oct 5, 1892, HL.
91 JEK to GEH, Oct 10, 1892, YOA; GEH to JEK, Oct 13, 1892, HL; GEH, *AA*, *11*, 822, 1892.
92 J. Ritchie, Jr., *Boston Commonwealth*, Oct 22, 1892; JEK, *AA*, *11*, 927, 1892.
93 GEH to JEK, May 4, 1893, HL; JEK to GEH, May 13, 1893, YOA.
94 W.W. Payne to GEH, May 14, 1892, YOA; GEH to JEK, Nov 15, 1892, W.W. Payne to JEK, Jan 5, Jan 11, 1893, HL.
95 JEK to GEH, Dec 20, 1892, YOA; GEH to JEK, Dec 23, 1892, HL.
96 W.W. Payne to GEH, Apr 19, 1893, YOA.
97 W.W. Payne to GEH, July 20, 1893, YOA.
98 W.W. Payne, *AA*, *12*, 376, 1893; *AA*, *12*, 470, 1893.
99 JEK, *PA*, *1*, 9, 1893; W.W. Payne to JEK, July 21, Oct 18, 1893, HL; W.W. Payne to GEH, Aug 5, 1893, YOA.
100 GEH to JEK, Oct 6, Oct 16, Oct 27, 1893, JEK to GEH, Oct 10, Oct 19, 1893, Aug 22, 1894 HL; JEK to F. Ellerman, Oct 27, 1893, W.W. Payne to GEH, Oct 9, 1893, YOA.
101 GEH to JEK, Nov 18, 1893, HL; GEH to WRH, Nov 23, 1893, RL.
102 GEH to JEK, Nov 25, 1893, HL; GEH to WRH, Jan 9, 1894, RL.

103 JEK to GEH, Dec 16, 1893, YOA; JEK to H. Crew, Nov 5, 1893, JEK to W.W. Payne, Nov 10, 1893, Jan 23, 1894, HL; *AA, 12,* 622, 860, 1893.

104 JEK to J.S. Ames, Apr 7, 1894, HL.

105 GEH to JEK, Mar 11, Apr 26, 1894, JEK to GEH, Mar 26, 1894, HL.

106 H.M. Paul to GEH, Oct 18, 1892, YOA.

107 GEH to JEK, Aug 12, Aug 14, 1894 (telegrams), HL.

108 GEH to JEK, Aug 21, Aug 24, 1894, HL.

109 GEH to JEK, Sep 1, 1894, HL.

110 H. Wright, op. cit., p. 115.

111 W.W. Payne to JEK, Sep 5, 1894, JEK to GEH, Sep 8, 1894, JEK to W.W. Payne, Sep 10, 1894, HL.

112 W.W. Payne to GEH, Sep 13, 1894, YOA.

113 W.W. Payne to GEH, Oct 10, Oct 18, 1894, YOA; GEH to JEK, Oct 14, 1894, HL.

114 JEK to GEH, Nov 30, 1894, W.W. Payne to GEH, Nov 14, 1894, YOA; GEH to JEK, Oct 17, Nov 10, 1894, HL.

115 W.W. Payne to GEH, Dec 3, Dec 4, Dec 5, Dec 13, 1894, YOA; W.W. Payne, *AA, 18,* 871, 1894.

116 GEH, *AA, 18,* 831, 1894; GEH, *Ap.J, 1,* 80, 1895.

117 GEH to JEK, Sep 29, 1894, JEK to GEH, Oct 2, 1894, HL.

118 WWC to GEH, Oct 9, 1894, YOA.

119 JEK to WWC, Oct 27, 1894, SLO.

120 WWC to JEK, Nov 5, 1894, HL; WWC to GEH, Nov 5, 1894, YOA.

121 WWC to GEH, Nov 16, Nov 19, 1894, YOA.

122 F.L.O. Wadsworth to WRH, June 2, 1894, RL; GEH to JEK, Oct 4, 1894, Mar 5, 1896, HL.

123 JEK to WWC, Nov 13, 1894; Minutes of Meeting of the Editorial Board, Nov 2, 1894, HL.

124 JEK to J. Boughton, Feb 23, May 7, 1893, HL; JEK to GEH, Nov 10, 1894, YOA.

125 *Ap.J, 1,* 1, 1895; JEK to GEH, Jan 8, Jan 11, Jan 16, 1895, YOA.

126 W.W. Payne to GEH, Jan 14, Jan 22, Jan 25, Jan 31, 1895, YOA.

127 GEH to JEK, Sep 18, 1894, Feb 9, 1895, HL.

128 JEK to GEH, Feb 12, 1895, YOA.

129 W.G. Hoyt, *Lowell and Mars.* Tucson, University of Arizona Press, 1976; S.C. Chandler to SN, May 26, 1897, LC.

130 GEH to JEK, Dec 19, 1894, Jan 6, Jan 21, 1895, JEK to GEH, Dec 27, 1894, Jan 4, Jan 17, Jan 18, 1895, F. Ellerman to JEK, Jan 16, 1895, JEK to P. Lowell, Jan 18, 1895, HL.

131 P. Lowell, *Ap.J, 1,* 393, 1895; JEK to GEH, May 11, June 27, 1895, GEH to JEK, July 5, 1895, HL; GEH to JEK, June 25, 1895, YOA.

132 GEH to WRH, Apr 23, Apr 25, May 5, 1895, RL; WRH to GEH, Apr 29, 1895, YOA.

133 JEK to GEH, Jan 31, 1896, YOA.

134 GEH to JEK, Mar 3, Mar 7, Mar 11, 1896, HL; JEK to GEH, Mar 6, Mar 11, Mar 15, 1896, YOA; B. Hasselberg, *Ap.J, 4,* 116, 212, 1894.

135 WRH to GEH, Jan 9, 1893, YOA; Storr, op. cit., p. 267.

136 GEH to WRH, Jan 15, 1893, T.C. Chamberlin to GEH, Jan 31, 1893, YOA; JEK to WRH, Mar 6, 1893, GEH to JEK, Apr 6, 1893, JEK to GEH, Apr 17, 1893, HL; GEH, *Ap.J, 5,* 164, 1897; A. Wolfmeyer and M.B. Gage, *Lake Geneva: Newport of the West 1870-1920.* Lake Geneva, Lake Geneva Historical Society, 1976.

137 SWB to GEH, Dec 9, 1893, YOA; GEH, *Ap.J, 5,* 254, 1897.

138 GEH to WRH, Sep 4, 1893, Oct 4, 1894, WRH to GEH, Dec 6, 1894, RL.

139 GEH to JEK, Mar 21, 1895, JEK to GEH, Apr 26, 1895, HL.

140 GEH to JEK, June 11, July 18, Aug 12, Sep 20, 1895, JEK to C. Burckhalter, Aug 23, 1895, HL; GEH, *Ap.J, 1,* 318, 1895.

141 JEK, *PLO, 3,* 161, 1894; JEK, *AA, 13,* 476, 1894; WWC, *AA, 13,* 384, 494, 1894.

142 GEH to JEK, Apr 6, June 26, 1895, HL; [GEH], *Ap.J, 1,* 439, 1895; GEH, *Ap.J, 2,* 165, 1895; E.B. Frost, *PASP, 7,* 317, 1895; C.A. Young, *PASP, 7,* 345, 1895.

143 A.G. Clark to GEH, Sep 23, 1895, YOA.

144 GEH to JEK, Aug 2, Aug 19, Sep 13, Sep 25, Oct 6, 1895, SWB to JEK, Sep 11, Oct 4, 1895, HL; E.C. Pickering to GEH, Sep 29, 1895, L.S. Pickering to GEH, Oct 7, [1895], JEK to GEH, Oct 9, 1895, YOA.

145 GEH, Announcement of Board of Editors Meeting [~Oct 4, 1895], JEK to GEH, Nov 1, 1895, JEK to P.Tacchini, Nov 1, 1895, HL.

146 GEH to JEK, Nov 7, Nov 11, Nov 25, 1895, JEK to GEH, Nov 14, 1895, HL; JEK, *Ap.J, 3,* 154, 1896.

147 GEH to WRH, Jan 28, 1894, RL.

148 T.J.J. See to Hon. Board of Curators of the State of Missouri, Mar 24, 1893, T.J.J. See to WRH, May 31 1894, RL.

149 T.J.J. See to WRH, Nov 15, 1895, WRH to T.J.J. See, Nov 18, 1895, RL.

150 T.J.J. See to WRH, Mar 4, 1898, RL.

151 W.G. Hoyt, op. cit., p. 124.

152 W. Huggins to GEH, July 10, 1898, HPM; WRH to SN, Nov 16, 1898, E.C. Pickering to SN, Jan 12, 1899, LC.

153 "Editor of Astronomical Life" [S.C. Chandler] to E.B. Frost, [~May 1, 1899], YOA; *Observatory, 22,*292, 1899.

154 G.C. Comstock to E.B. Frost, May 30, 1900, YOA.

155 J. Lankford, *JHA, 11,* 129, 1980. Many of the facts of See's career not otherwise referenced are from this interesting study, and from a longer, unpublished version of it.

156 GEH to WRH, Feb 24, Mar 4, Mar 27, 1896, RL; W. Huggins to GEH, Mar 10, 1896, HPM; GEH to JEK, Mar 21, Apr 6, 1896, HL.

157 GEH to JEK, May 24, June 2, June 19, July 10, July 17, Aug 19, 1896, W. Huggins to JEK, Feb 3, 1897, W. Huggins to GEH, Oct 14, 1896, HPM; GEH to JEK, Feb 13, 1897, YOA.

158 GEH to WRH, Oct 4, 1896, RL.

159 GEH to JEK, Nov 14, 1896, HL; GEH to JEK, Nov 24, 1896, YOA; EEB to GD, June 10, June 20, 1895, June 27, Nov 21, 1896, BL; EEB to H.H. Turner, Dec 2, 1896, Jan 27, 1897, RAS.

160 GEH to JEK, Dec 7, 1896, JEK to GEH, Dec 9, Dec 22, 1896, YOA.

161 D.C. Gilman to JEK, May 14, May 19, 1896, GEH to JEK, May 26, 1896, J.S. Ames to JEK, Jan 20, 1897, Program of Peabody Institute, 1896-1897, HL; JEK to GEH, May 28, 1896, YOA.

162 JEK to GEH, Nov 11, Nov 24, Dec 31, 1896, Jan 22, 1897, GEH to JEK, Jan 8, 1897, YOA.

163 GEH to JEK, Jan 20, 1897, YOA.

164 GEH to W.J. Holland, Mar 12, 1897, GEH to JEK, Mar 12, Mar 17, Mar 20, Mar 30, 1897, YOA.

165 GEH to JEK, May 24, 1897, HL; GEH to C.T. Yerkes, May 31, 1897, GEH to WRH, May 20, 1897, GEH to H.A. Rust, May 25, 1897, YOA; GEH to WRH, June 2, 1897, RL.

166 C.T. Yerkes to WRH, Mar 24, 1897, RL.

167 GEH to C.A. Young, June 1, 1897, GEH to JEK, June 1, 1897, C.T. Yerkes to GEH, June 2, 1897, GEH, Report of the Director for the Month Ending May 31, 1897, [June 7, 1897], YOA.

168 WRH to GEH, Apr 9, 1897, YOA; GEH to JEK, Apr 10, 1897, HL; GEH to WWC, Aug 9, 1897, SLO.

169 GEH to WRH, Dec 10, 1895, RL; GEH to W. Huggins, Apr 5, 1897, YOA; W. Huggins to GEH, July 17, Nov 23, 1896, Jan 16, Mar 7, Apr 21, 1897, HPM.

170 M.R. Cobb to GEH, July 22, 1897, YOA.

171 GEH to WRH, May 25, July 26, 1897, RL; C.A. Young to ESH, Jan 25, 1897, SLO; GEH to H.C. Vogel, May 29, 1897, SPL to GEH, Sep 1, 1897, R. Rathburn to GEH, Sep 1, 1897, YOA.

172 GEH to JEK, Sep 4, 1897, JEK to GEH, Sep 6, 1897 (telegram), YOA.

173 JEK to GEH, Sep 9, Sep 10, 1897, JAB to GEH, Sep 17, 1897, YOA.

174 GEH to JEK, July 14, Aug 27, Sep 26, Oct 1, Nov 2, 1897, JEK to GEH, July 21, Aug 23, Sep 15, 1897, F. Ellerman to GEH, Dec 28, 1898, YOA.

175 *Chicago Tribune*, October 18-22, 1897. The *Tribune* covered the Yerkes dedication activities both in Chicago and at Williams Bay in depth and is the source for most of the material not otherwise cited. See also *University of Chicago Weekly*, 6, No. 4, October 21, 1897; *Lake Geneva Herald*, Oct 22, 1897.

176 GEH to M.B. Hale, Aug 4, 1897, HPM; GEH to E.C. Pickering, Oct 13, 1897, GEH to J.M. Van Vleck, Oct 13, 1897, GEH to F.W. Shepardson, Oct 13, 1897, YOA; EEB to SN, Oct 10, Oct 14, Oct 25, 1897, LC.

177 JEK, *Ap.J*, 6, 423, 1897.

178 WWC to GEH, Sep 21, 1897, YOA.

179 *University Record*, Chicago, 2, 235, 1897, gives the complete program of the dedication.

180 JEK, *Ap.J*, 6, 271, 1897.

181 *University Record*, Chicago, 2, 246, 1897, gives the complete texts of the speeches of Yerkes, Ryerson and Harper.

182 B.W. Snow to GEH, Oct 17, 1897, J.D. Butler to GEH, Oct 17, 1897, YOA; J.D. Butler to WRH, Nov 4, 1897, RL.

183 GEH, *Ap.J*, 6, 353, 1897.

184 W.S. Adams, *Science*, 106, 196, 1947.

185 SN, *North American Review*, 122, 88, 1876; W. Alvord, *PASP*, 10, 49, 1898; WWC, *MNAS*, 17, 1, 1924.

186 S. Newcomb, *Ap.J*, 6, 289, 1897.

187 GEH to H.A. Rust, Oct 26, 1897, C.H. Rockwell to GEH, Oct 26, 1897, JEK to GEH, Nov 1, 1897, YOA; JEK to F.W. Very, Nov 4, 1897, HL.

188 GEH to WWC, Aug 25, 1897, SLO; WWC to GEH, Aug 10, Sep 9, 1897, C.D. Perrine to GEH, Sep 2, 1897, A.L. Colton to GEH, Sep 6, 1897, YOA. O. Struve, *Science, 106,* 217, 1947. Colton appears in the group picture taken at the Yerkes dedication on Oct 10, 1897 and published in the last reference.

Chapter 8

1 C.B. Hill to GD, Oct 14, 1892, BL; SWB to EEB, Aug 19, 1892, C.B. Hill to EEB, Dec 31, 1892, VU; H. Crew to C.A. Young, Dec 10, 1891, DCL; JAB to GD, Nov 12, 1896, JAB to RSF, Jan 20, 1889, BL; GD to EEB, June 7, 1897, VU; EEB to GD, May 10, Nov 24, 1894, BL; D.S. Jordan to WJH, May 2, [1897], BHL; EEB to GD, Oct 29, 1897, BL.

2 *SN, The Reminiscences of an Astronomer.* Boston, Houghton Mifflin & Co., 1903. Many of the actions and opinions attributed to Newcomb throughout this chapter are from this book.

3 S.E. Tillman, *Annual Report of the Association of Graduates of the U.S. Military Academy,* Annual Reunion, June 11, 1915, p. 52. See also W.C. Winlock, *Popular Science Monthly,* November 1886, p. 114; WWC, *BMNAS, 8,* 346, 1916. Facts of Holden's life and career not otherwise referenced in this chapter come from these three biographical articles.

4 Minutes, University of Wisconsin Board of Regents, Jan 18, 1881, UW; *Wisconsin State Journal,* Jan 19, 1881; ESH to RSF, Feb 10, 1881, SLO.

5 JEK to ESH, June 4, June 30, Sep 21, 1886, Sep 4, Sep 12, 1887, etc., SLO.

6 ESH to H.E. Mathews, May 1, 1888, ESH, "The Lick Trustees to E.S. Holden Dr. For expert services rendered ...", May 1, 1888, SLO.

7 SN to RSF, Apr 7, 1887, July 14, 1888, SN to H.E. Mathews, Aug 10, 1888, SLO; H.E. Mathews to SN, Aug 17, 1888, RSF to E.B. Mastick, Sep 8, 1888, LC.

8 E.B. Mastick to C.L. Floyd, June 2, June 7, 1888, C.L. Floyd to E.B. Mastick, June 1, 1888, H.E. Mathews to RSF, Feb 13, 1889, BL; H.E. Mathews to SN, Aug 17, Sep 5, 1888, C.L. Floyd to SN, Oct 16, 1888, LC; H.E. Mathews, *Reminiscences of the James Lick Trust,* 1920, SLO.

9 D.E. Osterbrock, *JHA,* in press, 1984 gives the details of these espisodes.

10 JEK to WWC, Sep 20, 1892, SLO.

11 JEK to ESH, June 28, 1892, ESH to T.G. Phelps, July 8, 1892, ESH to C.F. Crocker, Oct 5, 1892, SLO.

12 ESH to JEK, Sep 12, 1892, HL.

13 JEK to W.W. Payne, Sep 24, 1892, HL; *AA, 11,* 840, 1892.

14 ESH to JEK, Oct 28, Nov 14, 1892, SLO; ESH to JEK, Nov 15, Dec 1, 1892, Jan 7, Feb 11, Mar 16, 1893, HL.

15 EEB to GD, Nov 17, 1892, BL.

16 JAB to GD, Nov 18, 1892, BL.

17 ESH to JEK, May 17, 1893 (telegram), HL; ESH to A. Rodgers, May 3, 1892, ESH to M. Kellogg, Feb 20, May 10, 1893, Extract from Board of Regents Minutes, May 16, 1893, SLO.

18 JEK to ESH, May 17, 1893, JEK to E.W. Maunder, May 19, 1893, JEK to J.H.C. Bonté, Aug 1, 1893, HL; JEK to D.C. Gilman, May 19, 1893, FHA.

19 ESH to WJH, July 10, 1895 (telegram), ESH to Associated Press/United Press, July 18, 1895 (telegram), SLO.

20 R.P.S. Stone, *Sky & Telescope, 58,* 307, 1979; F.H.C., *MNRAS, 67,* 232, 1907; J. Gledhill to ESH, Mar 2, 1893, E. Crossley to ESH, Feb 22, 1896, SLO; J. Gledhill to A.C. Ranyard, Jan 22, 1880, J. Gledhill to W.H. Wesley, Jan 21, 1892, RAS.

21 ESH to J. Gledhill, Feb 10, 1893, J. Gledhill to ESH, Mar 2, Mar 3 (telegram), 1893, ESH to A. Carnegie, Mar 30, 1893, ESH to D.O. Mills, Apr 21, 1893, E. Crossley to ESH, Oct 21, 1893, Jan 22, 1894, ESH to A.S. Hallidie, Apr 20, 1893, SLO.

22 A. Carnegie to ESH, Apr 12, 1893, D.O. Mills to ESH, May 17, 1893, ESH to E. Crossley, Oct 9, 1893, Feb 13, Feb 17, 1894, SLO.

23 J. Gledhill to ESH, Jan 7, Feb 20, 1895; ESH to J. Gledhill, Mar 8, 1895; ESH to E. Crossley, Mar 8, 1895, SLO.

24 E. Crossley to ESH, Apr 6, May 19, Aug 22, Sep 28, 1895, SLO; ESH to A.S. Hallidie, Apr 17, Apr 23, 1895, ESH to E. Crossley, Apr 23, 1895, UCA; ESH, *PASP, 7,* 197, 1895.

25 EEB to GD, May 26, 1897, BL; JEK, *Ap.J., 11,* 325, 1900.

26 JEK to GEH, Nov 1, 1897, YOA.

27 JEK to G.W. Myers, May 8, 1896, JEK to T.H. Foulkes, July 1, 1897, JEK to W.W. Payne, July 3, 1897, JEK to GEH, July 10, 1897, JEK to A.E. Lyon, Aug 10, 1897, HL.

28 JEK to WWC, June 10, Aug 22, Sep 7, 1895, May 25, 1896, SLO; WWC to JEK, Sep 16, 1895, June 1, 1896, HL.

29 F. Ellerman to GEH, Jan 14, 1898, GEH to JEK, Feb 24, 1898, YOA.

30 JEK to GEH, Apr 9, 1895, JEK to Harley & Emmert, Feb 24, 1896, JEK to E. Lane, July 24, 1897, HL; JEK, "L.O. Salary List," [Aug 1, 1898], SLO; WRH to EEB, Jan 10, 1895, VU; SN to L.C. Seelye [~May 22, 1888], LC.

31 S. Lorant, *Pittsburgh, the Story of an American City.* Garden City, Doubleday & Co., 1964. JEK to ESH, May 16, 1892, HL.

32 JEK to ESH, Oct 23, 1892, JEK to V. Schumann, Nov 28, 1892, JEK to GEH, July 11, 1894, JAB to A. Carnegie, Feb 24, 1898, HL.

33 JEK to WWC, July 13, 1893, JEK to F.L. Sheppard, Jan 16, 1894, JEK to AOL, Mar 12, 1894, JEK to S.J. Cunningham, Nov 19, 1894, HL.

34 Observatory Committee Minutes, Mar 16, 1895, Councils of Allegheny, File of Select Councils, No. 84, Ordinance, July 26, 1895, AO; JAB to C. Lockhart, Feb 4, 1898, HL.

35 JEK to WWC, May 25, 1895, JEK to D.P. Todd, Jan 3, 1896, JEK to G.E. Lumsden, Apr 13, 1896, JEK to H.F. Griffiths, Nov 26, 1896, JAB to A. Carnegie, Mar 8, 1898, HL.

36 JEK, "Sketch of a Design for an Astro-Physical Observatory," 1892, JEK, "Plan of a Proposed Astronomical Observatory," 1895, AO; JAB, "Estimated Cost of Proposed Observatory in Riverview Park," [~Feb 4, 1898], HL.

37 JAB to EEB, May 6, 1897, VU; JAB to GD, May 27, 1897, BL.

38 GEH to JEK, Mar 11, Apr 26, 1894, HL; JEK to GEH, Mar 26, 1894, YOA.

39 GEH to WRH, Oct 24, 1894, RL.

40 WRH to GEH, Dec 6, 1894, GEH to WRH, Dec 5, 1895, Jan 16, 1896, RL.

41 GEH to WRH, Nov 21, 1896, RL.

42 JEK to GEH, Jan 27, 1896, GEH to JEK, Nov 24, 1896, July 14, Sep 4, Sep 26, 1897, YOA.

43 JEK to D.C. Gilman, Oct 20, 1897, MEL.

44 WWC, *PASP, 1,* 228, 1894.

45 W. Huggins, *Ap.J., 1,* 193, 1895; H.C. Vogel, *Ap.J., 1,* 196, 1895.

46 L.E. Jewell, *Ap.J., 1,* 311, 1895.

47 WWC, *Ap.J., 2,* 28, 1895, *Ap.J., 4,* 79, 1896; L.E. Jewell, *Ap.J., 3,* 255, 1896.

48 JEK, *Ap.J., 4,* 137, 1896.

49 JEK, *Ap.J., 5,* 328, 1897.

50 L.D. Kaplan, G. Münch and H. Spinrad, *Ap.J., 139, 1,* 1964.

51 JEK to J.S. Ames, Feb 14, 1896, HL.

52 JEK to H.H. Langell, Mar 4, 1896, JEK to D.C. Miller, Apr 6, 1896, JEK to M.I.
    Pupin, Apr 1, Apr 13, 1896, JEK to J.G. Ogden, Apr 13, 1896, D.C. Miller to
    JEK, Apr 10, 1896, M.I. Pupin to JEK, Apr 11, Apr 14, 1896, HL.

53 JEK to M.I. Pupin, Apr 17, 1896, JEK to F.W. Very, Apr 27, 1896, HL.

54 JEK to A.W. Goodspeed, July 1, 1896, JEK to F. Nicola, July 17, 1896, JEK to
    G.H. Clapp, Aug 25, 1896, HL; JAB to GD, Nov 12, 1896, BL.

55 G.H. Clapp to JEK, Feb 4, Apr 19, Apr 28, 1897, HL.

56 JEK to J.S. Ames, Dec 9, 1896, HL.

57 J.R. Collins to JEK, Jan 17, 1896, JEK to J.R. Collins, Jan 20, 1896, JEK to J.
    Whitehead, June 13, 1896, HL.

58 JEK to ESH, Dec 20, 1897, HL; ESH to GEH, Dec 6, 1897, GEH to ESH, Dec
    10, 1897, YOA.

59 WJH, *PASP, 36,* 309, 1924; RHT, *PASP, 36,* 312, 1924.

60 E.F. Hussey, "Mens Aequa in Arduis" - "Diary" Apr 1 - Sep 18, 1897, BHL;
    JMS to F.L.O. Wadsworth, Oct 14, Dec 26, 1897, Jan 10, 1898, AO; F.L.O.
    Wadsworth to JMS, [~Dec 20, 1897], Jan 2, 1898, RHT to MAT, Nov 5, 1893,
    Feb 18, 1895, Apr 4, 1898, JEK to AOL, Mar 10, 1899, SLO.

61 JAB to GD, Oct 30, 1897, BL. *San Francisco Examiner,* Oct 7, Oct 8, Dec 7,
    1897.

62 O. Lewis, *George Davidson. Pioneer West Coast Scientist,* Berkeley, University
    of California Press, 1954. EEB to GD, July 2, 1895, JAB to GD, [~July 28,
    1895], BL.

63 GD to EEB, Apr 7, 1898, VU; SWB to GD, Dec 28, 1897, EEB to GD, Jan 25,
    1898, BL.

64 JEK to R.W. Wright, Aug 19, 1892, JEK to H.E. Mathews, Dec 20, 1892, July
    18, 1895, JEK to A. Friant, Jan 30, 1894, HL; JEK to GD, June 28, 1895, BL;
    JAB to GD, Aug 19, 1894, Feb 26, May 31, June 8, Sep 28, 1897, BL.

65 WJH to D.S. Jordan, June 9, 1897, BHL; WJH to D.S. Jordan, Dec 7, 1897,
    SUA; WWC to *San Francisco Chronicle,* Sep 30, 1897, WWC to M. Kellogg, Oct
    7, 1897, WWC, RHT, WJH, C.D. Perrine, R.G. Aitken to M. Kellogg, Oct 15,
    1897, UCA; WWC to C. Burckhalter, Sep 30, Oct 7, 1897, CO; RHT to MAT,
    Dec 13, 1897, SLO.

66 JEK to C.D. Perrine, Nov 3, 1897, HL.

67 GEH to WRH, Oct 28, 1897, YOA.

68 WRH to GEH, Nov 4, Nov 26, 1897, GEH to WRH, Nov 23, 1897, YOA.

69 E.C. Pickering to GEH, Nov 6, 1897, W. Huggins to GEH, Nov 15, 1897, HPM;
    GEH to SN, Oct 26, 1897, GEH to E.C. Pickering, Nov 3, 1897, GEH to WRH,
    Nov 17, 1897, YOA.

70 M. Kellogg to GEH, Nov 15, 1897, HPM; JEK to GEH, Nov 24, 1897, YOA.

71 GEH to M. Kellogg, Nov 27, 1897, UCA.

72 M.H. Newcomb to J. LeConte, July 14, 1897, C.D. Perrine to M. Kellogg, Dec 3, 1897, RHT, WJH, R.G. Aitken, C.D. Perrine to M. Kellogg, Dec 3, 1897, J. LeConte to M. Kellogg, Dec 13, 1897, UCA; GD to EEB, Nov 8, 1897, VU; JEK to GEH, Dec 16, Dec 27, 1897, YOA.

73 M. Kellogg to D.C. Gilman, Dec 11, 1897, MEL; D.C. Gilman to M. Kellogg, Dec 19, 1897, UCA; RHT to MAT, Dec 18, 1897, SLO.

74 C.T. Yerkes to WRH, Apr 24, 1897, W.E. Hale to WRH, Nov 24, 1897, RL; GEH to C.T. Yerkes, Oct 16, 1897, YOA.

75 W.W. Payne, *PA, 8,* 235, 1900.

76 EEB to WRH, July 26, July 28, Aug 11, 1897, RL.

77 GEH to JEK, Nov 17, 1897, GEH to C.W. Bruce, Nov 22, 1897, JEK to GEH, Jan 6, 1898, YOA; GEH to F.L.O. Wadsworth, Dec 19, 1897, AO.

78 M.W. Bruce to GEH, Dec 17, Dec 28, Dec 31, 1897, Jan 3, Jan 5 (letter and telegram), 1898; WRH to GEH, Jan 4 (telegram), Jan 5, 1898; E.C. Pickering to GEH, Jan 2, 1898, YOA; GEH to WRH, Jan 4, 1898 (letter and telegram), RL; GEH to F.L.O. Wadsworth, Dec 28, 1897, AO.

79 GEH to WRH, Jan 6, Jan 9, 1898 (telegrams), WRH to GEH, Jan 8, 1898 (telegram), GEH to JEK, Jan 9, 1898 (telegram), YOA; GEH to WRH, Jan 15, [Jan] 22 (telegram) 1898, RL; JEK to F.L.O. Wadsworth, Jan 13, 1898, AO.

80 GEH to F.L.O. Wadsworth, Jan 7, [1898], AO; WRH to M. Ryerson, Jan 27, 1898, WRH to JEK, Jan 29, 1898, RL; JEK to GEH, Jan 31, 1898, YOA.

81 EEB to GEH, Dec 23, Dec 28, 1897, YOA.

82 JAB to Warner & Swasey, Jan 30, 1898, JAB to A. Carnegie, Feb 7, 1898, JAB to H. Phipps, Feb 12, 1898, JAB to C.L. Magee, Feb 22, 1898, HL.

83 JAB, *The Autobiography of a Man Who Loved the Stars.* Boston, Houghton Mifflin Co., 1925.

84 JEK to C.H. McLeod, Jan 4, 1898 (two letters), JEK to J.M. Cattell, Mar 8, 1898, HL.

85 RHT to MAT, Jan 3, Feb 28, Mar 6, 1898, SLO; W.H. Wright to F.L.O. Wadsworth, Feb 10, 1898, AO; J.P. Hall to GEH, Jan 22, 1898, GEH to JEK, Feb 1, Feb 15, 1898, JEK to GEH, Feb 17, 1898, YOA.

86 V.A. Stadtman, *The University of California 1868-1968.* New York, McGraw Hill, 1970. See especially pp. 116-119.

87 E.C. Pickering to M. Kellogg, Nov 21, 1897, M.L. Elkins to M. Kellogg, Nov 27, 1897, SN to M. Kellogg, Nov 21, 1897, UCA; SPL to P.A. Hearst, Jan 5, 1898, BL.

88 GD, "Memo of personal interviews with Rodgers, Stoss, Niebaum, Reinstein and Gen Houghton about Directorship of the Lick Obs[ervator]y," Mar 6, 1898, BL.

89 *San Francisco Chronicle,* Mar 9, 1898, *San Francisco Examiner,* Mar 9, 1898; Tally sheet of votes for Director of Lick Observatory, Mar 8, 1898, UCA.

90 RHT to MAT, Mar 8, Mar 14, 1898, SLO; WJH to D.S. Jordan, Mar 8, 1898, WJH to J.C. Branner, Mar 8, 1898, WJH to WWC, Mar 10, 1898, BHL.

91 JEK to GEH, Mar 9, Mar 17, 1898, E.B. Frost to GEH, Feb 22, 1898, YOA; JEK to M. Kellogg, Mar 9, Mar 15, 1898 (both telegrams), JEK to J.S. Ames, Mar 9, 1898, JAB to C. Lockhart, Mar 14, 1898, L. Weinek to JEK, Apr 9, 1898, HL; JEK to D.C. Gilman, Mar 23, 1898, MEL; *Allegheny Evening Record,* Mar 17, 1898.

92 JAB to E. O'Neil, Mar 17, 1898, JAB to G. Westinghouse, Jr., Mar 17, Mar 24, 1898, JAB to A. Carnegie, Mar 24, 1898, JAB to JEK [~Mar 31, 1898], HL.

93 JAB to Mrs. W. Thaw, Apr 4, 1898, HL.

94 JEK to M. Kellogg, Apr 2, 1898 (telegram), HL.

95 GEH to JEK, Mar 10, Mar 12, Apr 6, 1898, YOA.

96 JEK to A.D. Keeler, Apr 4, 1898, SLO; C. Burckhalter to EEB, [~Apr 5, 1898], VU.

97 EEB to AOL, Oct 2, 1897, Apr 18, 1898, BL; JEK to EEB, Mar 19, Dec 10, 1898, Oct [~10], 1899, VU.

98 ESH to JEK, Apr 5, 1898, SLO.

99 GD to JEK, Mar 5, Dec 20, 1893, Feb 3, 1897, JAB to GD, Apr 23, 1898, HL; JEK to GD, Apr 7, 1893, BL; GD to EEB, Apr 18, 1898, VU.

100 RHT to MAT, Nov 15, 1897, Jan 10, Mar 21, Apr 11, 1898, JEK to JMS, Apr 2, 1898, JMS to JEK, Apr 11, 1898, SLO; JMS to M. Kellogg, Mar 10, Apr 15, May 23, 1898, JMS to the Honorable Board of Regents, Mar 10, 1898, Regents' Resolution, Apr 12, 1898, UCA.

101 GEH to W.H. Wright, Oct 12, 1897, GEH to JEK, Mar 22, 1898, YOA; W.H. Wright to ESH, May 11, 1895, Aug 24, 1896, Mar 30, July 7, 1897, W.H. Wright to WWC, Apr 19, Sep 5, 1896, ESH to W.H. Wright, Mar 8, Apr 6, 1897, E.W. Davis to ESH, July 15, 1897, ESH to C.W. Slack, July 29, 1897, SLO.

102 GEH to JEK, May 2, May 6, 1898, JEK to GEH, Apr 18, 1898, YOA; W.A. McKowen to R.G. Aitken, May 28, 1897, SLO; W.H. Wright to AOL, July 10, 1895, BL; *San Francisco Chronicle,* May 12, 1897.

103 GEH to C. Runge, Feb 24, Mar 10 (cablegram), Mar 11, 1898, YOA.

104 GEH to JEK, Mar 21, 1898, JEK to GEH, Mar 28, 1898, GEH to M.W. Bruce, Mar 21, 1898, M.W. Bruce to GEH, Apr 7, 1898, GEH to E.B. Frost, Mar 11, Apr 19, Apr 26, Apr 28, 1898, E.B. Frost to GEH, Mar 15, Apr 5, Apr 11, May 2, 1898, GEH to C.W. Bruce, Aug 29, 1899, YOA; GEH to WRH, Mar 29, June 27, 1898, RL.

105 GEH to WRH, Feb 8, 1898, WRH to GEH, Feb 21, 1898, YOA; T.C. Mendenhall to WRH, Jan 31, 1898, F.L.O. Wadsworth to WRH, May 28, July 11, July 27, 1898, F.L.O. Wadsworth to T.W. Goodspeed, June 23, 1898, RL.

106 JEK to GEH, Mar 24, Apr 2, 1898, GEH to JEK, Mar 26, 1898, YOA; JAB to JEK, May 19, 1899, SLO.

107 *Proc. Am. Acad. Arts & Sci., 34,* 626, 683, 1898; *PASP, 10,* 123, 1898; GEH to JEK, June 3, 1898, JEK to GEH, June 22, 1898, SLO.

108 JEK to GEH, May 5, May 10, May 20 (telegram), 1898, JEK to GEH, May 14, 1898, YOA; JAB to J.K. Rees, May 20, 1898, HL.

109 JEK to AOL, Apr 20, May 10, 1898, BL; *San Francisco Chronicle,* May 26, May 31, 1898; *San Francisco Examiner,* May 28, May 31, 1898; P. Herget, *BMNAS, 49,* 129, 1978.

110 *PASP, 10,* 113, 1898; JEK to EEB, June 3, 1898, VU.

111 WWC, *PASP, 10,* 127, 1898; WWC, *Ap.J, 10,* 186, 1899; E.B. Campbell, *In the Shadow of the Moon* (undated manuscript), SLO.

Chapter 9

1 JMS to E.W. Davis, May 31, 1898, JMS to C. Burckhalter, May 31, 1898, JEK to E.W. Davis, June 1, 1898, JEK to C.A. Young, June 3, 1898, RHT to MAT, Mar 1, Mar 19, 1900, SLO.

2 JEK to JMS, Apr 2, 1898, JEK to A.D. Keeler, Apr 4, 1898, JMS to JEK, Apr 11, 1898, JEK to GEH, June 22, 1898, SLO; JEK to EEB, June 3, 1898, VU.

3 *San Jose Mercury,* Sep 19, 1910, *San Francisco Chronicle,* Sep 22, 1910, SLO.

4 JEK, "L.O. Salary List", Aug 9, 1898, JEK to W.H.L. Barnes, Aug 30, 1899, SLO.

5 RHT to P.T. Templeton, Aug 24, 1893, RHT to MAT, Aug 27, 1893, May 8, Dec 25, 1898, Jan 16, Apr 3, 1899, JEK to B.I. Wheeler, Apr 7, 1900, SLO; B.I. Wheeler to JEK, Apr 9, 1900, UCA.

6 RHT to MAT, Dec 27, 1897, Jan 31, Dec 12, 1898, Aug 28, Oct 2, Oct 8, Nov 27, Dec 4, Dec 18, 1899, June 11, 1900, SLO.

7 JEK to H.A.L. Floyd, Aug 2, 1898, SLO.

8 M. Kellogg to JEK, Apr 17, 1898, HL; JEK to AOL, Apr 5, 1898, BL; SDT to JEK, June 21, 1898, JEK to SDT, June 28, 1898, SLO.

9 ESH to SFT, Apr 11, 1892, SDT to ESH, May 21, 1892, ESH to P.A. Hearst, Jan 30, 1893, P.A. Hearst to ESH, Mar 16, 1896, ESH to H.C. Lord, Mar 25, 1897, E.F. Coddington to ESH, Apr 3, July 2, 1897, ESH to E.F. Coddington, July 9, 1897 (telegram), SLO.

10 ESH to C.W. Slack, Mar 23, 1895, ESH, "Report Made to the Joint Committee of the Regents on the Organization and Courses in the Department of Astronomy", Apr 15, 1895, ESH to T.G. Phelps, July 31, 1895, SLO.

11 AOL, *PASP, 16,* 68, 1904.

12 JEK to M. Kellogg, June 2, June 28, 1898, M. Kellogg to JEK, June 16, 1898, JEK to AOL, June 18, 1898, AOL to JEK, June 20, June 21, 1898, SLO.

13 JEK to T.G. Phelps, June 8, 1898, SLO. *San Francisco Chronicle,* June 15, 1898.

14 JEK to T.G. Phelps, June 28, July 6, 1898, T.G. Phelps to JEK, July 3, 1898, JEK to AOL, July 6, 1898, SLO.

15 W.A. McKowen to JEK, July 13, Aug 11, 1898, JEK to T.G. Phelps, July 15, Aug 4, 1898, JEK to AOL, July 15, 1898, T.G. Phelps to JEK, July 17, 1898, AOL to JEK, July 20, Aug 12, 1898, SLO.

16 H.K. Palmer to JEK, June 16, 1898, JEK to AOL, June 28, 1898, JEK to M. Kellogg, July 8, July 16, 1898, R.T. Crawford to JEK, July 18, July 29, 1898, JEK to F.E. Ross, Aug 1, 1898, SLO.

17 ESH to A. Rodgers, May 1, 1891, H.K. Palmer to the Graduate Council, Oct 23, 1899, R.T. Crawford to the Graduate Council, Oct 23, 1899, SLO.

18 JEK to M. Kellogg, Dec 7, 1898, WWC to GEH, Aug 21, 1901, SLO.

19 JEK to M. Kellogg, May 5, 1899, JEK to AOL, Nov 6, 1899. A.H. Joy, *BMNAS, 39,* 417, 1967.

20 JEK to JAB, June 7, 1898, JEK to H. Harrer, Sep 28, 1898, Feb 8, Mar 2, Sep 4, 1899, JEK to GEH, Sep 25, 1899, SLO. H. Harrer, *Ap.J., 10,* 290, 1899.

21 JEK to M. Kellogg, Aug 15, 1898, JEK to AOL, Feb 1, 1899, JEK to E. Brown, Feb 10, 1899, JEK to Secretary, University of California, Feb 20, 1899, WWC to Secretary, UC, Mar 10, 1899, RHT to MAT, Oct 4, 1897, Feb 12, Feb 20, Mar 5, Mar 12, Mar 27, 1899, SLO.

22 JEK to B.I. Wheeler, Jan 3, 1900, RHT to MAT, Feb 18, Feb 26, 1900, SLO; B.I. Wheeler to JEK, Jan 5, 1900, UCA.

23 JEK to Secretary, University of California, Sep 30, 1898, SLO; A.L. Colton to Special Committee of the Regents, [~Oct 9, 1897], BHL.

24 JEK to F.G. Hesse, Oct 26, 1898, May 1, 1899, JEK to AOL, Aug 21, 1899, Mar 31, 1900, SLO.

25 JEK, "Polaris a Triple Stellar System," Sep 8, 1898, WWC to E. Bradford, Sep 11, 1898 (telegram), SLO.

26 ESH to T.G. Phelps, Apr 13, 1892, M. Merriman to RHT, June 20, 1895, RHT to J.A. Tucker, July 14, 1895, SLO.

27 M. Kellogg to JEK, Dec 5, 1898, JEK to M. Kellogg, Dec 8, 1898, Mar 9, Mar 11, 1899, M. Kellogg to WWC, Feb 10, 1899, WWC to M. Kellogg, Mar 14, 1899, SLO; WWC to M. Kellogg, Feb 6, 1899, WJH to M. Kellogg, Mar 20, 1899, UCA.

28 E.F. Hussey, "Diary", Apr 1 - Sep 18; 1897, BHL; JMS to T.G. Phelps, Sep 24, Sep 25, 1897, JEK to ESH, Aug 11, Sep 27, 1898, JEK to H. Eichelbaum, Sep 28, 1898, JEK to A.G. Dixon, Nov 28, 1898, JEK to R. Lehmann-Fihles, Jan 4, 1899, JEK to G.E. Lumsden, Jan 5, 1899, SLO; JEK, *Ap.J., 5*, 150, 1897.

29 JEK to J. Perrotin, Apr 3, 1900, SLO.

30 JAB to JEK, June 15, 1898, JEK to JAB, June 29, 1898, May 4, 1899, JEK to C. Zeiss, Sep 6, 1898, May 6, 1899, SLO.

31 JEK, *Ap.J., 6*, 423, 1897; JEK to GEH, Aug 11, 1898, GEH to JEK, Aug 29, 1898, YOA.

32 JEK to J.W. Benham, June 3, 1898, SLO.

33 JEK to T.R. Bruce, Nov 27, 1899, SLO.

34 JEK to M. Disney, Aug 20, 1898, JEK to M. Sievers, Apr 22, 1899, JEK to S. York, Feb 13, 1900, JEK to D. Satterwaite, Dec 22, 1898, SLO.

35 EEB, Memo, Feb 3, 1891, VU; JEK to H.G. Haskell, Mar 30, 1900, JEK to C. Burckhalter, Mar 30, 1900, SLO.

36 JEK to J.B. Mhoon, June 16, July 4, 1899, JEK to C. Roelling, July 3, 1899, JEK to J. Bernal, July 3, 1899, JEK to B.I. Wheeler, Dec 21, 1899, JEK to W.H.L. Barnes, Jan 10, 1900, SLO.

37 JEK to R.J. Holway, Sep 3, 1898, JEK to C.D. Perrine, Nov 29, 1898, JEK to B.R. Baumgardt, Dec 9, 1898, JEK to W.H. Knight, June 26, July 21, 1900, SLO.

38 J.H. Montgomery to JEK, May 9, 1900, JEK to J.H. Montgomery, June 5, 1900, JEK to H.C. Lord, Jan 6, 1899, JEK to J.S. Riscard, July 25, Dec 20, 1898, SLO.

39 JEK to A. Cichi, Nov 11, 1898, SLO.

40 JEK to S. Bowers, Nov 24, 1899, SLO.

41 JEK to D.T. Day, Apr 19, 1899, D.T. Day to JEK, Jan 11, 1900, SLO.

42 JEK, *Ap.J., 8*, 113, 1898.

43 M.L. Huggins, *Ap.J., 8*, 54, 1898.

44 WWC, *Ap.J., 6*, 363, 1897. WWC to GEH, Sep 21, 1897, YOA.

45 JEK to V. Schumann, Dec 9, 1891, JEK to E.B. Frost, Dec 14, 1892, HL.

46 JEK, *PASP, 10*, 141, 1898.

47 GEH, *Science, 12*, 353, 1900.

48 JEK to H.C. Vogel, Jan 4, 1899, SLO.

49 WJH, "Crossley 36″ Telescope," undated memo [~June 1897], BHL.

50 WJH to E.A. Lyman, Oct 10, 1898, BHL.

51 WJH, *PASP, 36*, 309, 1924.

52 JEK to D.S. Jordan, Apr 2, 1898, SUA.

53 D. Gill to RSF, Sep 18, 1876, SLO; H. Grubb, *Proc. Royal Dublin Society, 1*, 1, 1877; GEH, *Ap.J., 5*, 119, 1897.

54 *University Record (The University of Chicago), 2*, 235, 1897; G.W. Ritchey, *Ap.J., 5*, 143, 1897; *Smithsonian Contributions to Knowledge, 24*, 1, 1904.

55 C.D. Perrine, *PA, 8*, 409, 1900.

56 M.W. Bruce to ESH, May 15, 1895, SLO; ESH to J.H.C. Bonté, Sep 19, 1895, UCA.

57 ESH to JAB, Mar 6, Mar 27, Apr 22, 1896, JAB to ESH, Mar 18, Mar 26 (telegram), Apr 6, 1896, ESH to T.G. Phelps, Mar 27, 1896, ESH to A.H. Babcock, Mar 27, 1896, SLO.

58 ESH to M. Urmy, Mar 27, 1896, ESH to P. Urmy, Apr 3, 1896, ESH to JAB, Feb 19, Mar 6, 1896, SLO.

59 JAB to WWC, July 2, 1896, Jan 28, Feb 25, 1897, SLO; JAB to GD, Nov 12, 1896, Feb 26, Apr 15, 1897, BL.

60 JEK, *Ap.J., 11*, 325, 1900. (Reprinted as JEK, *PLO, 8*, 1, 1908.)

61 JEK, Crossley Reflector Observing Book 1, June 1, 1898 - Apr 10, 1899, LO. Many of the statements about Keeler's observing, not otherwise referenced, are based on this book, and the later books of refernces 77 and 104.

62 ESH to WJH, Dec 3, Dec 6, 1895, BL; WJH to ESH, Dec 4, 1895, SLO.

63 JEK to F.L.O Wadsworth, July 18, 1898, AO.

64 JEK to JAB, Nov 18, 1898, SLO.

65 JEK, *Ap.J., 8*, 1989; JEK, *AJ, 19*, 451, 1898.

66 JEK, *PASP, 10*, 245, 1898; JEK to EEB, Dec 10, 1898, VU.

67 JEK to EEB, Jan 5, 1899, VU.

68 JEK to GEH, Apr 19, 1899, YOA.

69 V.M. Slipher, *Lowell Obs. Bull., 55*, 2, 1913.

70 JEK, *PASP, 11*, 39, 1899.

71 JEK to J.R. Eastman, Apr 17, 1899, SLO.

72 JEK to J. Carbutt, Nov 15, Dec 27, 1898, Feb 20, Mar 9, 1899, SLO.

73 JEK, *Ap.J., 9*, 133, 1899; JEK, *PASP, 11*, 70, 1899.

74 JEK to J.P Hall, Mar 16, 1899, SLO.

75 GEH to JEK, Mar 23, 1899, YOA.

76 J. Scheiner, *Ap.J., 10*, 164, 1899; JEK, *Ap.J., 10*, 167, 1899.

77 JEK, Crossley Reflector Observing Book 2, Apr 11 - Nov 29, 1899, LO.

78 JEK, *PASP, 12*, 89, 1900.

79 JEK to GEH, June 5, 1899, YOA.

80 GEH to JEK, June 12, 1899, SLO.

81 JEK, *Ap.J., 10*, 193, 1899.

82 GEH to JEK, Sep 11, 1897, JEK to GEH, Sep 15, 1897, J.S. Ames to GEH, Mar 19, 1898, T.C. Chamberlin to GEH, Mar 26, 1898, GEH to E.C. Pickering, Mar 26, May 3, 1898, GEH to SN, May 5, 1898, YOA; JEK to E.C. Pickering, June 21, 1898, SLO.

83 GEH, *Ap.J., 8*, 193, 1898.

84 GEH, *Ap.J., 10*, 211, 1899.

85 G.C Comstock to GEH, Oct 6, Oct 13, 1898, GEH to E.C. Pickering, Sep 9, 1899, YOA.

86 GEH to JEK, June 26, Aug 16, 1899, SLO; JEK to GEH, June 16, Aug 7, Aug 24, 1899, GEH to WWC, July 24, Aug 16, 1899, Oct 27, 1900, YOA.

87 GEH to WWC, Sep 28, 1898, Jan 24, Aug 29, 1899, WWC to E.B. frost, July 26, 1898, Oct 27, 1900, YOA.

88 JEK to G.C. Comstock, July 26, 1899, JEK to GEH, Aug 21, Aug 25, 1899, SLO.

89 GEH to JEK, Oct 6, 1899, UCA.

90 JAB to WWC, Nov 25, 1898, SLO.

91 JEK to B.I. Wheeler, Oct 12, 1899, UCA; JEK to J.F. Houghton, Nov 6, 1899, SLO.

92 F.P. Whitman to JEK, Sep 13, 1899, JEK to G.C. Comstock, Sep 15, 1899, JEK to F.P. Whitman, Sep 21, 1899, SLO.

93 JEK to ESH, Dec 13, 1898, JEK to J.K. Rees, Apr 11, 1899, JEK to C.L. Doolittle, Apr 17, 1899, JEK to GEH, July 28, Sep 4, 1899, SLO.

94 JEK to the Secretary of the Royal Astronomical Society, Sep 9, 1899, J. Gledhill to W.H. Wesley, Mar 14, 1900, RAS; E. Crossley to JEK, Nov 2, 1899, F.L.O. Wadsworth to JEK, May 5, 1899, SLO.

95 JEK to C.A. Young, Mar 2, Apr 2, 1900, DCL; H.K. Palmer to WWC, Sep 11, 1906, SLO.

96 *San Francisco Examiner,* June 12, 1899; JEK, *PASP, 11,* 164, 1899; JEK to A.S. Hallidie, June 20, 1899, UCA.

97 V.A. Stadtman, *The University of California,* 1868 - 1968. New York, McGraw Hill Book Co., 1970. AOL to JEK, May 19, May 27, 1899, JEK to AOL, May 30, 1899, SLO; JEK to A.S. Hallidie, July 24, 1899, JEK to M. Kellogg, May [~22], 1899, UCA.

98 F.E. Ross to JEK, Mar 30, Apr 24, May 23, 1899, JEK to AOL, Apr 3, 1899, JEK to C.W. Slack, July 6, 1899, SLO; JEK to AOL, Apr 14, 1899, BL.

99 C.M. Keeler to P.A. Hearst, Aug 12, [1898], BL; JEK to P.A. Hearst, July 17, 1899, P.A. Hearst to JEK, Aug 10, 1899, JEK to C.W. Slack, July 4, 1899, SLO.

100 JEK to D.O. Mills, Feb 1, 1900, D.O. Mills to JEK, Mar 12, 1900, SLO.

101 D.G. Gilman, *The Launching of a University.* New York, Dodd, Mead & Co., 1906; JEK to B.I. Wheeler, July 20, 1899, RHT to MAT, Oct 22, Oct 29, 1899, SN to JEK, Nov 30, 1899, SLO; JEK to SN, Nov 24, 1899, LC; JEK to D.C. Gilman, Nov 27, 1899, MEL.

102 JEK to K. Kostersitz, June 6, July 17, 1899, JEK to Chairman of the North Austrian Landes-Ausschuss, Sep 18, 1899, SLO.

103 JEK, *PASP, 11,* 199, 1899.

104 JEK, Crossley Observing Book 3, Dec 1, 1899 - July 24, 1900, LO.

105 EK, *PASP, 12,* 126, 1900; JEK, *AN, 152,* 173, 1900; HK. Palmer, *AN, 152,* 173, 1899.

106 WWC to C.M. Keeler, Jan 4, Jan 31, 1907, C.M. Keeler to WWC, Jan 9, 1907, SLO.

107 International Astronomical Union Circular No. 3595, 1981; Minor Planet Circular 5975, 1981; P. Herget to author, Mar 27, 1978; B.G. Marsden to author, July 16, 1981.

108 JEK to A.M. Clerke, Aug 21, 1899, SLO.

109 JEK, *MNRAS, 59,* 537, 1899.

110 JEK, *MNRAS, 60,* 128, 1899.

111 JEK, *AN, 151,* 2, 1899.

112 JEK, *Ap.J., 12,* 1, 1900.

113 W.W Payne to JEK, Dec 3, Dec 30, 1898, [Dec 17, 1899] (telegram), JEK to W.W. Payne, Dec 22, 1898, Dec 18, 1899 (telegram), Jan 15, Jan 29, 1900, JEK to H.C. Wilson, Dec 12, 1899, Jan 19, 1900, SLO; W.W. Payne, *PA, 8,* 4, 1900.

114 R.T. Chamberlin, *BMNAS, 15,* 307, 1932; S.G. Brush, *JHA, 9,* 1, 1978.

115 T.C. Chamberlin to JEK, Jan 30, 1900, SLO.

116 JEK to T.C. Chamberlin, Feb 3, 1900, T.C. Chamberlin to JEK, Feb 28, 1900, T.C. Chamberlin to WWC, Oct 18, 1901, SLO.

117 T.J.J. See to JEK, Nov 30, Dec 20, 1899, JEK to T.J.J. See, Dec 8, 1899, SLO.

118 A.C. Maury to JEK, Dec 12, Dec 29, 1899, Mar 12, 1900, JEK to A.C. Maury, Dec 20, 1899, Feb 19, 1900, SLO; A.C. Maury, *Annals Astron. Obs Harvard College, 28,* 1, 1897; B.L. Welther, *Bull. American Astronomical Society, 13,* 861, 1981.

119 JEK to W.H. Wesley, Dec 29, 1898, RAS; H.H. T[urner], *MNRAS, 61,* 197, 1901.

120 JAB to A.D. Keeler, Apr 30, [1899], SLO.

121 GEH to JEK, May 10, May 31, 1899, JEK to GEH, May 16, 1899, YOA.

122 C.A. Young to SN, Apr 15, 1897, Dec 22, 1898, LC; SN to JEK, Feb 10, 1900, JEK to I.Remsen, Feb 20, 1899, SLO.

123 A.J. Cannon, *BMNAS, 15,* 193, 1934; S.I. Bailey, *BMNAS, 15,* 169, 1932.

124 JEK to E.C. Pickering, Feb 13, [~Feb 16], May 24, June 13, June 26, 1900, E.C. Pickering to JEK, June 6, 1900, JEK to S.I. Bailey, June 26, 1900, S.I. Bailey to JEK, June 20, July 23, 1900, SLO.

125 JEK to SPL, Feb 1, 1900, JEK to D. Gill, Nov 8, 1899, Apr 3, 1900, H.S. Budd to the Chief Computer, Jan 7, 1900, JEK to H.S. Budd, Jan 10, 1900, SLO.

126 RHT to MAT, Dec 31, 1899, SLO.

127 EEB to GD, Mar 26, 1900, BL.

128 JEK to E.H. Hills, Dec 16, 1898, Jan 7, 1899, JEK to W. Moore, Jan 7, 1899, SLO.

129 JEK to GEH, Feb 21 1899, YOA.

130 C. Burckhalter to GEH, Feb 17, 1899, H.S. Pritchett to GEH, Mar 7, 1899, AOL to GEH, July 18, 1899, WWC to GEH, Aug 16, 1899, UOA.

131 GEH to SPL, Feb 25, 1900, YOA.

132 SPL to WWC, Jan 22, Feb 21, 1900, SPL to JEK, Feb 21, 1900, JEK to SPL, Feb 12, 1900, WWC to SPL, Mar 14, 1900, JEK to SPL, Feb 12, 1900, WWC to SPL, Mar 14, 1900, JEK to GEH, Mar 19, 1900, SLO.

133 C.A. Young, *Ap.J., 12,* 77, 1900; J. Stebbins to WWC, Apr 11, 1901, SLO.

134 JEK to S. J. Brown, Feb 17, 1900, JEK to H.C. Lord, Apr 4, 1900, SLO.

135 WWC to JEK, Apr 24, May 16, June 4, June 5, 1900, JEK to WWC, May 21, May 28, 1900, SLO; WWC and C.D. Perrine, *PASP, 12,* 175, 1900.

136 JEK to WWC, June 14, 1900, SLO.

137 JEK to C. Burckhalter, July 21, 1900, SLO; C.G. Abbot, *Ap.J., 12,* 69, 1900; C. Burckhalter, *PASP, 12,* 169, 1900.

138 J.S. Ames to H. Crew, June 2, 1900, R.W. Wood to H. Crew, June 9, 1900, GEH to H. Crew, June 9, 1900, AIP; WWC to JEK, June 16, 1900, SLO; S.J. Brown, *Ap.J., 12,* 58, 1900; GEH, *Ap.J., 12,* 80, 1900.

139 JEK to J.K. Rees, Apr 10, 1900, JEK to G.C. Comstock, May 12, 1900, SLO.
140 JEK to H.C. Vogel, Mar 3, 1900, H.C. Vogel to JEK [~Apr 23, 1900], JEK to B.I. Wheeler, May 23, 1900, SLO.
141 JEK to GEH, Mar 7, Mar 23, Apr 20, 1900, SLO.
142 JEK to GEH, Apr 20, 1900, SLO.
143 WWC to GEH, Aug 18, 1900, YOA; WWC to C.A. Young, Aug 24, 1900, DCL; RHT to MAT, Aug 16, 1900, SLO; JAB to C. Burckhalter, Aug 24, 1900, CO; JAB to GD, Oct 1, 1900, BL. Descriptions of Keeler's symptoms are from these five letters; further details of his illness and death not otherwise referenced are from the two by Campbell.
144 RAT to MAT, June 25, 1900, JEK to A. Bearda, July 5, 1900, R.T. Crawford to Secretary of the University, July 6, 1900, R.T. Crawford to E.F. Coddington, July 11, 1900, RHT to Boschen Hardware Co., July 17, 1900, JEK to J.P. Jarman, July 19, 1900, SLO.
145 JEK, "Biography", June 20, 1900, JEK to J.A. Parkhurst, [July] 24, 1900, SLO; JEK to GEH, July 24, 1900, YOA; JEK to AOL, July 27, 1900, BL.
146 JEK to F.W. Gardam, July 30, 1900, WWC to Secretary, University of California, July 31, 1900, SLO.
147 JEK to B.I. Wheeler, Aug 2, 1900, UCA; WWC to J.E. Budd, Aug 13, 1900, SLO.
148 *San Francisco Call, San Francisco Bulletin, San Francisco Examiner*, all Aug 13, 1900.
149 WWC to Secretary of the Royal Astronomical Society, Sep 15, 1900, RAS.

Chapter 10

1 *San Francisco Examiner, San Francisco Call, San Francisco Chronicle*, all Aug 16, 1900; JAB to C.M. Schwab, Aug 17, 1900, HL; RHT to MAT, Aug 22, 1900, JAB to WWC, Aug 23, 1900, SLO; C.A. Young to SN, Aug 25, 1900, LC; GEH to C.A. Young, Aug 28, 1900, YOA.
2 WWC to C.M. Keeler, Oct. 19, Oct 26, Dec 14, 1900, C.M. Keeler to WWC, Oct 13, Oct 14, Dec 27, 1900, Feb 11, 1901, SLO.
3 F.L.O. Wadsworth, Annual Reports of the Director, Apr 1, 1900, Jan 31, 1901, HL; JAB to GEH, Mar 25, 1899, Nov 10, 1902, HPM.
4 *Dedication of the New Allegheny Observatory*, Aug 28, 1912 (pamphlet); N.E. Wagman, *Pitt, 10*, 9, 1942; T.M. Lauterbach, *Pitt, 46*, 21, 1952.
5 Report of Observatory Committee, June 4, 1906, AO; W.W. Payne, *PA, 14*, 11, 1906.
6 RHT to MAT, Aug 27, Sep 10, 1900, C.M. Keeler to WWC, Mar 13, 1901, Apr 8, 1905, SLO; C.M. Keeler to C. Burckhalter, Sep 30, 1900, CO; C.M. Keeler to AOL, Oct 16, 1900, BL.
7 JAB to GD, Oct 1, 1900, C.M. Keeler to AOL, Sep 20, Oct 12, 1900, BL; *San Francisco Post*, Oct 23, 1900; *San Francisco Examiner*, Dec 12, 1900.
8 C.M. Keeler to P.A. Hearst, Jan 9, 1903, Oct 1, 1909, Aug 27, Sep 9, Oct 19, Nov 8, 1910, BL; WWC to JAB, Aug 3, 1910, SLO.
9 WWC to JAB, July 6, Nov 21, 1910, JAB to WWC, July 23, Dec 27, 1910, Jan 11, 1911, WWC to C.M. Keeler, Oct 18, Nov 21, 1910, SLO.

10 W.P. Field to R.A. Franks, Jan 12, June 12, 1911, Sep 21, 1912, A.A. Hamer-schlag to C.M. Keeler, Apr 21, 1911, J.H. Leete to H.B. Keeler, July 6, 1911, J.H. Leete to JAB, Jan 30, 1914, A.A. Hamerschlag to A.T. Doremus, Oct 22, 1914, W.T. Field to A.T. Doremus, Oct 26, 1914, Carnegie-Mellon University.

11 C.M. Keeler to WWC, Dec 17, 1917, SLO; M.T. Smith to C.M. Keeler, Nov 8, 1918, N.N. Babcock to C.M. Keeler, Nov 8, 1918, Carnegie-Mellon University.

12 C.M. Keeler to P.A. Hearst, 'Tuesday'' [~Oct 1, 1912], Jan 6, 1919, BL; W.H. Wright to J.E. Johnson, Oct 4, 1939, C.K. Moore to H. Crawford, Feb 28, 1961, SLO; C.K. Moore to author, June 9, 1977.

13 JAB, *PA, 8,* 476, 1900; WWC, *PASP, 12,* 139, 1900, *Ap.J., 12,* 239, 1900; GEH, *Science, 12,* 353, 1900.

14 C.S. Hastings, *AJS, 10,* 325, 1900, *BMNAS, 5,* 231, 1903; C.D. Perrine, *PA, 8,* 439, 1900; RHT, *AN, 153,* 99, 1900.

15 JEK, *PASP, 12,* 146, 1900, *PLO, 8,* 1, 1908.

16 M.W. Bruce to ESH, May 15, [1897], SLO; ESH, *PASP, 9,* 104, 168, 1897.

17 JEK to F.R. Ziel, Oct 27, 1898, Oct 31, 1899, SLO; GEH to F.R. Ziel, Oct 26, 1897, Oct 11, 1898, Sep 11, 1899, F.R. Ziel to GEH, Oct 15, 1900, YOA.

18 K. Campbell, *Life on Mount Hamilton 1899-1913,* Interviewed and Edited by Elizabeth Spedding Calciano, Santa Cruz, 1971, SLO.

19 J. Stebbins to G.C. Comstock, Aug 18, 1901, UW.

20 J.H. Moore, *PASP, 50,* 189, 1938; C.D. Shane, *PASP, 76,* 77, 1964; C.D. Shane, interview with author, July 14, 1978.

21 W.H. Wright to W.F. Kellogg, Nov 9, 1922, SLO.

22 C.D. Perrine and H.K. Palmer, Crossley Reflector Observing Book 4, Sep 16 - Oct 24, 1900, LO.

23 B.W. Snow to JEK, July 2, 1900, JEK to B.W. Snow, July 23, 1900, R.W. Wood to JEK, July 22, 1900, JEK to R.W. Wood, July 24, 1900, R.W.Wood to WWC, July 15, Aug 23, 1900, Mar 17, 1901, WWC to R.W. Wood, Aug 21, 1900, SLO; R.W. Wood, *Ap.J., 12,* 281, 1900.

24 B.I. Wheeler to GEH, Aug 17, 1900, YOA; B.I. Wheeler to SN, Aug 17, 1900, LC.

25 WWC to GEH, Aug 24, 1900, E.C. Pickering to GEH, Sep 8, 1900, GEH to E.C. Pickering, Sep 27, 1900, YOA; RHT to MAT, Sep 3, 1900, SN to WWC, Sep 26, 1900, JAB to WWC, Oct 31, Nov 13, 1900, SLO; WWC to C. Burckhalter, Aug 24, Sep 22, 1900, CO.

26 WWC to SN, Oct 8, 1900, LC; WWC to GEH, Sep 15, Oct 19, 1900, YOA; B.I. Wheeler to WWC, Dec 11, 1900, WWC to B.I. Wheeler, Dec 11, 1900 (both telegrams), SLO.

27 JEK to SN, Dec 7, 1899, SN to WWC, Oct 15, Nov 13, 1900, Feb 12, 1901, SLO.

28 R.G. Aitken, *PASP, 50,* 204, 1938; W.H. Wright, *BMNAS, 25,* 35, 1949.

29 Interviews with and letters from three former Lick staff members who knew Campbell and Wright personally, 1978-1981.

30 JAB to H. Crew, July 4, 1908, AIP; S.E. Tillman, *Annual Report of the Association of Graduates of the U.S. Military Academy, Annual Reunion,* June 11, 1915, pp. 52-74.

31 WWC to ESH, Jan 29, 1907, Sep 19, 1910, Jan 10, 1911, ESH to WWC, Sep 26, Dec 15, 1910, Jan 18, 1911, M.H. Rosen to WWC, Mar 18, 1914, WWC to M.H. Rosen, Mar 26, 1914, SLO.

32 JAB to JEK, Nov 23, 1899, H. Harrer to WWC, Dec 26, 1900, SLO; H. Harrer to
    SN, Aug 13, 1898, LC.

33 SN to H.J. Furber, Sep 4, 1895, LC; E.F. Coddington to JEK, Apr 20, 1900, JEK
    to B.I. Wheeler, Apr 26, 1900, RAT to MAT, June 4, 1900, E.F. Coddington to
    JEK, July 24, 1900, E.F. Coddington to WWC, Jan 26, Sep 29, 1901, Feb 2, 1902,
    SLO; Ohio State University Board of Trustees' Minutes, Feb 12, 1951.

34 JEK to A. Raumm, June 3, 1898, JEK to M. Kellogg, June 3, June 11, 1898,
    A.M. Mattoon to JEK, June 26, 1900, JEK to A.M. Mattoon, July 4, 1900, SLO.

35 F.P. Leavenworth to JEK, Mar 15, 1900, JEK to B.I. Wheeler, Mar 21, 1900,
    A.F. Maxwell to JEK, Apr 21, June 1, 1900, JEK to A.F. Maxwell, Apr 26, 1900,
    SLO.

36 JEK to H. Bingham, Jr., July 5, July 28, 1899, H. Bingham, Jr., to JEK, July 14,
    July 31, 1899, SLO.

37 JEK to J.S. Ames, Apr 4, 1900, H.M. Reese to JEK, Apr 23, June 28, 1900, JEK
    to H.M. Reese, May 1, 1900, JEK to B.I. Wheeler, July 2, July 3, 1900, SLO.

38 H. M. Reese to WWC, Sep 4, Sep 24, 1900, WWC to H.M. Reese, Sep 14, 1900
    (telegram), SLO.

39 R.R. McMath, *PASP, 54,* 69, 1942; R.G. Aitken, *BMNAS, 22,* 275, 1943.

40 H.D. Curtis to ESH, Sep 27, 1896, Feb 18, May 15, 1897, ESH to H.D. Curtis,
    Sep 30, 1896, SLO.

41 J. Sutton to JMS, Feb 1, 1898, H.D. Curtis to JMS, Feb 6, Apr 10, Apr 13 (two
    letters), 1898, H.D. Curtis to JEK, Dec 11, 1898, Oct 16, 1899, JEK to H.D.
    Curtis, Oct 21, 1899, SLO.

42 H.D. Curtis to JEK, Feb 15, Mar 24, 1900, JEK to H.D. Curtis, Feb 19, Mar 26,
    Mar 31, 1900, H.D. Curtis to WWC, Mar 29, 1900, SLO.

43 WWC to H.D. Curtis, Aug 21, 1900, SLO; WWC and C.D. Perrine, *PASP, 12,*
    175, 1900.

44 JEK to GEH, July 4, 1900, GEH to JEK, July 11, 1900, SLO.

45 E. Wylie to Lick Observatory Director, July 19, 1900, JEK to E. Wylie, July 24,
    1900, JEK to J.S. Ames, July 24, 1900, SLO.

46 AOL to GEH, Dec 26, 1900, YOA; *PASP, 51,* 63, 1939.

47 F.E. Ross, Final Examination for PhD, May 3, 1901 (printed program), SLO;
    S.B. Nicholson, *PASP, 73,* 182, 1961; W.W. Morgan, *BMNAS, 39,* 391, 1967.

48 R.T. Crawford, Final Examination for PhD, May 3, 1901 (printed program),
    SLO; M.W. Makemson, *PASP, 71,* 503, 1959.

49 J. Stebbins to GEH, Dec 17, 1900, YOA; AOL to WWC, Dec 27, 1900, J.
    Stebbins to WWC, Apr 11, 1901, WWC to AOL, [Jan 9, 1903], AOL to WWC,
    Jan 22, 1903, SLO.

50 WWC to D.O. Mills, Nov 17, Dec 12, 1900, D.O. Mills to WWC, Dec 1, 1900,
    SLO; WWC to C. Burckhalter, Dec 15, 1900, CO; WWC, *PASP, 17,* 64, 1905.

51 WWC, *PASP, 15,* 70, 1903.

52 H.K. Palmer to WWC, Oct 14, Nov 15, 1901, June 23, 1902, Feb 12, Feb 15, Feb
    28, 1903, WWC to H.K. Palmer, Sep 11, Oct 19, Dec 9, 1901, Feb 14, Feb 19,
    1903, AOL to WWC, Feb 12, Feb 20, Mar 7, 1903, WWC to AOL, Feb 24, 1903,
    H.K. Palmer, Final Examination for PhD, Feb 27, 1903 (printed program), SLO.

53 WWC, *PASP, 15,* 244, 1903, *PASP, 40,* 249, 1928; C.D. Shane, *BMNAS, 50,*
    377, 1979.

54 GEH to WRH, Jan 16, 1896, RL; GEH to W.E. Hale, Nov 17, 1897, YOA;
    GEH, *Ap.J., 6,* 353, 1897.

55 G.W. Ritchey, Bill to the University of Chicago, July 16, 1898, YOA; G.W. Ritchey, *Ap.J.*, *14*, 217, 1901, *Pub. Yerkes Obs.*, *2*, 387, 1904.
56 GEH to WRH, Sep 19, 1898, RL; SWB to GEH, Nov 17, 1898, WRH to GEH, Nov 18, 1898, HPM.
57 GEH to H.M. Goodwin, Nov 23, 1898, HHL; JAB to JEK, Nov 26, 1898, SLO; JEK to GEH, Nov 25, 1898, SWB to GEH, July 13, 1899, HPM.
58 GEH to N.B. Ream, Jan 19, 1899, GEH to JEK, Mar 23, May 10, 1899, JEK to GEH, Apr 19, 1899, GEH to WRH, May 9, 1899, YOA.
59 WRH to GEH, Mar 19, Nov 15, 1898, GEH to WRH, Sep 20, 1899, YOA.
60 GEH, Diary, 1901. See especially entries for Oct 3-15. This diary is in the possession of W.L.W. Sargent.
61 GEH to H.M. Goodwin, Dec 24, 1903, Feb 28, Aug 4, 1904, HHL.
62 H.K. Palmer to WWC, Mar 29, June 25, 1905, June 12, Oct 6, 1906, Feb 4, 1912, WWC to H.K. Palmer, Feb 15, Apr 19, 1905, H.K. Palmer to B.I. Wheeler, May 14, 1906, H.K. Palmer to W.H. Wright, Feb 28, 1943, Apr 23, 1951, SLO.
63 G.W. Ritchey, *Ap.J.*, *5*, 143, 1897, *Smithsonian Contributions to Knowledge, 34,* 1, 1904.
64 Helen Wright, *Explorer of the Universe: A Biography of George Ellery Hale.* New York, E.P. Dutton & Co., 1966. This excellent book should be consulted for Hale's complete life and career.
65 WWC to JAB, May 18, 1895, ESH to JAB, Mar 6, Mar 27, 1896, SLO.
66 JEK to JAB, Sep 12, Sep 30, Dec 13, 1898, Jan 20, 1899, JAB to JEK, Dec 1, 1898, Feb 21, 1899, SLO.
67 JEK to H.C. Vogel, Mar 18, 1899, SLO.
68 JEK, *Ap.J.*, *10*, 193, 1899; JEK, Crossley Observing Book 2, Apr 11 - Nov 29, 1899, LO; JEK to JAB, July 17, 1899, SLO.
69 JEK to E.C. Pickering, June 13, 1900, HCO.
70 H.K. Palmer, *LOB, 2,* 46, 1903.
71 JEK, *Ap.J.*, *10*, 266, 1899, *PASP, 11,* 177, 1899.
72 J. Scheiner, *AN, 148*, 325, 1899.
73 JEK to T.J.J. See, July 3, 1899, SLO.
74 JAB to JEK, July 24, 1899, SLO.
75 JEK to JAB, Nov 25, Nov 30, Dec 18, 1899, Jan 31, Feb 13, Mar 2, 1900, JAB to JEK, Dec 2, Dec 28, 1899, Feb 22, 1900, SLO; JEK, Crossley Reflector Observing Book 3, Dec 1, 1899 - July 24, 1900, LO.
76 JAB to JEK, Mar 31, Apr 17, June 13, June 22, June 25, 1900, JEK to JAB, Apr 12, Apr 24, 1900, SLO.
77 J. Stebbins, *LOB, 1,* 57, 1901.
78 K. Burns, *LOB, 6,* 92, 1910, *PASP, 23,* 33, 1911.
79 E.A. Fath, *LOB, 5,* 71, 1909.
80 H.D. Curtis, *PLO, 13,* 9, 45, 57, 1918.
81 H.D. Curtis, *Journal of the Washington Academy of Sciences, 9,* 217, 1919.
82 K. Lundmark, *PASP, 33,* 324, 1921.
83 W.G. Hoyt, *BMNAS, 52,* 411, 1980.
84 N.U. Mayall, *BMNAS, 41,* 175, 1970.
85 N.U. Mayall, *PASP, 49,* 42, 1937.

Bibliography

Brashear, John A. *The Autobiography of a Man Who Loved the Stars*. Boston & New York, Houghton Mifflin Co., 1925.

Campbell, W.W. *Stellar Motions*. New Haven, Yale University Press, 1913.

Canning, Jeff and Buxton, Wally. *History of the Tarrytowns, Westchester County, N.Y. from Ancient Times to the Present*. Harrison, N.Y., Harbor Hills Books, 1975.

Clerke, Agnes, M. *A Popular History of Astronomy during the Nineteenth Century*. London, Adam & Charles Black, third edition 1893.

Cordasco, Francesco. *Daniel Coit Gilman and the Protean Ph.D.* Leiden, E.J. Brill, 1960.

de Vaucouleurs, Gerard. *Astronomical Photography from the Daguerretype to the Electron Camera*. New York, Macmillan Co., 1961.

Elliott, Orrin Leslie. *Stanford University - The First Twenty-Five Years*. Stanford, Stanford University Press, 1937.

Ferrier, William Warren. *Ninety Years of Education in California*. Berkeley, Sather Gate Book Shop, 1937.

Flexner, Abraham. *Daniel Coit Gilman: Creator of the American Type of University*. New York, Harcourt Brace and Co., 1946.

Frost, Edwin B. *An Astronomer's Life*. Boston and New York, Houghton Mifflin Co., 1933.

Gilman, Daniel Coit. *The Launching of a University*. New York, Dodd, Mead & Co., 1906.

Goodspeed, Thomas W. *William Rainey Harper: First President of the University of Chicago*. Chicago, University of Chicago Press, 1928.

Hawkins, Hugh. *Pioneer: A History of the Johns Hopkins University 1874 - 1889*. Ithaca, Cornell University Press, 1960.

Hoyt, William Graves. *Lowell and Mars*. Tucson, University of Arizona Press, 1976.

Jones, B.Z. and Boyd, L.G. *The Harvard College Observatory - The First Four Directors 1839 - 1919*. Cambridge, Harvard University Press, 1971.

Keeler, Charles Augustus. *San Francisco and Thereabout*. San Francisco, A.M. Robertson, 1906.

Lewis, Oscar. *George Davidson. Pioneer West Coast Scientist*. Berkeley, University of California Press, 1954.

Lick, Rosemary. *Generous Miser. The Story of James Lick of California*. Palo Alto, Ward Ritchie Press, 1967.

Livingston, Dorothy Michelson. *The Master of Light: A Biography of Albert A. Michelson.* Chicago, University of Chicago Press, 1967.

Lorant, Stefan. *Pittsburgh - The Story of an American City.* Garden City, N.Y., Doubleday & Co., 1964.

Lubove, Roy (ed.) *Pittsburgh.* New York, New Viewpoints, 1976.

Miller, Howard S. *Dollars for Research. Science and Its Nineteenth Century Patrons.* Seattle, University of Washington Press, 1970.

Mirrielees, Edith R. *Stanford. The Story of a University.* New York, G.P. Putnams Sons, 1959.

Newcomb, Simon. *The Reminiscences of an Astronomer.* Boston & New York, Houghton Mifflin Co., 1903.

Prescott, Samuel C. *When M.I.T. Was "Boston Tech" 1861 - 1916.* Cambridge, The Technology Press, 1954.

Seabrook, William. *Doctor Wood - Modern Wizard of the Laboratory.* New York, Harcourt, Brace and Co., 1941.

Stadtman, Verne A. *The University of California 1868 - 1968.* New York, McGraw Hill Book Co., 1970.

Starrett, Agnes L. *Through One Hundred and Fifty Years, The University of Pittsburgh.* Pittsburgh, University of Pittsburgh Press, 1932.

Storr, Richard J. *Harper's University: The Beginnings. A History of the University of Chicago.* Chicago, University of Chicago Press, 1966.

Webb, W.L. *Brief Biography and Popular Account of the Unparalleled Discoveries of T.J.J. See.* Lynn, Mass., Thos. B. Nichols & Son Co., 1913.

Wolfmeyer, Ann and Gage, Mary Burns. *Lake Geneva: Newport of the West 1870 - 1920.* Lake Geneva, Wis., Lake Geneva Historical Society, 1976.

Wright, Helen. *Explorer of the Universe. A Biography of George Ellery Hale.* New York, E.P. Dutton, 1966.

Index